Mobile Communications

IFIP – The International Federation for Information Processing

IFIP was founded in 1960 under the auspices of UNESCO, following the First World Computer Congress held in Paris the previous year. An umbrella organization for societies working in information processing, IFIP's aim is two-fold: to support information processing within its member countries and to encourage technology transfer to developing nations. As its mission statement clearly states,

> IFIP's mission is to be the leading, truly international, apolitical organization which encourages and assists in the development, exploitation and application of information technology for the benefit of all people.

IFIP is a non-profitmaking organization, run almost solely by 2500 volunteers. It operates through a number of technical committees, which organize events and publications. IFIP's events range from an international congress to local seminars, but the most important are:

- the IFIP World Computer Congress, held every second year;
- open conferences;
- working conferences.

The flagship event is the IFIP World Computer Congress, at which both invited and contributed papers are presented. Contributed papers are rigorously refereed and the rejection rate is high.

As with the Congress, participation in the open conferences is open to all and papers may be invited or submitted. Again, submitted papers are stringently refereed.

The working conferences are structured differently. They are usually run by a working group and attendance is small and by invitation only. Their purpose is to create an atmosphere conducive to innovation and development. Refereeing is less rigorous and papers are subjected to extensive group discussion.

Publications arising from IFIP events vary. The papers presented at the IFIP World Computer Congress and at open conferences are published as conference proceedings, while the results of the working conferences are often published as collections of selected and edited papers.

Any national society whose primary activity is in information may apply to become a full member of IFIP, although full membership is restricted to one society per country. Full members are entitled to vote at the annual General Assembly, National societies preferring a less committed involvement may apply for associate or corresponding membership. Associate members enjoy the same benefits as full members, but without voting rights. Corresponding members are not represented in IFIP bodies. Affiliated membership is open to non-national societies, and individual and honorary membership schemes are also offered.

Mobile Communications

Technology, tools, applications, authentication and security

IFIP World Conference on
Mobile Communications
2 - 6 September 1996,
Canberra, Australia

Edited by

José L. Encarnação
Fraunhofer - IGD
Darmstadt
Germany

and

Jan M. Rabaey
EECS Department
University of California
Berkeley
USA

Published by Chapman & Hall on behalf of the
International Federation for Information Processing (IFIP)

CHAPMAN & HALL
London · Weinheim · New York · Tokyo · Melbourne · Madras

Published by Chapman & Hall, 2–6 Boundary Row, London SE1 8HN, UK

Chapman & Hall, 2–6 Boundary Row, London SE1 8HN, UK
Chapman & Hall GmbH, Pappelallee 3, 69469 Weinheim, Germany
Chapman & Hall USA, 115 Fifth Avenue, New York, NY 10003, USA
Chapman & Hall Japan, ITP-Japan, Kyowa Building, 3F, 2-2-1 Hirakawacho, Chiyoda-ku, Tokyo 102, Japan
Chapman & Hall Australia, 102 Dodds Street, South Melbourne, Victoria 3205, Australia
Chapman & Hall India, R. Seshadri, 32 Second Main Road, CIT East, Madras 600 035, India

First edition 1996

© 1996 IFIP

Printed in Great Britain by Hartnolls, Bodmin, Cornwall

ISBN 0 412 75580 7

Apart from any fair dealing for the purposes of research or private study, or criticism or review, as permitted under the UK Copyright Designs and Patents Act, 1988, this publication may not be reproduced, stored, or transmitted, in any form or by any means, without the prior permission in writing of the publishers, or in the case of reprographic reproduction only in accordance with the terms of the licences issued by the Copyright Licensing Agency in the UK, or in accordance with the terms of licences issued by the appropriate Reproduction Rights Organization outside the UK. Enquiries concerning reproduction outside the terms stated here should be sent to the publishers at the London address printed on this page.
  The publisher makes no representation, express or implied, with regard to the accuracy of the information contained in this book and cannot accept any legal responsibility or liability for any errors or omissions that may be made.

A catalogue record for this book is available from the British Library

∞ Printed on permanent acid-free text paper, manufactured in accordance with ANSI/NISO Z39.48-1992 and ANSI/NISO Z39.48-1984 (Permanence of Paper).

Foreword

The IFIP 1996 World Conference on Mobile Communications was prepared by an International Programme Committee (IPC) that consisted of 25 members, international top experts for the area and coming from 10 countries. The IPC published a Call for Papers that was internationally distributed on a large scale. All papers submitted were reviewed by at least three international experts. Based on these reviews the IPC developed a very high-quality conference programme in two meetings (one in Providence, RI, USA, and the other one in Darmstadt, Germany). The programme is divided into two tracks:

Track 1: Mobile Technology, Tools and Applications
Track 2: Trusting in Technology, Authentification, Security

Track 1 presents 21 papers and includes four invited papers in eight technical sessions. Track 2 presents 12 papers and includes two invited papers and one industrial state-of-the-art report in four technical sessions. These proceedings publish the revised versions of all accepted papers.

We would like to express our thanks to the IPC members for all their contributions and all the support in putting this programme together, which we believe is of a very high technical quality and includes major contributions in science, technology, applications, and trends for the area of Mobile Communications. Thanks also to all the organizers of IFIP'96, especially to Professor Hörbst for all his personal advice. I would also like to express my gratitude to my Co-Chairman Professor Jan Rabaey, University of California at Berkeley, USA, for sharing with me the effort needed to develop this technical, scientific programme for the 1996 World Conference on Mobile Communications in Canberra, Australia.

Darmstadt, Germany   José L. Encarnação
June 1996

IFIP PRESIDENT'S INTRODUCTION

The most important single event in the IFIP program of activities is the World Computer Congress, currently held every two years. Thirteen World Computer Congresses have been organized by IFIP. In 1980, the 8th IFIP Computer Congress was jointly held in Japan and Australia. The 13th IFIP Congress was held in Hamburg in September 1994. The 15th Congress is scheduled to take place in Vienna and Budapest in September 1998 and the venue of Congress 2000 will be Beijing, China.

IFIP is delighted to come back to Australia which is also an appreciation and recognition of the contributions Australia and the other countries from the Pacific Region have made to the development of information processing.

The name "Canberra" comes from an ancient Aboriginal word which means "Meeting Place". From 2 - 6 September 1996, this modern and beautiful city will host the **14th IFIP World Computer Congress** and will provide a meeting place for many academics and practitioners coming from all regions of the world to discuss achievements, interests and future developments in the field of information processing.

Congress '96 and its three specialized conferences are focused on the latest developments in multimedia, information highways, intelligent systems, mobile communications, use of computer and communication technologies in teaching and learning. The Congress format offers a unique opportunity to all participants to discuss and contribute to subject areas with a critical impact on the use and application of IT in the future.

The organization of an IFIP Congress is not an easy task. Since 1991, when the IFIP General Assembly selected Canberra as the site for Congress '96, preparations have been under way with contributions from many IFIP volunteers, technical committees and IFIP as a whole. We are hopeful that these efforts will be reflected in a very successful Congress.

I would like to take this opportunity to thank everyone who has contributed to the organization of Congress '96. May I express IFIP's grateful thanks to the International Program Committee and the hosting organization, The Australian Computer Society. Last but not least, I extend IFIP's thanks to all Congress '96 participants, wishing them an enjoyable and professionally successful stay in Canberra, and hope to see many of them in Vienna and Budapest in 1998.

Prof. Kurt Bauknecht
President
International Federation for Information Processing

CONTENTS

Foreword v

Introduction vi

Keynote Paper
1 Trends of flat displays in the multimedia age
  H. Sasaki 3

TRACK 1   MOBILE TECHNOLOGY, TOOLS AND APPLICATIONS

PART ONE   System Support for Mobile Communications 7

Invited Paper
2 System support and applications for mobile computing
  N. Diehl 9

3 Developments in mobile data system technology - particularly GSM
  J. Leske 11

PART TWO   Caching and Replication for Mobile Communications 21

4 Replication-support for advanced mobile applications
  D. Kottmann 23

5 Caching data over a broadcast channel
  H.V. Leong, A.Si, B.Y.L. Chan 31

PART THREE   Basic Architectures for Mobile Information Systems 39

6 Mobile frames: a pragmatic approach to automatic application partitioning based on an end-user data model
  T. Kirste 41

7 Generic personal communications support for open service environments
  T. Eckardt, T. Magedanz, C. Ulbricht, R. Popescu-Zeletin 50

PART FOUR   Mobile Agents and Multimedia Applications 67

Invited Paper
8 Supporting user mobility
  M. Brown 69

9 Designing secure agents with O.O. technologies for user's mobility
  D. Carlier, P. Trane 78

10 The idea: integrating authoring concepts for mobile agents into an authoring tool for active multimedia mail
  J. Schirmer, T. Kirste 86

11 Animation within mobile multimedia on-line services
C. Belz, M. Bergold, H. Häckelmann, R. Strack ... 98

PART FIVE Networking and Protocols for Mobile Communication ... 109

12 Theoretical analyses of data communications integrated into cordless voice channels
R. Canchi, Y. Akaiwa ... 111

13 Random access, reservation and polling multiaccess protocol for wireless data systems
T. Buot ... 119

PART SIX Methods and Algorithms for Mobile Information Access ... 127

Invited Paper
14 Challenges in mobile information systems and services
R. Strack ... 129

15 A buffer overhead minimization method for multicast-based handoff in picocellular networks
E. Ha, Y. Choi, C. Kim ... 132

16 Impact of mobility in mobile communication systems
M. Zonoozi, P. Dassanayake ... 141

PART SEVEN Mobile Communication Architectures ... 149

17 Mobile computing based on GSM: the Mowgli approach
T. Alanko, M. Kojo, H. Laamanen, K. Raatikainen, M. Tienari ... 151

18 An adaptive data distribution system for mobile environments
S. Kümmel, A. Schill, K. Schumann, T. Ziegert ... 159

19 Object oriented system architecture and strategies for the exchange of structured multimedia data with mobile hosts
J. Bönigk, U. von Lukas ... 167

PART EIGHT QoS-Management and Resource Discovery for Mobile Communication ... 179

Invited Paper
20 Agent skills and their roles in mobile computing and personal communications
M. Mendes, W. Loyolla, T. Magedanz, F.M. Assis Silva, S. Krause ... 181

21 A global QoS management for wireless network
M.T. Le, J. Rabaey ... 205

22 Resource discovery protocol for mobile computing
C. Perkins, H. Harjono ... 219

TRACK 2 TRUSTING IN TECHNOLOGY, AUTHENTICATION, SECURITY

Invited Paper
23 The future of smart cards: technology and application
V. Cordonnier ... 239

Industrial State-of-the-Art Report
24 Use of smart cards for security applications by Deutsche Telekom
   B. Kowalski — 245

PART ONE   Protocols for Authentication, Secure Communication and Payment — 247

25 An authentication and security protocol for mobile computing
   Y. Zheng — 249

26 Design of secure end-to-end protocols for mobile systems
   V. Varadharajan, Y. Mu — 258

27 Yet another simple internet electronic payment system
   J. Zhao, C. Dong, E. Koch — 267

PART TWO   General Security Aspects in Mobile Communication — 275

28 Difficulties in achieving security in mobile communications
   I. Nurkic — 277

29 GSM digital cellular telephone system: a case study of encryption algorithms
   T. Smith — 285

PART THREE   New Security Algorithms and Methods — 299

30 A new algorithm for smart cards
   C. Marco, P. Morillo — 301

31 A new approach to integrity of digital images
   D. Storck — 309

Invited Paper
32 How to protect multimedia applications
   E. Koch — 317

PART FOUR   Secure Mobile Applications — 321

33 Phone card application and authentication in wireless communications
   C.H. Lee, M.S. Hwang, W.P. Yang — 323

34 Real-time mobile EFTPOS: challenges and implications of a world first application
   S.R. Elliot — 330

Index of Contributors — 339

Keyword Index — 341

Keynote Paper

1

Trends of Flat Displays in the Multimedia Age

H. Sasaki
NEC Corporation
7-1, Shiba 5-Chome, Minato-ku, Tokyo 108-01 Japan
Tel:+81-3-3454-1111   Fax:+81-3-3798-6541
e-mail:hajime_sasaki@hq-exec.ccgw.nec.co.jp

Abstract
Today, in a multimedia age, electronic displays are the essential key devices involved in the relation between multimedia equipment and human beings. Flat panel displays are overcoming the restrictions of traditional CRTs (Cathode Ray Tubes) regarding portability and large screen capability, and they are ready to expand their multimedia application fields. TFT-LCDs (Thin Film Transistor-Liquid Crystal Displays) in particular have advantages such as light weight, low profile, low power consumption, and motion picture display, and importantly contribute the development of multimedia. The color PDP (Plasma Display Panel) features emissive display, a large screen, and a low profile, and mass production aimed at implementing space-saving, and wall-hung HDTV (High Definition Television) is about to be started. Since further development of multimedia is expected in the twenty-first century, the scale of this market is immeasurable. At the same time, the flat panel display market will also expand. Wide variety of flat panel displays will gain wider use as they secure their respective markets through their particular advantages.

Keywords
Multimedia, Device technology, Electronic display, Flat panel display, LCD, TFT-LCD, Color PDP

1 Introduction

The multimedia society has come. Multimedia is not simply a combination of various media. Multimedia contribute to the creation of a sophisticated social life through new forms of creation, presentation, understanding, education, and entertainment activities in which the human intellect as well as sensibility are supported by computers.

As multimedia are based on C&C technologies such as computers, networks, databases, and hypermedia, the development of device technologies including semiconductors, batteries, and electronic displays is the key to implement these technologies.

This article covers the trends of flat panel displays while focusing on TFT-LCDs and color PDPs.

2 Development of Electronic Devices Supporting Multimedia

Semiconductor devices have progressed at a remarkable rate for a broad range of applications such as data processing, transmission, and storage. Microprocessors are expected to achieve the remarkably high performance levels ranging from 1000 MIPS to 5000 MIPS. Memory devices are also entering the gigabit generation. The development of ASIC ICs which support the elementary multimedia technologies will directly provide high performance, high functionality and reduced size for multimedia equipment.

The dramatic evolution of semiconductor devices allows image-oriented high-volume data to easily be processed. Much higher levels of multimedia system performance are expected in future.

Electronic displays are essential key devices in the relation between these multimedia equipment and human beings. Among the various electronic displays, traditional CRTs offer general purpose display performance. However, they have limitations regarding portability and large screen. Flat panel displays are overcoming these disadvantages of CRTs, and are about to expand the application field of multimedia. Currently, available flat panel displays are VFDs (Vacuum Fluorescent Displays) in the small size display area, LCDs and EL (Electro-Luminescent) displays in the medium size area, and LED (Light Emitting Diode) array displays and color PDPs in the large size area.

Among these flat panel displays, the evolution of LCDs, in particular of TFT-LCDs, is contributing to the development of multimedia. In addition, color PDPs are ready for full production. Since color PDPs can easily be produced in large screen size, they are expected to grow mainly in the fields of HDTV equipment, space-saving large display, etc.

As secondary batteries having a higher energy density, such as lithium-ion batteries, are increasingly being used as the energy source for the personal and portable equipment, the application fields of flat panel displays will also expand.

3 Trends of Color TFT-LCDs

The first LCDs to be introduced were monochrome LCDs for calculators and watches. STN-LCD evolved into large screen sizes, but could not afford high contrast motion picture displays. Higher contrast, higher speed, and full-color flat panel displays were realized by TFT-LCDs which have TFT switches on every pixels. Then, in the early 90's, the large-scale production of color TFT-LCDs started to apply to notebook PCs.

TFT-LCDs can support the wide area of media, since it can display motion pictures as well as characters, and still images. LCDs, which offer the advantages of lightweight, low profile, and low power consumption, are increasingly being used for applications linked with multimedia networks. These applications are centered on notebook computers and workstations.

The first generation of TFT-LCDs for notebook computers were for 8 to 9-inch VGA resolution (approx. 0.3 million pixels), while current models are 12-inch SVGA or XGA resolution (0.5 to 0.8 million pixels). For PC monitors and workstations, new products with 13 to 17-inch screens and SXGA resolution (approx. 1.3 million pixels) are being introduced. Highly finished 20-inch class models have already been developed.

Mainstream of full-color notebook computer LCDs support 0.26 million colors (with 64-level gray scale) by using the multi-bit digital signals. Further progressed full color capability has been achieved with 16.7 million colors (with 256-level gray scale) or with analog signal processing which provides infinite colors similar to CRTs.

The multi-scan resolution expansion technology, which expands the display format from VGA to SXGA, has also been put to practical use. In addition, the adoption of new LCD modes or the development of viewing-angle-expanded-film enables wider viewing angles equivalent to the CRT level.

Portability of a TFT-LCD has been greatly improved. Weight and thickness of TFT-LCDs were reduced to approximately half the initial value. Power consumption has decreased to 1/4 the initial level. Reflection type LCDs without a backlight also offer improved display performance. They are expected to come into wide use for portable data terminal applications.

While current TFT-LCDs use a-Si TFTs (amorphous silicon TFTs), higher performance is being realized by shifting to low-temperature-processed p-Si TFTs (polycrystalline silicon TFTs). Low-temperature-processed p-Si TFTs on glass substrates have high electron mobility, and raise TFT-LCD functionality as integration of associated circuits on the substrate glass, opening up the "System on Glass" world.

4 Trends of Color PDPs

CRTs or LCDs present manufacturing and practical use restrictions that limit the maximum screen size to approximately 30 - 40 inches. On the other hand, color PDPs offer the same advantages of emissive display and wider viewing angle as CRTs. They also have the advantage that the large flat panel display size more than 40 inches can be produced through a relatively simple manufacturing process.

Comparing with traditional CRTs, the thickness of color PDPs is estimated to be about 1/10, and their weight less than 1/5. In the multimedia age, color PDPs are expected to be the optimal displays for wide-screen flat panel displays, especially for HDTV. In particular, we expect a new market to be created through the implementation of space-saving, large-size wall-hung TVs, something which has been a long-cherished dream.

As the result, the 40-inch class with 850 x 480 x RGB pixels, 16.7 million colors, and 200 to 350 cd/m^2 level of luminance has been achieved.

At present, we are pursuing aggressive investment and further improvement for color PDPs as we are preparing for mass production, with the goal of introducing 40 to 60-inch products. To realize a true HDTV, a reduced pixel pitch and faster drive will be important to support an increased number of pixels. The key to meeting these challenges is to obtain higher luminance by improving fluorescent materials and driving method, to optimize manufacturing processes for reduced pixel pitch, and to reduce power consumption.

5 Future Prospect of Flat Panel Displays

In the society of twenty-first century, multimedia will grow dramatically and the infrastructure supporting social life will be based in large part on multimedia systems. Along with the development of multimedia, the application of electronic displays will also increasingly expand in the twenty-first century.

Current applications of LCDs focus on personal computers and workstations. In the future, however, the main features of LCDs, i.e. portability, and display performance, will be further improved. Thus, LCDs will further evolve as the indispensable display that can be used anytime, anywhere, and by anybody in the multimedia age, with one display for each person. Also in the monitor market, LCDs will steadily replace CRTs.

Color PDPs, featuring emissive display, large screen size and low profile, will adorn every home as the essential " wall-hung TV ". Color PDPs will also steadily grow as they penetrate the public marketplace use.

Wide variety of other flat panel displays will gain wider use as they secure their respective markets through their particular advantages.

TRACK 1: MOBILE TECHNOLOGY, TOOLS AND APPLICATIONS

PART ONE

System Support for Mobile Communications

2
System Support and Applications for Mobile Computing
- Extended Abstract -

Norbert Diehl, Albert Held
Daimler-Benz AG, Research Center Ulm, Information Technology
Wilhelm-Runge-Straße 11, P.O. Box 2360, 89013 Ulm, Germany
Phone: +49 731 505 2132/2831, Fax: + 49 731 505 4218
e-mail: {diehl, held}@dbag.ulm.DaimlerBenz.COM

1. INTRODUCTION: DISTRIBUTED, MOBILE APPLICATIONS

In future networks, mobile computers will become increasingly important. New portable computers and wireless communications technologies particularly enable mobile computing. Distributed, mobile applications allow information access anywhere, anytime. Mobility (in the sense of location independence) and adequate information support are no longer opposites but are supported simultaneously.

The possibility to support mobility and thus changing locations (i.e. network access points) while being connected offers new possibilities for future applications, but also introduces new problems. Some of them are:
- A basic but far-reaching fact is that the users are moving and thus the topology of the system is constantly changing. This also includes location dependent information and therefore the need for new addressing schemes.
- The mobile terminals will always be less powerful than the stationary computers. Thus we have a performance disparity between the mobile and stationary computers, that has a strong influence on the system design, e.g., workload balance, etc.
- The data rates of wireless connections will always be clearly lower than those of wired connections. Wireless connections are not as reliable as wired. There is also the need to support interworking in heterogeneous networks.

Additional problems due to mobility and the specific characteristics of portable computers and wireless communications are dynamic configuration, moving resources, reachability, data and function consistency. All these problems have to be treated in more than one of the communication layers. A full system architecture is necessary. This includes an additional „mobility layer" above the OSI transport layer.

We present a system called MOBI-DICK that addresses the requirements of mobile computing and we discuss two fields of applications in more detail: mobile services in the industrial environment and traffic-telematics.

2. MOBI-DICK

MOBI-DICK is a system platform to support distributed mobile applications and addresses basic aspects of mobile computing:
- terminal mobility -user mobility - service mobility
- location dependent services - location independent services
- ad-hoc networking

- autonomous work during disconnection
- IT-security (authentication, authorization, ...)
- adaption and support of differnet wireless and wired communication systems

MOBI-DICK includes an application manager for mobile applications and a service directory and service trading mechanisms that allow location transparency and location awareness. Basic security services for authentication and access control are integrated in MOBI-DICK

3. MOBILE APPLICATIONS

Mobile Services in the Industrial Environment

Mobile applications in industrial environments add special requirements. Typical scenarios:
- plant operatives are mobile: For example, the set-up or maintenance staff, equipped with mobile terminals, is working at different locations and having full multi-media information support.
- products are mobile: Within a large production plant the products, e.g. cars, are moved from one manufacturing location to another. To allow the easy identification of the products and the concrete work to be done as well as the status of work carried out, the product should be equipped with escort memory systems (ems) that can be read and written via wireless communication.
- machinery or production systems are mobile: Within h ghly flexible production plants the machinery may be reconfigured and moved relatively often. Problems such as recabling, installing and interconnecting new sensors and control units have to be considered.
- transportation systems or platforms in a manufacturing plant are mobile: To achieve an optimal material flow, transportation systems should be equipped with sufficient computing facilities and access to the plant's communication network.

For all types, wireless communications and new multi-media mobility services are needed. Thus we will both look at the requirements for the new high performance radio network and the application support services for mobility, multi-media, and interworking.

Traffic Telematics

Traffic Telematic Services such as Guidance-Systems, Tourist- and Travel-Information, Traffic-Information, Trip-Planning, Fleet Management, Mobile Office, etc. become more and more important. These services are used from stationary hosts (e.g., PC at home) as well as from mobile hosts (in vehicles).

In addition to global service providers, a large number of smaller service providers with local services (e.g., hotel reservation, local entertainment) or specialized services (e.g., stock market information) will coexist. This leads to dynamic systems with various distributed information sources. While driving from one city to another, several service providers are involved. A seamless service handover with adopted service trading protocols is needed.

Other topics of interest are: flexibility and scalability of the service system, efficient information acquisition, filtering and presentation, optimized use of the available communication systems, location dependency.

We show how these requirement can be handled and present solutions based on the MOBI-DICK system.

3
Developments in mobile data system technology - particularly GSM

J. Leske
Centre for Telecommunications Information Networking (CTIN)
33 Queen St, Thebarton, SA 5031, Australia
Phone +61-8-303 3222, Fax +61-8-303 4405
Email: J.Leske@CTIN.adelaide.edu.au

Abstract
This paper details the current and near future developments in mobile data systems. It looks in particular detail at the work currently being undertaken to provide advanced data capabilities in the GSM digital mobile phone system, particularly packet and high speed circuit switched data. This paper also details other current developments in mobile data systems, from two-way paging systems through to high speed wireless LANs.

Keywords
mobile data, wireless, GSM, GPRS, HSCSD

1 INTRODUCTION

According to the Wireless LAN Alliance and the Yankee Group (Chambers, 1996), in North America wireless local area network (WLAN) sales tripled from 1993 to 1995 (US$157 million). Of the cellular phone users in the USA, approxiamately 2% use some form of mobile data communications (Data Communications, 1995). In Europe the use of data over cellular phone networks is conservatively forecast to grow to 3% by 1998 and to 18% by 2003 (Yankee Group, 1996). Other forms of wireless data access exhibit strong growth, with enthusiastic forecasts for their future.

What are the reasons for such strong growth? Why go Wireless?
Customer expectation and work practices have changed to require responsiveness, immediacy, and accuracy. While half of typical USA mobility is within the company building, the remainder comprises: within a campus area (14%), in metropolitan area (18%) within the country (12%) or out of the country (1%) (Pryor, 1996). Staying connected allows access to central information storage at the point of service, for consistency, efficiency and security. The

resurgence in centralised information storage can be seen in the use of client/server systems and the network computer concept. Typical mobile data applications include: field service; fleet management; public safety; field sales; EFTPOS; and the mobile professional seeking data base access and information sharing.

Wireline remote access can meet some of these needs, especially where daily access or updates are sufficient. For broadband style applications, such as video conferencing, a mobile solution cannot meet the bandwidth demand of 1.5 Mbit/s or higher. But workgroup applications, such as Lotus Notes, rely on large central information stores where a typical database size is 500MB. A dial-in update will often take half an hour, so continual mobile access to the central database has many advantages.

With the many advantages wireless can offer, why isn't it used more frequently?

What's so hard about not having a connecting wire?

The choices for wireless connection are InfraRed (IR) or radio. IR is limited by line-of-sight issues and so is restricted to localised coverage. IR requires multiple sources to cover a complex geometry like the inside of a partitioned office, and multiple receivers to be in sight of each terminal unit. Interference problems are fairly minimal, however direct sunlight can saturate receivers. Range is quite limited, typically being 3m for high speed (>1Mbit/s) and 50m for low speed IR systems.

Radio is the best current solution for mobility. With appropriate frequencies radio can operate around corners and penetrate inside most buildings and vehicles. However the electromagnetic spectrum is a limited resource which must be shared between multiple uses (TV, radio, RADAR, etc) and multiple users (subscribers). Interference can be from other radio transmitters nearby on the same system; from transmitters on adjacent channels of a different system; intermodulation results from separate frequencies; or noise such as clock signals from devices not intended to be radio transmitters. Signal attenuation is another difficulty: radio signals have losses proportional to distance squared and distance to the fourth power; buildings and trees cause additional losses; at higher frequencies atmospheric effects such as rain also significantly attenuate the signal. Finally, the multiple paths by which a radio signal travels cause fading, which is worse with high frequencies and motion.

Hence the major limitations in radio are that to increase range higher power is needed, with attendant interference issues; and to increase data rate a larger portion of the limited spectrum is required. These difficulties are continuously being addressed to provide and improve mobile data facilities. The following sections of this paper investigate current systems which offer mobile data. capability.

2 GLOBAL SYSTEM FOR MOBILES (GSM)

The development of the second generation (digital) cellular mobile phone system, GSM, commenced in Europe in 1982. Commercial operation of GSM, in the 890 to 960 MHz band, subsequently commenced in 1992.

Using TDMA (Time Division Multiple Access), each 200kHz channel is divided into 8 timeslots. Primarily designed for voice communications, GSM also supports data, facsimile and short pager-like text messages.

2.1 Circuit switched data

The primary method of computer to computer data communication in GSM at present is the circuit-switched data call. GSM currently supports up to 9600 bit/s asynchronous or synchronous communication, and can also incorporate RF error correction. The raw RF bit rate is 33.8 kbit/s. Channel overheads reduce this to 22.8 kbit/s. Interleaving and some minor forward error correction designed specifically for voice reduce this to 13 kbit/s. Additional error correction for data transmission reduces the available user data rate to 9.6 kbit/s.

Radio Link Protocol (RLP) is the specialist layer 2 error correction and data burst retransmission scheme for GSM. Without RLP, the residual average bit error rate (BER) from the sole use of forward error correction (FEC) schemes is 10^{-3} *. When using RLP, the data stream is divided into blocks of 200 bits, which can be selectively retransmitted if they are corrupted by the radio transmission. This reduces the typical bit error rate to 10^{-6}. The penalty of achieving this is increased end-to-end delay. Without RLP, GSM introduces a maximum delay in transmission from GSM handset to network connection of 330ms. With RLP, the delay is dependant on the quality of the radio link at the time, with interference causing retransmissions and hence longer delays.

The GSM specification defines links for data transmission to the telephone network via modem bank, to ISDN, and direct access to X.25. Modem connection has been available since November 1994, and all GSM operators in Australia currently support this. ISDN and X.25 access are to be offered soon.

The data rate of 9.6 kbit/s may seem slow compared to 28.8k kbit/s now common over the telephone network. Additionally these V.34 modems cannot be used with GSM while they can be used over the analogue cellular network. Because GSM is a fully digital system, instead of plugging a modem into a computer, a "data card" is used instead, which converts the data stream and adds the GSM transmission protocols to it. However V.34 is not suitable for the noisy mobile environment. Typically over the analogue network the reliable throughput obtainable is 2400 bit/s. This is being addressed by special analogue mobile error correction protocols like ETC and MNP-10 introduced last year, which by performing a similar function to RLP bring the reliable throughput up to 9.6 kbit/s, or potentially 14.4 kbit/s for a stationary phone in a high signal strength area.

Data compression standard V.42bis has been approved for use with GSM. This is currently being implemented in GSM networks around the world. For compressible data, such as text, V.42bis can offer a 2:1 or better improvement in throughput.

GSM data communications are currently used for facsimile services (transmit mostly rather than receive), dial-in access to hosts, for Email, and file access. The limited data rate means that applications like LAN extension or large file transfers (such as multimedia or the World Wide Web) are quite limited over GSM. Additionally, the connection orientated nature means that long duration calls become very expensive. Staying connected to the office LAN all day is unrealistic, and even spending an hour connected to central office while conducting an interview is prohibitive in cost.

Where applications have not been designed for mobile operation, the overhead of data transmission added to the occasional delay of RLP, can make interactive operation frustrating. An Email application which echoes keystrokes onto the screen from the central computer can lag typing by a number of characters.

---

* GSM network design calls for a 12dB carrier to interference ratio. Where the quality of the radio link is worse, the error rate will be correspondingly higher.

Applications currently running over GSM circuit switched data include field service, sales, fleet telemetry, Email, and remote host access.

2.2 Short Message Service (SMS)

SMS messages are 140 byte data packets sent to and from mobile stations. These messages are sent on control channels, thus allowing simultaneous voice conversation. The intended usage for this facility is notification messaging: pager-like 160 character alphanumeric messages. The messages are sent using a store and forward system via a central SMS centre. This guarantees delivery of messages to turned off or out of range phones once they become available again.

SMS is typically used to provide operator transcribed messages and automatic notification of voicemail. It is also used for limited email, and some telemetry applications. In the UK one application developed is as a data link for police offers sending in their reports from Apple Newtons. Enhancements to the GSM standard will provide for chaining of SMS messages together, allowing full email for example, and message overwriting, which would allow stock quotes or weather reports to be sent regularly without overflowing the message storage.

While the performance of a standard SMS system is not that fast, especially because of the store and forward system, a network can accommodate more than one SMS centre, allowing certain centres to be optimised for use as data links, which makes traffic telemetry and similar real-time applications viable.

2.3 Packet Data over Signalling (PDS)

PDS was originally proposed by DeTeMobil (the national German mobile telephone network) to provide a data transport mechanism for real-time traffic telemetry. Research is being conducted by the European Union in a project called 'SOCRATES' to design a comprehensive traffic monitoring, toll paying, and driver assistance network for Europe. As DeTeMobil is heavily involved in this, they proposed a solution where the existing cellular network, common throughout Europe and soon to be covering most motorways, could be used as the communications link.

The data rates required are quite moderate: 2400 bit/s broadcast downlink to cars, and 100 bit/s uplink from cars to the network. Since GPRS, the fully featured packet data system, was considered to be progressing quite slowly, a simpler solution was proposed explicitly for this purpose. The design borrowed heavily from SMS, while also taking on concepts being developed in GPRS. However the time frame for prototype trials has not been met, and it seems as though PDS while being standardised, will not reach prototype stage and be dropped in favour of GPRS once it becomes available.

2.4 General Packet Radio Service (GPRS)

GPRS is the main effort for the development of a general purpose packet data capability for GSM, as opposed to the specific capability which SMS provides. The aim is to provide a capability to handle everything from large numbers of telemetry users with small data packets, through to high speed packet data transmission with low latency for individual users up to speeds of 80kbit/s.

The discussion for developing GPRS has been underway since 1994 at ETSI (European Telecommunications Standards Institute). The broad standard is expected to be finalised by the end of this year, followed by detailed design in 1997, with prototypes developed in 1998 (optimistically). The two main drives for the design of GPRS are the 'SOCRATES' road telemetry project, and the European Railways Organisation (UIC) requirements for a unified railway communications and telemetry system.

The UIC desires a system that will not only provide voice access across Europe for train controllers and drivers, but will also provide the data communications link for centralised train monitoring and control, with a particular emphasis on the planned 500km/h next generation of very fast trains (VFT). In this application GPRS would have to provide 99.9% confidence maximum message delay of 0.5 seconds for a 250 byte message every 2 seconds. This presents some interesting challenges, which should be able to be met.

The basic design of GPRS is constrained to use, as far as practical, the existing GSM system. The intention is to ease the cost of adding GPRS capability to an existing network, to assist in the widespread introduction of the service. Software changes to existing network infrastructure are achieved by downloads from the central operations and maintenance (OAM) system. Physically going to every BTS to change hardware is an order of magnitude more expensive. GPRS will require additional equipment: intelligent network (IN) style 'GPRS support nodes' (GSN) to manage the packet data sessions, located at the base station controller (BSC) or mobile switch (MSC) level.

In contrast consider the approach necessary for cellular digital packet data (CDPD) in North America. To add packet data to the analogue cellular network an entirely separate network is parasitically added to the AMPS network. This involves significant capital and installation costs, with CDPD radio transmitters and receivers added at every base station. After an expenditure of nearly US$1billion to set up the capability, there is still some doubt as to the viability of the network.

GPRS operates by using the radio channel to transmit short burst containing packets of data. Once a packet is sent, the timeslot used is then available for other GPRS data packets from other users, or for voice calls. In this way a large number of low volume users can be accommodated. To achieve high data rates adjacent timeslots are used, subject to availability.

With a true high speed connectionless data communications link, the possibilities for applications are wide: day-long LAN extension, web browsing, telemetry, multimedia, broadcast services.

2.5 High Speed Circuit Switched Data (HSCSD)

HSCSD is a recent (since 1995) effort to quickly provide a capability for high speed data transmission. By extending the current circuit switched data transmission system, it aims to provide data rates of up to 80 kbit/s without requiring major extensions to the existing GSM system. It will use multiple circuit switched timeslots identical to the current circuit switched data transmission system, sending data over them simultaneously. The aim is to support applications requiring high data rates, such as file transfer, high speed fax, video and multimedia.

The use of multiple timeslots will be restricted to adjacent timeslots. One of the main reasons for this is that GSM handsets presently use a half duplex radio element. Adjacent timeslots allows a half duplex radio element to be used for up to 30 kbit/s. To achieve the highest data rate a handset would require 2 receiver and 1 transmitter radio elements. Handsets supporting the highest data rates will be more expensive than standard GSM handsets because of the additional circuitry and the lower volumes of production. Of course this also applies to GPRS handsets which support the highest data rates.

Of particular importance is the ability to set up and tear down circuit switched data calls rapidly, enabling additional timeslots beyond the first to be used in an on demand basis. This should allow HSCSD to share frequencies with voice calls, and hence be available at a reasonable cost to customers. For network operators HSCSD is intended to be able to be implemented without excessive network redesign, and only software changes to the network infrastructure.

The broad HSCSD standard is expected to be finalised by the end of 1996, leading to prototypes in 1997. Nokia displayed a laboratory concept demonstrator at Telecom '95 in Geneva, operating on two timeslots.

One interesting possibility for operators in the Australian environment would be the use of high speed data services in the soon to be released spectrum at 1.8 GHz for DCS1800 systems (GSM at 1800MHz). With the greater spectrum likely to be available to each operator, high speed data services could be offered without competing so heavily with voice.

3 MOBILE DATA AROUND THE WORLD

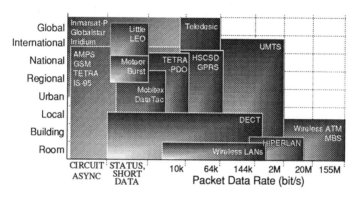

Figure 1. Mobile data comparison chart, contrasting area coverage against data rate.

3.1 Satellites

Satellites can be divided into three broad categories: Geostationary (GEO); Low Earth Orbit (LEO) and Little LEO. While the primary satellite focus is on voice, nearly all systems support data communications. Satellite communications systems are not well suited to urban environments because of tall building shadowing.

Geostationary satellite systems (eg. Imarsat) provide broad coverage with a small number of satellites. The distance of the satellite from earth requires the use of large directional antennas and an antenna guidance system if the phone is moving. The maximum data rate supported is 2400 bit/s, and calls are very expensive at around A$9 per minute. A regional spot satellite such as Optus MobilSat charges A$2.50 per minute.

LEO systems place a large number of small satellites in close orbit, with communications being handed over from one satellite to the next as they move over the horizon. The closer orbit means lower power transmitters can be used, leading to the expectation of pocket sized handsets. Data rates supported are from 2.4 to 7.2 kbit/s, depending on the system. The proposed Teledesic system has a specific focus on data communications and claims its 840 satellites will support data rates from 16 kbit/s and higher. (One of Teledesic's backers is Microsoft.) LEO systems aim to charge their calls at a rate similar to cellular phone calls. Launches are planned to begin in 1998.

Little LEOs are store and forward data only systems, usually also incorporating a position location function. By not needing to provide uninterrupted coverage, far fewer satellites need to be launched. ORBCOMM began its launches in 1995, starting with coverage of North America. Target applications include telemetry, tracking, messaging and emergency services.

3.2 Two way paging

Two way paging is just beginning to be implemented, mostly as extensions of current paging technology, incorporating a return channel. It is characterised by low data rates (around 1kbit/s), small message size (a few hundred bytes), wide area of coverage and usually a very low speed return path. To be a viable consumer product it must offer lowest cost service, smallest size, excellent coverage and building penetration, and long battery life.

Systems either under trial, or soon to be tested include FLEX, pACT, Nexus and Glenayre. The European paging standard ERMES is also considering adding a return path. The greatest difficulty in such systems is providing enough transmit power from the mobile unit to be received. This means bigger antennae, bigger batteries, multiple receiver aerials, or all three.

3.3 Low speed

The Low Speed category includes networks such as DataTAC, MobiTex, Cognito and Paknet. These networks are characterised by low speed data rates (typically under the 10kbit/s), moderate typical message size, moderate to large latency (usually from being a store and forward system), wide area coverage, and typically symmetric uplink and downlink channels.

This category also includes fleet management and vehicular tracking systems such as QuikTRAK, Fleetcomm, Pinpoint, Datatrak and Teletrak. These typically require large aerials for their return transmissions (often via HF radio), have a very low data rates and small packet size. One interesting system, which is wireless though not mobile, is meteor burst communications, where small (30 byte) data packets are transmitted up to 1000 miles using VHF radio bounced off the ionosphere.

Low speed systems are usually packet based rather than circuit switched, and are mainly used in specific industrial applications, such as telemetry, EFTPOS, fleet management, stores and inventory, or customised sales applications. The developments in this field are new specialised mobile radio systems (SMR) being introduced with both voice and data capability, such as TETRA, iDEN, Geotek FHMA, and TMR-MAP27.

3.4 Cellular

Cellular systems are characterised by moderate data rates, few limits on message size, low latency, cellular coverage (and so high infrastructure cost) and often circuit switched, asynchronous connectivity. The existing analogue mobile phone networks AMPS, TACS, and NMT support data rates of 2400 bit/s reliably, though this can be improved with special protocols in the modems at each end, as mentioned earlier. The digital networks such as GSM, DAMPS and soon IS-95 CDMA support data speeds of around 9.6 kbit/s.

Next generation Personal Communications Systems (PCS) such as PHS, PACS Edge and Ricochet can support higher data rates of 28.8 kbit/s because of the smaller cell sizes and greater bandwidth allocated. DECT, the new European cordless standard, can support data rates of up to 552 kbit/s in certain configurations.

The European design for third generation mobile phones, UMTS, calls for data rates inversely related to the coverage area currently occupied. When in microcells within buildings, 2 Mbit/s would be available. When in medium sized urban cells, 385 kbit/s. In large cells 144 kbit/s, and in very large cells 16 kbit/s. UMTS has a 2002/2005 timeframe for introduction.

3.5 High speed

This category contains the high speed wireless data systems, which typically are wireless LANs. Existing wireless LANs include AIRLAN, WaveLAN and other similar systems. The 2Mbit/s raw bit rate provides 600-800 kbit/s user data rate, which is available to be shared between all the terminals in the area. Range is limited to 50m indoors and 300m in an optimal outdoors environment. These systems have a capacity for large packets and provide low latency. InfraRed is also a viable option indoors in such limited ranges, and a number of systems are in use.

An increasing use of these systems is to provide point to point links with directional antennas, to provide a LAN link into nearby buildings. In certain cases and where line of sight is available, a range of 10km can be achieved. IEEE 802.11 is being developed in an attempt to standardise these systems for both wireless LAN and point to point usage.

In addition to providing a wireless link to the office LAN, these systems have found use in large stores for check-out counters, where not having additional wiring and being able to relocate systems is an advantage.

Future developments primarily concentrate on providing higher data rates, supporting new applications like video, and increasing size of files transferred. The European HIPERLAN standard is being finalised this year, which will support 20Mbit/s data with a 50m range, or 1Mbit/s data with an 800m range. It is also expressly designed to support slowly moving (36km/h) terminals. The Japanese are developing a similar standard supporting 10Mbit/s, and there are North American proposals to build on HIPERLAN to develop a local standard (Apple's NII and WinForum's SUPERNET, which have recently combined).

Wireless versions of ATM are being considered and proposed in a number of forms. Certainly any future wireless LAN system will have ATM connection capability. Olivetti are testing in laboratory a system using ATM packets sent over the air. The goal for such systems is to be able to support broadband applications such as HDTV (6Mbit/s). Mobile Broadband System (MBS) is a European project, complimentary to UMTS in telephony, to support data rates from 2Mbit/s up to 155Mbit/s (B-ISDN compatible). Since it is planned to operate in the 60 GHz frequency range, it will be several years before the equipment is available and affordable.

4 CONCLUSION

Any new wireless system being designed will support data communications to meet the demand of the growing market. Indeed, current wireless systems in many cases are enhancing their current support for data communications. GSM in particular will be enhanced in ways which will allow the widespread use of data communications, with improved performance and more attractive costs. However, the limitations which wireless imposes still call for innovative software design and careful systems integration - and convergence between telecommunications and computing.

5 REFERENCES

Alanko T., Kojo M., Laamanen I., Liljeberg M., Moilanen M., Raatikainen K. (1994) Measured Performance of Data Transmission Over Cellular Telephone Networks. *University of Helsinki Department of Computer Science*, Report C-1994-53.

Chambers, P.(ed.) (April 8th, 1996) Wireless LAN Market forecasts. *Wireless Office Newsletter*, Telecomeuropa.

Data Communications (March 21, 1995) Special Report / Wireless Data Services. Data Communications.

ETSI SMG2 and SMG3 GPRS working groups. (1994-96) GSM 02.60, GSM 03.60, GSM 04.60 and associated working documents. *ETSI*.

GSM 04.22 (1992) Radio Link Protocol (RLP) for data and telematic services on the MS-BSS and BSS-MSC interface. *ETSI*.

Gilchrist, P. (1996) One stop data shop FROM GSM's GPRS. *Mobile Communications International*, **Issue 29**, 62–64.

Hämäläinen, J. (1996) GSM's support of HIGH SPEED DATA. *Mobile Communications International*, **Issue 27**, 72–78.

Mouley, M. and Pautet, B. (1993) The GSM System for Mobile Communications. Published by the authors, ISBN 2-9507190-0-7.

Pryor, R. (1996) DECT - From Cordless to Wireless to Tetherless. *Proceedings of Mobiles '96*, IIR Conferences

Rune, T. (1995) Wireless Local Area Networks. *Proceedings of Telecom '95*, ITU Geneva.

Yankee Group (April 1996) Sending data through the air, *d.Comm*. The Economist.

6 BIOGRAPHY

John Leske graduated from the University of Adelaide in 1987 with an Honours degree in Electrical and Electronic Engineering. He worked for a number of years in signal processing and high speed computer design before heading off to see why his ancestors had left Europe. After designing safety systems for oil platforms in the UK, he decided that despite there being some very nice places in Europe, perhaps some of his ancestor's reasons were valid. He joined CTIN in 1994 as a Research Engineer, concentrating on mobile data.

PART TWO

Caching and Replication for Mobile Communications

4
Replication–Support for Advanced Mobile Applications[1]

D. A. Kottmann
Institut für Telematik, Universität Karlsruhe
Kaiserstr. 12, 76128 Karlsruhe, Germany
Tel.: (+49) 721 608-4022, Fax: (+49) 721 388097
e-mail: kottmann@telematik.informatik.uni-karlsruhe.de

Abstract
Applications in mobile computing have to live with massive fluctuations of connectivity, bandwidth, latency, and cost in the underlying communication system. These fluctuations must be masked in order to provide the end user with predictable expenses and functionality. This can be achieved in employing cheap high bandwidth connections to set up a long term usage capability in replicating objects. This capability is afterwards exploited when communication cost increases or its quality decreases. However, off the shelf replication systems that are unaware of the applications they support, often fail to produce sufficient results. In this paper we present an approach for object based replication–support systems for mobile computing which can be tailored to specific applications. After discussing the basic concepts, we present our current prototype system MISTRAL that opens this flexibility to C++–based applications.

Keywords
Mobile computing, application–support systems, replication, advanced mobile applications

1 INTRODUCTION

Untethered communication promises to alter information processing in a way comparable to the emergence of computers from a terminal pool to the desktop (Bagrodia et al, 1995). The core of this paradigm shift towards mobile computing is that applications have to cope with massive fluctuations of connectivity, bandwidth, latency, and cost in the underlying communication system. The level of service can range from a bandwidth of 0 in coverage

---

[1] This work was partly supported by the German Research Council (Deutsche Forschungsgemeinschaft DFG) under grant SFB 346–A6.

blackspots to broadband access in the order of several hundred Mbps when directly connected to a fixed network. Permitting users to operate as effectively as possible under these challenging conditions is the aim of advanced mobile applications. The key success factor for masking these variations is to replicate information between mobile and stationary devices and to employ available communication facilities to manage the replicas (Bagrodia et al, 1995), (Satyanarayanan, 1996). In this paper we present techniques that provide flexible replication–support for these kind of applications.

The paper is organized as follows. Section 2 gives a brief overview about the prospects of replication–support for mobile applications. Then we discuss techniques for tailoring a replication-support system to object oriented advanced mobile applications in section 3. How these techniques can be realized is shown in section 4 that presents our current prototype MISTRAL. Finally, section 5 concludes the paper.

2 THE CASE FOR REPLICATION IN MOBILE COMPUTING

Applications in mobile computing can be classified into three types (Davies, 1996): stand–alone applications, existing distributed applications, and advanced mobile applications.

Stand–alone applications, like textprocessors, are existing applications that only interact with other applications in exchanging data over a file-system or a database. These are the places where one can introduce replication–support. As the application itself exists, the support has to be transparent. Hence, issues that are specific for the mobility context must be handled completely inside the support system.

Most systems that employ replication for mobile computing today operate at the file system level and have been designed with stand–alone applications in mind. Systems like Coda (Kistler and Satyanarayanan, 1993) or LITTLE WORK (Honeyman and Huston, 1995) allow mobile applications arbitrary operations on each replicated file. This so called **optimistic update** philosophy is complemented with protocols for validating consistency of the replicas such that consistency degrades gracefully with vanishing communication abilities. As soon as the available service becomes too poor or too expensive (just think of a 24 hour GSM–connection) updates are performed locally and reintegrated when service quality and cost improves. Such independent updates bear the risk of conflicts. Sometimes conflicts can be resolved automatically by facilities like ASR (Kumar and Satyanarayanan, 1993), but often manual intervention is indispensable. As long as only stand–alone applications are used, the conflict rates are tolerable. An extensive study in Bagrodia, et al. (1995) reports rates below 0.3 % for applications like software development. On the other hand, the observed conflict rate for commercial applications peaked at 25 %. Considering specific applications, like appointment calendars or physical stocktaking, conflicts are no longer the exception. Hence, the conventional approach to replication–support is insufficient when one wants to use those applications in the mobile field.

It is nowadays a widely accepted fact that for these kinds of applications one should no longer try to keep all effects of mobility completely transparent at the system support layer (Davies, 1996), (Satyanarayanan, 1996). However, those advanced mobile applications, which have been designed with mobility in mind, still have to replicate information to compensate for the above mentioned variations in the underlying communication system. Doing this in an application specific way is a tedious task when one has to start from scratch. Generic replication–support systems for advanced mobile applications have to offer an interface that allows the provision of application specific information. The sketched approach has already been the primary design philosophy of Bayou (Demers et al, 1994) which provides replication–support for SQL-based advanced mobile applications.

The third application type, existing distributed applications like WWW, lies in between the two discussed cases. The application itself should not be reimplemented completely, but it is possible to interfere the interactions with remote components, e.g. to satisfy requests in consulting cached information. The cache itself can either be taken from a generic support–system for mobile computing or implemented in an application specific way as in the wireless WWW system from Kaashoek et al. (1994). In the latter case, we can once again employ a replication–support system that has the capability to be tailored to specific application needs.

3 A REPLICATION–SUPPORT TECHNOLOGY FOR ADVANCED MOBILE APPLICATIONS

To ease the task of developing advanced mobile applications that employ replication, we devised a replication–support technology for objects like appointment calendars which is especially tailored for the use of mobile communication services.

3.1 Basic operations of replication support systems

First of all, we take a look at the basic operations that have to be offered by a replication–support system in mobile computing. As it is expensive or sometimes even impossible to reach other replicas, the system normally has to operate in disconnected mode (Kistler and Satyanarayanan, 1993). All cited approaches like Code, LITTLE WORK or Bayou stick to this paradigm. The replication system is employed when new replicas are created, when replicas are reintegrated or when two replicas are mutually updated. All these operations are two–party operations on exactly two replicas, as depicted in figure 1.

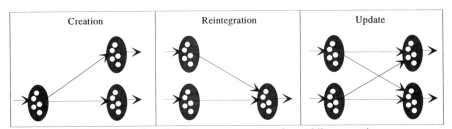

Figure 1 Basic operations of replication–support systems for mobile computing

We use figure 1 to structure replication protocols for mobile computing. We call the ellipsis **virtual partitions**. Inside a virtual partition, operations are performed disconnectedly, unless the replication protocol decides that it has to contact other replicas for reasons of consistency.

3.2 Handling the semantics of objects

A support system has to know what kind of replication–support the application requires. Our approach is to use a description of the objects' semantics via a **serial–dependency–relation** (Herlihy, 1990) and a lean classification of application–specific consistency requirements. Informally, a serial–dependency–relation tells what subset of operations executed on all replicas has to be known when one wants to be sure that another operation returns a consistent value. We discuss the idea behind this relation for the case of a simple stock–item object. The relation is depicted in table 1.

Table 1 Serial–dependency–relation for the stock–item object.

$(op_1, op_2) \in R$	$op_2 = \mathbf{onStock}() \rightarrow i_3$	$op_2 = \mathbf{addToStock}(i_4) \rightarrow \varepsilon$
$op_1 = \mathbf{onStock}() \rightarrow i_1$	false	false
$op_1 = \mathbf{addToStock}(i_2) \rightarrow \varepsilon$	$i_2 \neq 0$	false

The interpretation of the table is that the predicate in a cell tells when one has to know the operation in the row in order to be sure that the execution of the operation in the column returns a globally consistent value. Hence, one can execute arbitrary modifying **addToStock()**–operations without consulting other replicas, while one needs to know all **addToStock(i)**–operations with $i \neq 0$, executed on any replica to get a consistent result from an **onStock()** call. Thus, it is possible to give several replicas the right to increase the amount of a stock–item. How many replicas have to be contacted before executing is a matter of the consistency requirement. Due to the limited place in this paper we only allow to weaken the requirements for operations that do not alter the state of the object and we only tell the case of consistent operations from operations that return a value based on the state at the time of disconnection and the operations executed since then on that replica. Note, that we can provide two different **onStock()**–operations. One **cOnStock()** that is consistent and is used when physical stocktaking has completed to inquire the overall number of items and one **iOnStock()**, used to inquire intermediate results. Even under these simplified conditions it is possible to perform physical stocktaking in multiple stocks.

When a replication–support system knows the serial–dependency–relations R of all objects and it knows what operations are up to be performed in a virtual partition, it has enough information to test whether it can pessimistically guarantee that all operations are allowed to be performed disconnectedly. When operation op_1 is allowed in virtual partition p_1, and there is an operation op_2 allowed in virtual partition p_2 with $(op_1, op_2) \in R$, then the support system can only guarantee that one can execute op_2 disconnectedly, when there is no op^2 allowed in p_2 and op^1 allowed in p_1 with $(op^2, op^1) \in R$. When the system avoids such cycles, that can also span multiple virtual partitions, it can guarantee that the requested operations can be executed disconnectedly. This approach relies on the repeatedly reported observation that mobile users approximately know in advance what operations they need on a replica (Satyanarayanan et al, 1993); i.e. that they can tell the replication–support system what operations they disconnectedly like to execute in a virtual partition.

We use two special locks to detect such cycles efficiently: a **past**-lock for operations that have no successor in the serial–dependency–relation R, a **future**-lock for operations that have no predecessor in R, and both locks for the remaining operations. Then we say that a virtual partition p_1 **precedes** a virtual partition p_2 when there is an object with a past–lock in p_1 and a future–lock in p_2. It is easy to verify that when we avoid cyclic precedes between partitions, also no of the above discussed cycles can result. Note, that although the lock–representation is only a coarse approximation of the serial–dependency–relation, it still suffices to give several replicas the right to increase the amount of a stock–item.

3.3 Alternatives in case of insufficient guarantees

The so far discussed approach allows to perform operations on the local replica as long as the allocated rights (i.e. locks) are sufficient. Thus, it keeps the illusion to work with a single object without the necessity of permanent expensive access to a single copy allocated a server. When an operation beyond the allocated right is requested, we can offer the application the choice among four alternatives. The superiority of one of the alternatives depends on the currently available communication system. The alternatives are:

Permanent reconciliation: Consistency is preserved via contacting other replicas. This is appropriate for high–bandwidth connections or to perform a bulk reconciliation over wireless services that charge for connection time, regardless of the amount of data that is actually transmitted.
Reassigning rights: Rights are redistributed among the replicas. This requires only minimal data transfer in the order of 20 Bytes per object. Hence it is suited for transfer over wireless links that charge the amount of transmitted data.
Switch to optimism: The operation is performed optimistically. This alternative is appropriate for coverage blackspots, when the conflict rate is low, or when automatic reconciliation is possible. We discuss below how optimistic operations are seamlessly integrated.
Defer: Try the operation later. This should be chosen when one is in a small coverage blackspot, as in a railway tunnel, or when conflict rates are high and only manual conflict resolution is possible.

Our support–technology allows applications to choose the right alternative for each situation. As an example, permanent reconciliation is the right choice as long as the mobile device is directly connected to an Ethernet or via an ISDN link.

3.4 Integrating optimism

A novel aspect of our approach is that anomalies that might arise from optimistic operations are only visible for the initiating user who in performing the operation beyond the allocated rights has accepted the risk of failure in the first place. All operations that are performed without violating the allocated rights are guaranteed to converge into a common state, even if optimistic operations have been performed on some replicas. We call this seamless integration of pessimistic guarantees and optimism **mixed–mode–replication**.
Mixed–mode–replication is achieved in logging operations that have been performed optimistically and in postponing reconciliation until the replica can contact another one so

that the sum of the guarantees covers all logged operations. On reintegration, a fine–grained conflict analysis is performed. It uses the serial–dependency–relation to verify whether an operation on the other replica may invalidate one of the optimistically performed ones. In this case, the optimistically executed operation has to be rolled back. As other optimistically performed operations might depend on the deleted operation, we use a **forward–dependency–relation** to detect these conflicts. Informally, a forward–dependency–relation tells what operations must not be deleted in order to guarantee that another operation stays valid. To see the difference between those two relations, consider an account whose balance is not allowed to drop beyond zero. Here we have to know all **credit()**–operations to make sure that we can apply another **credit()**. Hence **credit()**–operations are related by the serial–dependency relation. On the other hand, we can roll back arbitrary many credit() operations without compromising the validity of a **credit()**. Hence **credit()**–operations are not related by the forward–dependency–relation.

We can employ this fine–grained analysis to perform a flexible conflict classification. Take an appointment calendar as an example and consider the case that the same appointment is deleted independently on two replicas. Note, that on the application level the conflict is harmless and can be ignored, as both users who issued the deletions get what they wanted when a single deletion is reflected in the reintegrated object. Conventional conflict detection algorithms that are unaware of the semantics would detect an update conflict and invoke some kind of repair action. Our replication–support system detects that the conflict was caused by two deletions of the same appointment and infers from the conflict classification that no specific repair action is required.

4 REALIZATION

The discussed replication-support approach has been realized in the system **MISTRAL**. MISTRAL replicates C++–objects. The standard implementation of a C++–class has to be complemented by a specification of the internal structure to enable MISTRAL to create new objects and specifications of the serial–dependency–relation and the forward–dependency–relation. A preprocessor generates a wrapper along the lines of DC++ (Schill and Mock, 1993) to trap and log invocations. Complementing a conventional implementation of a stock–item object comprises 8 lines of description. 7 lines are necessary to build the wrapper for logging invocations and creating replicas. Only one line describes the semantics of the object. Based on this description, MISTRAL allows to dynamically create replicas on new mobile devices, to join replicas, and to assign flexible rights — e.g. to allow several replicas to increase the quantity on stock. The implementation uses XDR (RFC 1014, 1987) to code all exchanged information and currently rests on TCP/IP as a lean communication platform.

The application development process in MISTRAL is depicted in figure 3. It shows that the inputs that have to provided by the programmer consist of the implementation of the C++–class (UserClass), the specification of the used parameters in XDR (UserXDR.x), and the discussed descriptions for MISTRAL (User.def).

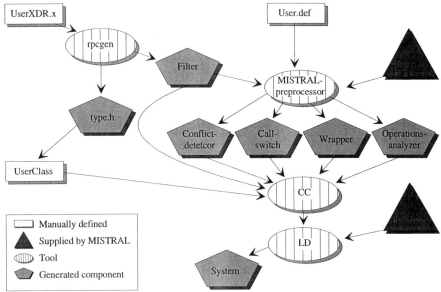

Figure 3 Application development in MISTRAL.

MISTRAL has successfully replicated objects among DECstations, DEC–AXPs, and LINUX–PCs. Figure 4 gives the performance of reintegration. The results have been obtained for the stock-item replicated on two AXP 3000/500 connected via a lightly loaded Ethernet.

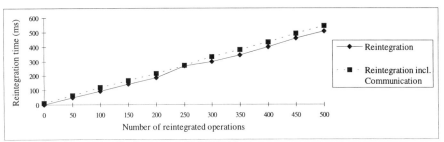

Figure 4 Reintegration Performance of MISTRAL.

5 CONCLUSIONS

In this paper we presented an approach to provide generic replication–support for advanced mobile applications. The approach exploits the specific semantics of objects, the approximate knowledge of mobile users about what they are up to, and offers four different reaction alternatives for the case that the preallocated guarantees are insufficient. Those alternatives are seamlessly integrated into the concept of mixed–mode–replication. Those features ease the task of developing replication strategies that exploit the specifics of objects, they result in a system that can make best use of the cost structure of the currently available communication link, and they provide the mobile user with the flexibility to trade between consistency and cost while he is on the move.

REFERENCES

Bagrodia R., Chu W. W., Kleinrock L., Popek G. (1995): Vision, Issues, and Architecture for Nomadic Computing}, *IEEE Personal Communications*, 2(6), Dec. 1995, 14–27

Davies N. (1996): The Impact of Mobility on Distributed Platforms, *in International Conference on Distributed Platforms*, Dresden, Feb. 1996, 18–25

Demers A., Petersen K., Spreitzer M., Terry D., Theimer M., Welche B. (1994): The Bayou Architecture: Support for Data Sharing among Mobile Users, *Workshop on Mobile Computing Systems and Applications,* Santa Cruz, Dec. 1994, 2–7

Herlihy M. P. (1990): Concurrency and Availability as Dual Properties of Replicated Atomic Data, *Journal of the ACM*, 37(2), April 1990, 257–278

Honeyman P., Huston L.B. (1995): Communications and Consistency in Mobile File Systems, *IEEE Personal Communications*, 2(6), Dec. 1995, 44–48

Kaashoek M. F., Pinckney T., Tauber J. A.: Dynamic Documents: Mobile Wireless Access to WWW, *Workshop on Mobile Computing Systems and Applications,* Santa Cruz, Dec. 1994, 179–184

Kistler J. J., Satyanarayanan M. (1992): Disconnected Operation in the Coda File System, *ACM Transactions on Computer Systems*, 10(1), Feb. 1992, 3–25

Kumar P., Satyanarayanan M. (1993): Supporting Application–Specific Resolution in an Optimistically Replicated File System, in *Fourth Workshop on Workstation Operating Systems*, Napa, CA, Oct. 1993, 66–70

RFC–1014 (1987), Sun Microsystems, XDR: External Data Representation, June 1987

Satyanarayanan M. (1996): Mobile Information Access, *IEEE Personal Communications*, 3(1), Feb. 1996, 26–33

Satyanarayanan M., Kistler J.J., Mummert L. B., Ebling M. R., Kumar P., Lu Q. (1993): Experiences with Disconnected Operations in a Mobile Computing Environment, in *USENIX Symposium on Mobile and Location–Independent Computing*, Cambridge, Aug. 1993, 11–28

Schill A.B., Mock M. U. (1993): DC++: Distributed Object–Oriented System Support on Top of OSF DCE, *Distributed Systems Engineering Journal*, 1(2), 1993

BIOGRAPHY

Dietmar Kottmann graduated from Karlsruhe in 1992. Since then he has been a research assistant at the institute of telematics at the computer science department of the university of Karlsruhe. He is working in the interdisciplinary research group SFB 346 and is currently pursuing his Ph.D. His research interests include mobile computing, mechanisms for distributed platforms and distributed object–oriented databases.

5
Caching Data Over a Broadcast Channel

H. V. Leong   A. Si   B. Y. L. Chan
Department of Computing,
The Hong Kong Polytechnic University,
Hung Hom, Hong Kong.
Telephone: (852)-2766-7243.
Fax: (852)-2774-0842.
Email: {cshleong,csasi,csylchan}@comp.polyu.edu.hk

Abstract
We consider an environment in which a collection of mobile clients interacts with a stationary database server. Due to the low bandwidth of wireless communication channels, it is necessary to broadcast highly popular data items and deliver other data items on a demand basis to the mobile clients. This broadcast wireless media can be considered as an extra layer of cache storage, which we term air-storage. We investigate several mechanisms in selecting the data items to be placed over this new layer of air-storage, and illustrate their effectiveness by means of simulation.

Keywords
Mobile database, data broadcasting, data caching

1 INTRODUCTION

In a mobile computing environment, users carrying portable computers, referred to as mobile clients, will be able to access information via wireless channels to a stationary database server (Imielinski and Badrinath, 1994). In such a mobile environment, the broadcast paradigm is effective for disseminating database items from the server to multiple clients since the downstream communication bandwidth from the server to the clients is usually much larger than the upstream communication bandwidth from the clients back to the server (Acharya *et al.*, 1995). Each client then, picks its own items of interest selectively from the broadcast database (Imielinski *et al.*, 1994).

In general, the performance of a query over the broadcast database will largely depend on the length of a broadcast cycle (Imielinski *et al.*, 1994; Leong and Si, 1995). which indicates the amount of time required to broadcast one copy of all database items. Figure 1 illustrates this idea on a database with five relations.

In Figure 1, the average response time for 100 selection queries against a broadcast

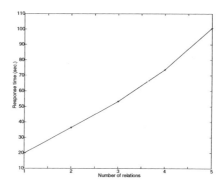

Figure 1 Response time *versus* number of relations.

database over a wireless channel at a bandwidth of 19.2 Kbps is measured. A broadcast cycle is composed of 1 to 5 relations. Each relation contains 640 tuples with each tuple occupying 128 bytes. The access probabilities of the five relations are 0.5, 0.2, 0.15, 0.1, and 0.05 respectively. As shown in Figure 1, the response time of a query increases if all database items are broadcast uniformly regardless of their access frequencies.

To optimize database queries against broadcast database, the server should only broadcast database items that would be frequently accessed by a lot of mobile clients. These database items are usually called "hot items". "Cold items" which are accessed much less frequently should be delivered on an "as needed" basis. In this paper, we propose schemes to identify the database items suitable to be broadcast over the wireless channel. We conduct simulated experiments to study the effect of various proposed schemes on the availability of a requested database item over the broadcast channel. This paper is organized as follows. In Section 2, we offer a brief description on our model and propose several schemes in identifying those hot candidates to be broadcast. Section 3 describes our experiments and illustrates some preliminary results, along with explanations to the results. We conclude this paper with a brief discussion on our current research directions.

2 ENERGY EFFICIENT CACHING SCHEMES

When a database is periodically broadcast over a wireless channel, the channel in effect acts as a storage layer which we term **air-storage**. The database server can thus be regarded as a tertiary storage and the broadcast database items can be considered to be cached into this air-storage. The size of this air-storage is characterized by the length of a broadcast cycle as it indicates the amount of database items that could be accommodated in the air-storage. This forms a hierarchy of storage systems with increasing bandwidths: database items frequently requested by most clients are cached over the air-storage with a low bandwidth; database items frequently requested by a particular mobile client could be cached into its local disk storage with a much higher bandwidth (Barbara and Imielinski, 1994); and finally, currently queried database items will be cached into a client's main memory buffer with the highest bandwidth.

By treating the broadcast channel as an air-storage, its management becomes similar in spirit to the management of cache memory. In particular, strategies to determine which database items should be broadcast over the channel or refrain from being broadcast will be similar to the cache replacement strategies. However, due to the different natures of the air-storage and conventional cache memory, four issues need to be readdressed.

First, the size of the air-storage, i.e., the length of a broadcast cycle, could be dynamically changed. This is in contrast to conventional cache memory whose size is static. Obviously, increasing the size of the air-storage might penalize the performance of certain data retrievals, but it might benefit the overall performance for the whole collection of mobile clients globally. Critical here is the mechanism to determine the size of the air-storage for each broadcast while striking a balance between local and global optimizations. Owing to space limitation, we are not able to discuss this issue further here.

Second, database items stored in conventional cache memory are accessed in a random manner; by contrast, database items stored over the air-storage could only be accessed sequentially. Techniques tailored to the sequential nature of the air-storage for retrieving the useful items are in demand and have been addressed to some extent in Acharya et al. (1995), Imielinski et al. (1994), and Leong and Si (1995).

Third, conventional caching mechanisms are usually performed on a per-page basis due to the principle of locality that the page containing the database items currently being accessed will have a high probability of being accessed in the near future. Since the bandwidth of the air-storage is much lower than that of conventional cache memory, caching the whole page of data from the server into the air-storage will be too expensive. A smaller granularity for caching is needed. In our research, we are experimenting in using an entity as a caching unit for the air-storage, i.e., the database is broadcast on an entity basis. Each entity will roughly correspond to an object in an object-oriented database and will correspond to a tuple in a relational database.

Finally, with the size of the air-storage being fixed for a particular broadcast, we need to identify the hot items (entities) suitable to be cached into the air-storage, i.e., being broadcast. During subsequent broadcast cycles, additional entities might be qualified as hot. Since hot items are those accessed by many clients, the access pattern of each client on the entities must be propagated from the clients where the accesses are originated, back to the server where the statistics will be compiled and caching strategy is being implemented. If the air-storage is exhausted, replacement strategy is needed to replace aged cold database items being broadcast currently with new hot items. While caching is performed on an entity basis, the spatial locality information among the different cached items in database applications is lost. This is amplified by the time lag until the server is aware of the access patterns on database items issued by the clients. This in turn implies that conventional replacement strategies that perform good (such as least recently used or LRU) (Korth and Silberschatz, 1991) are no longer suitable in the new context since they are based on the principle of locality. New cache replacement strategies need to be introduced.

Without the locality information among the cached entities, it is natural to adopt the access probabilities of entities as an indicator for the necessity of being cached into or being replaced from the air-storage. The access probabilities could be collected at each mobile client and forwarded to the server at regular intervals. They may also be piggybacked in a request to the server, when there is an air-storage miss, i.e., the requested database entity is not being broadcast in the current broadcast cycle. We have experimented with several ways of estimating the access probabilities and their effectiveness in cache replacement.

The simplest way to keep track of the access probability for each entity is by measuring the cumulative access frequency. We call this the **mean** scheme. If the air-storage is not exhausted, an entity is cached if its cumulative access frequency is higher than a certain threshold τ. However, if the air-storage is exhausted, an entity is cached only if its accumulated access frequency is higher than the lowest access frequency of a cached entity. In other words, with an air-storage of size c, only those c items with the highest access frequencies are broadcast. Alternatively, a window for the access frequencies spanning several broadcast cycles could be maintained for each entity. The cached entity with the lowest average access frequency within the window is replaced. We call this scheme the **window** scheme. The effectiveness of the window scheme depends on the window size W. A problem for the window scheme is the amount of storage needed in maintaining the windows. To avoid the use of a moving window and to adapt quickly to changes in access patterns, our third scheme measures the exponentially weighted moving average of access frequencies. Access frequencies of current broadcast cycle have higher weights and the weights tail off as they become aged. This is called the **exponentially weighted moving average (EWMA)** scheme. A parameter to EWMA is the weight α.

The changing metrics in the schemes above can be computed incrementally. Assume that a new access frequency $f_{x,n+1}$ from time interval n to $n+1$ is collected for object x and the metric for the previous n access frequencies is $\overline{f}_{x,n}$. Here a typical time interval for the access frequencies is the broadcast cycle. At the end of a broadcast cycle, the server accumulates the required information used to compute the metrics at the beginning of a new broadcast cycle which in turn acts as a guideline in selecting the database items to be broadcast. In the mean scheme, the new metric $\overline{f}_{x,n+1}$ can be computed as $(n\overline{f}_{x,n} + f_{x,n+1})/(n+1)$. For the window scheme with a window size W_x for object x, when the window is not full (i.e., $n < W_x$), computing the new metric when adding a new access frequency is the same as that for the mean scheme. Otherwise, the new metric $\overline{f}_{x,n+1}^{(W_x)}$ is computed as $(W_x \overline{f}_{x,n}^{(W_x)} + f_{x,n+1} - f_{x,n-W_x+1})/W_x$. A total of W_x intermediate values need to be stored for object x. Finally, in EWMA with weight α_x, the weighted moving average $\overline{f}_{x,n+1}^{(\alpha_x)}$ is computed as $\alpha_x \overline{f}_{x,n}^{(\alpha_x)} + (1 - \alpha_x) f_{x,n+1}$. No intermediate value needs to be maintained.

It can be observed that the window scheme reduces to the mean scheme if an infinite window size W is used. When $\alpha = 0$, the EWMA scheme becomes identical to the window scheme with $W = 1$; if α tends towards 1, older values have larger weights and the scheme becomes similar to the mean scheme.

3 SIMULATION RESULTS

To illustrate the relative effectiveness of the three cache replacement schemes, namely, mean, window, and EWMA, we have conducted a simulation study. In this simulation, we assume that there are 1000 entities in the database server and the air-storage has a capacity to broadcast 300 entities. The size of each entity is set to 128 bytes. The experiments are also repeated with an entity size of 16 bytes. Two different access patterns on data entities are experimented with. The first experiment assumes that the "temperature"

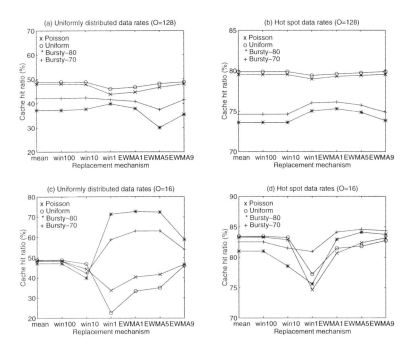

Figure 2 Effectiveness of cache replacement strategies.

of data entities are uniformly distributed, in that the access probabilities of the entities follow a uniform distribution. The second experiment is more realistic in reflecting the data affinity of data accesses. About 20% of the data entities have a very high access rate, constituting 80% of the operations; the remaining 80% of the entities have a much lower 20% access rate. This models a situation in which the access patterns exhibit some locality or skewed behavior (Franklin et al., 1992). We investigate different arrival patterns of the operations for accessing the data entities including the standard Poisson arrival following the exponential distribution, the simple uniformly distributed arrival events, and the bursty arrival showing some form of temporal locality. We experiment with two bursty arrival patterns: Bursty-80 and Bursty-70, meaning that 80% (and respectively 70%) of the operations arrive uniformly in 20% (and respectively 30%) of the time span and the remaining 20% (and respectively 30%) arrive uniformly in another 80% (and respectively 70%) of time span. A total of 100 000 read accesses are generated in each experiment. Each experiment is repeated 30 times, and the average over the 30 experiments is taken to be a data point.

As an indicator for the effectiveness of a replacement strategy, we measure the cache (air-storage) hit ratio which is the number of hits by the client on the air-storage divided by 100 000 operations, i.e., the number of times that a requested entity could be found on the broadcast channel. The average cache hit ratios of mean scheme, window schemes with $W = 100$ (win100), $W = 10$ (win10), $W = 1$ (win1), and EWMA schemes with

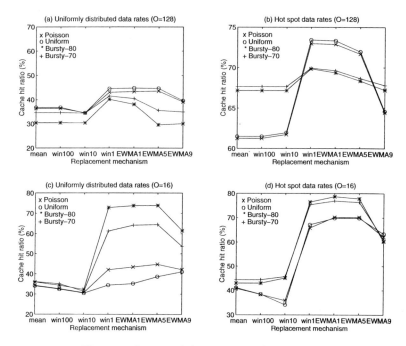

Figure 3 Impact of changes in access patterns.

$\alpha = 0.1$ (**EWMA1**), $\alpha = 0.5$ (**EWMA5**), and $\alpha = 0.9$ (**EWMA9**) over 30 repeated runs are measured. The cache hit ratios of the various replacement schemes for the uniform and skewed access patterns when each entity has a size of 128 bytes (**O=128**) are depicted in Figures 2a and 2b respectively. The corresponding hit ratios when the entity size is 16 bytes (**O=16**) are shown in Figures 2c and 2d.

It is obvious in Figure 2 that the performances of all schemes are better when there are hot spots in the database items. All the proposed schemes are capable of using access information to identify the hot entities and to broadcast them over the air-storage. Furthermore, we can observe that the performances of the window-based schemes are not stable and that the mean scheme performs relatively well. EWMA also seems to perform good, especially when the entity size becomes smaller (16 bytes). A smaller object size results in a shorter broadcast cycle such that changes in access patterns can be incorporated into the scheme in a more timely fashion and be reflected in the next broadcast cycle. When compared with using EWMA for local caching which exhibits a stable performance (Leong and Si, 1996), EWMA in the management of air-storage does show some fluctuations in performance. This is mainly due to the lag-behind of the changes on broadcast database items to cope with the changes on access patterns.

EWMA schemes usually perform better than the mean scheme in reality, when we consider that access patterns may change over time. To demonstrate this feature, we conducted another set of experiments. This time, however, the data access rates to each database entity are randomly permuted after every five broadcast cycles; this has an effect of moving the hot spot randomly every few cycles. The result is depicted in Figure 3, in one-to-one correspondence with those in Figure 2. We observe that all hit ratios drop since the server now is not able to cope with the changes in access patterns very well. The mean scheme, in particular, suffers seriously from this change and performs very poorly, since the changes in patterns cannot be reflected rapidly in a change of the cumulative access frequencies. The window scheme with large window sizes also suffers from the same phenomenon. The EWMA scheme with a small or moderate value of α reacts very well to this change. This is due to the much stronger adaptive nature of EWMA than the mean scheme. One might argue that window scheme with $W = 1$ (**win1**) has the best performance and should always be used. It is not surprising that **win1** performs best in this case since it is able to completely forget about all past history, except the information of the previous cycle. Any change in the access patterns will instantaneously be reflected in the next broadcast cycle. However, its performance is highly unstable in most cases, by forgetting virtually everything. EWMA therefore appears to be a valuable compromise, with a good and yet relatively stable performance in all cases. Other statistical operators may also be used, to exploit the temporal and spatial information about the database entities accessed.

4 DISCUSSION

In this paper, we show that the problem of determining if a database item needs to be broadcast is similar to the cache management problem and propose a solution based on cache replacement strategy to the problem. Several new replacement strategies have been proposed and preliminary experimental results have also demonstrated that our proposed strategies are promising. We are currently conducting experiments to measure the effectiveness of the different replacement strategies under different conditions such as in an environment containing a group of mobile clients, each showing a different access pattern.

Another issue that we are currently working on is to characterize the factors that determine the length of a broadcast cycle, i.e., the size of the air-storage. This issue is very important as that affects how many database items could be broadcast over a cycle and hence the response time. We believe that in a typical mobile environment, it is best to employ a hybrid broadcast and on-demand communication paradigms so that the hot items will be broadcast while the cold items will be disseminated on-demand. The size of the air-storage will thus depend on the relative ratio of the response time between on-demand and broadcast queries. For instance, if the average response time for an on-demand query is t, the response time for queries against broadcast database should not exceed t or clients might as well access the data on demand. This will in turn create additional traffics over the on-demand channel and drive up the average response time t.

ACKNOWLEDGEMENT

This research is supported in part by the Hong Kong Polytechnic University Central Grant numbers 353/029 and 350/570.

REFERENCES

S. Acharya, R. Alonso, M. Franklin, and S. Zdonik (1995). Broadcast Disks: Data Management for Asymmetric Communication Environments. In *Proceedings of the ACM SIGMOD International Conference on Management of Data*, pages 199–210.

D. Barbara and T. Imielinski (1994). Sleepers and Workaholics: Caching Strategies in Mobile Environments. In *Proceedings of the ACM SIGMOD International Conference on Management of Data*, pages 1–12.

M. Franklin, M. Carey, and M. Livny (1992). Global Memory Management in Client-Server DBMS Architectures. In *Proceedings of International Conference on Very Large Databases*, pages 596–609.

T. Imielinski and B. Badrinath (1994). Mobile Wireless Computing: Challenges in Data Management. *Communications of the ACM*, 37(10):18–28.

T. Imielinski, S. Viswanathan, and B. Badrinath (1994). Power Efficient Filtering of Data on Air. In *Proceedings of EDBT*, pages 245–58.

H. F. Korth and A. Silberschatz (1991). *Database System Concepts*. McGraw-Hill.

H. V. Leong and A. Si (1995). Data Broadcasting Strategies over Multiple Unreliable Wireless Channels. In *Proceedings of ACM International Conference on Information and Knowledge Management*, pages 96–104.

H. V. Leong and A. Si (1996). Semantic Caching in a Mobile Environment: Model and Evaluation. Submitted for publication.

BIOGRAPHICAL NOTES

H. V. Leong was graduated from the University of California at Santa Barbara, and is currently an assistant professor at the Hong Kong Polytechnic University. He received his undergraduate and first graduate degree from the Chinese University of Hong Kong in Computer Science. Since then, he had been working as teaching and research assistants throughout his graduate career and held several scholarships and fellowships, and has published over a dozen of research papers. His research interests are in distributed systems, distributed databases and mobile computing.

A. Si received his PhD. degree in Computer Science from University of Southern California. He is currently an assistant professor of the Department of Computing at the Hong Kong Polytechnic University. He has served as external reviewer for a number of international conferences and journals, and as program committee member for a number of international conferences. His research interests include heterogeneous and federated databases, mobile computing, distributed and parallel databases. He is a member of ACM, IEEE Computer Society, and Usenix Association.

B. Y. L. Chan is a graduate student at the Hong Kong Polytechnic University. He received his undergraduate degree from the University, before committing himself to the research degree. His current research focuses on distributed systems and mobile computing.

PART THREE

Basic Architectures for Mobile Information Systems

6
Mobile Frames: A pragmatic approach to automatic application partitioning based on an end-user data model

Thomas Kirste
Interactive Graphics Systems Group
Darmstadt Technical University, Department of Computer Science,
Wilhelminenstr. 7, D-64283 Darmstadt, Email: kirste@igd.fhg.de

Abstract
One important optimization technique for coping with resource constrained mobile hosts is *application partitioning*, the dynamic, resource-dependent distribution of application functionality across the available network nodes. This paper proposes a very pragmatic approach to this concept based on a specific *end-user data model*, the *mobile frame model*, which is an extension of the conventional frame model.

Keywords
Mobile computing, application partitioning, distributed systems, data models, frame models

1 INTRODUCTION

The introduction of mobile components (*e.g.*, mobile computers, wireless networks, mobile users) into a distributed system infrastructure creates an application execution environment that is highly dynamic with respect to quantity and quality of the available resources. Applications that want to provide optimal performance in such an environment have to be *resource aware*. This includes the ability to *adapt* to the current resource configuration.

A substantial amount of research in the field of mobile computing is devoted to making applications adaptive and resource aware. This includes the definition and tracking of acceptable resource states as well as the introduction of suitable reconciliation strategies (such as substituting a monochrome still image for a high-definition color video stream) – see for instance (Davies et al., 1994; Satyanarayanan et al., 1995).

One important adaption technique is *application partitioning*, the dynamic, resource dependent distribution of application functionality across the available network nodes. A fundamental prerequisite for a successful generalized use of this technique is the definition of a strategy for computing a suitable partitioning scheme *dynamically* at application startup, so that the currently available execution environment is optimally utilized. Currently, there is no suitable concept available.

The point of this paper is to propose a very pragmatic solution to this problem based on a specific data model, the *mobile frame model*, which is an extension of the conventional frame model. The interesting fact is that support for application partitioning results as *inherent consequence* from the introduction of this end-user data model.

The contributions of the work presented in this paper to the field of mobile computing are:

- To realize the general importance of an expressive data model for mobile computing.
- The proposal of a suitable model, integrating both end-user data management and definition of interactive applications.
- The observation that fully dynamic, automatic application partitioning can be provided by a straightforward extension of this model.

(It is worth noting that the claim of this paper is not "mobile frames are *the* solution to application partitioning", but rather "the data model needed anyway is *also* able to solve the partitioning problem for a small additional price".)

The paper is further structured as follows: Section 2 gives a detailed discussion of the "application partitioning" technique. Section 3 discusses the need for an expressive data model and proposes the frame model as a suitable alternative. Section 4 describes, how application partitioning can be achieved by a simple extension of the frame model, the *mobile* frame model. Conclusions are given in Section 5.

2 APPLICATION PARTITIONING

The basic idea behind "application partitioning" is to divide an application into individual component processes, which have specific requirements in terms of computation and communication resources. Upon application startup, the component processes are dynamically allocated to the network nodes based on available resources (*startup adaptivity*). As the resources change during application execution (*e.g.*, drop of bandwidth of a wireless communication link), the application components may dynamically be reallocated to other nodes (*runtime adaptivity*).

The basic concept is described, for instance, in (Watson, 1994), in (Joseph et al., 1995) (using the model of *Relocatable Dynamic Objects*), in (Cardelli, 1994) (within the definition of the language *Obliq*), and in (Kirste, 1995) (based on a combination of *Object Fragmentation* and *Pro-*

cess Migration). However, the important problem of *how* to partition an application is unsolved (Joseph et al., 1995; Kirste, 1995). In the current proposals, application partitioning schemes are individually designed on a per-application base by a programmer; the partitioning scheme is fixed at *implementation time*. The partitioning is usually even tailored towards a specific resource configuration, so that application partitioning today is very similar to conventional client/server programming – with the additional feature of client migration at startup (as provided, *e.g.*, by Java (SUN Microsystems, 1995) through the down-loading of applets).

But in order to support startup and runtime adaptivity, the partitioning scheme must not be fixed at implementation time. It must be dynamically computed from the application definition based on the currently available execution environment.

In the next section, a seemingly quite unrelated problem is discussed, the question of a data model for a specific class of mobile applications. The interesting point is that this data model provides a solution to the problem outlined here!

3 DATA MODELS FOR MOBILE INFORMATION SYSTEMS

One of the visions of mobile computing is the creation of an integrated information environment, providing unrestricted access to both public and private data – anytime, anywhere. (A discussion of such a scenario and a prototype system is given in (Kirste, 1995).)

Besides location transparency, this environment should provide a *service transparent* data access. This means, the operations available for a data entity are determined by the entity type, and not by the service used for accessing this entity. (Today, access to, *e.g.*, a spreadsheet, is not service transparent: access through the user's file system using its native application allows much more operations than when accessing it, say, as HTML-page through the World-Wide Web.) This requires an *expressive* data model, in which arbitrary data types and operations can be represented, in order to make them available for easy access by the user in a distributed environment.

Also, because the integrated information environment has to be open towards arbitrary future ehancements and extensions of its data set with repect to properties, types, and functionality, it needs a data model supporting *incremental type and data definition*.

One data model integrating incremental data definition, ease-of-use, expressiveness, and the ability for representing arbitrary interactive applications, is the "frame" model.

Originally, frames have been developed as a means for knowledge representation in artificial intelligence applications (Minsky, 1975). Powerful frame representation languages have been developed as early as 1977 (*e.g.*, FRL (Roberts and Goldstein, 1977)). From this application area, frames have inherited the ability to cope with structured, fast changing information – as it also exists in the above scenario. Incidentally, other application areas for frame models, besides knowledge representation, have been Hypermedia-based information management systems (*e.g.*, (Shibata and Katsumoto, 1993) and Oval (Malone et al., 1995)) and personal information management systems, such as Apple's MessagePad (Smith, 1994).

Therefore, it seems appropriate to assume a frame-based model as underlying data model for an integrated information environment, rather than starting with the less expressive data model provided by the current incarnation of this scenario, the World-Wide Web. (See also (Kirste, 1996a) for a more detailed discussion of this point.)

4 THE MOBILE FRAME MODEL

4.1 Basic aspects of frames

The specific frame model underlying this work is basically the one provided by the MessagePad's object system (Smith, 1994), which in some aspects is a simplified version of the object system built into the language *Self* (Ungar and Randall, 1987).

As far as this paper is concerned, a frame is – quite similar to an object – an entity with a unique identity that contains a set of name/value pairs ("slots"). Among conventional data types, frame references may be slot values, so that frame structures can be created. Also, slot values can be *functions*, which may be invoked by message passing. Finally, frames can use other frames as *prototypes*, from which they inherit.

As an example, consider the frames p and q, defined as follows[*]:

```
p := {a: 1, f: func() a+b};
q := {_proto: p, b: 2}
```

p is a frame with two slots, a, which contains an integer, and f, which contains a function. q is a frame with two slots, _proto, containing a frame reference (q's prototype), and b, containing an integer. So q inherits from p. As one would expect, q.b gives the value 2, q.a the value 1. The message invocation q:f() gives 3. Here, the values of f (and a, which is accessed in f) are inherited from p. The assignment q.c := 3 creates a new slot c with the value 3 in q (so, q.c now gives 3). The assignment q.a := 4 overrides the slot a inherited from p, so now q.a = 4, and q:f() = 6, but still p.a = 1.

It is also possible to define a frame

---

[*]The frame language syntax used throughout this document is based on NewtonScript (McKeehan and Rhodes, 1994), using the following conventions: The notation "$\{s_i : v_i, ...\}$" denotes a frame with slots s_i, which have values v_i. "$f.s$" denotes the access of f's slot s and "$f.s := v$" is the assignment of the value v to f's slot s. "func$(x_i, ...)$ *body*" denotes function with parameters x_i. "$f:m(a_i, ...)$" denotes the invocation of the function stored in $f.m$ with parameters a_i (i.e., a method invocation).

```
r := {f: func() a*b)}
```

and change q's prototype by assigning `q._proto := r`. So now $q:f() = 8$, because this time f is inherited from r.

The central differences to the conventional class/instance model of object-oriented programming (OOP) are the dynamic addition/removal of slots to/from frames, which enables incremental data definition and data reuse/data sharing, and the prototype-based dynamic inheritance mechanism that allows a frame to inherit slots from an arbitrary other frame.

Frames therefore provide a dynamic environment that allows the user to flexibly create, augment, and modify information structures to suit his personal needs. It is now very interesting that frame structures can be used for the definition of interactive applications. The most prominent example for this is probably NewtonScript. This means, frames integrate a powerful end-user data model *and* a means for defining non-trivial interactive applications. As a consequence, the *same* system services (*e.g.*, frame migration and replication, frame caching) can be used for managing the access of data and applications.

In the next section, it will be outlined how the basic frame-model can be extended to support automatic application partitioning.

4.2 Frame-based application partitioning

The fundamental idea for supporting application partitioning is to group the frames defining an application into *clusters*. These clusters may then be allocated to the available network nodes (each cluster representing an application partition). Initial cluster setup (startup adaptivity) and dynamic reclustering (runtime adaptivity) are supported by the fundamental frame migration / replication facilities needed for basic remote data access. Clustering itself is guided by estimations of a frame's requirements in terms of computation, communication, and storage resources.

A (very simple) example for frame-based application partitioning is given in Figure 1. The "application" – computation of a factorial – consists of a "front-end" frame (doing some pre-processing) and two "back-end" frames, doing the "number crunching". The application can be called, *e.g.*, using the expression `a:f(10)`, producing `3628800`.

4.3 The mobile frame model

In order to support the above application partitioning facility including runtime adaptivity, frames need to be *mobile* (ie., migratable). Also, frames must support the parallel execution of methods to allow for multi-user interaction. For achieving these goals, the following strategy has been used:

```
{a: {f: func(x)   b:g(abs(rnd(x)))},
 b: {g: func(y)   if y <= 1
                  then 1
                  else c:h(y)},
 c: {h: func(z)   z*b:g(z-1)}}
```

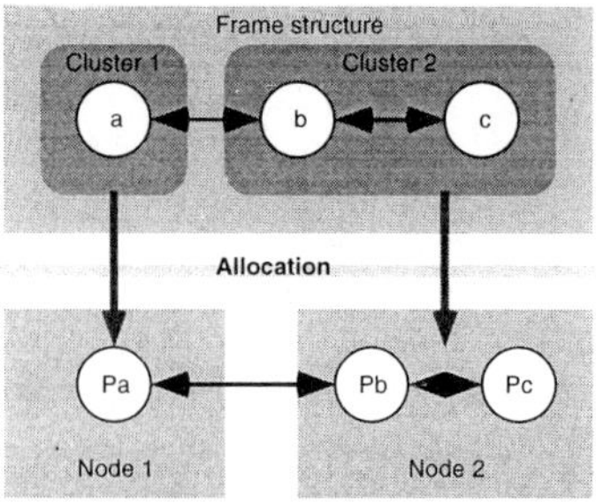

Figure 1 Example frame structure (left) and partitioning(right)

- Each frame is mapped to an individual process whose state also contains the frame's slots. Frame processes solely communicate by message passing (i.e., no hidden shared memory is introduced).
- Frame processes are designed to support recursive remote method calls (maintaining original frame execution semantics).
- Frame processes also support multi-threading – *e.g.*, lookup of slot values during method execution (enabling basic parallel execution). Specifically, the modell of *thread diffusion* is supported, providing complete location transparency for processing threads.
- The frame state completely describes the frame's process, so that copying the frame state implements migration of the frame process (which is required for runtime adaptivity).

The resulting model is called the *mobile frame model* (MFM). It provides the basic facilities for automatic application partitioning with startup- and runtime adaptivity in a multi-user environment. It is also quite easy to add basic support for "itinerant agents" by allowing frame processes to *actively* move between nodes.

The fundamental messages defined by the MFM for communcation between frame processes are summarized in Table 1.

Furthermore, the current definition of the MFM provides additional important aspects, such as:

- The choice between static inheritance ("cloning") and dynamic inheritance.
- The choice between remote message execution in calling frame or in called frame.
- A frame definition language (FDL) that allows to denote frames and frame structures. (The concrete syntax of the FDL adheres to the notation used in this paper.) The FDL language semantics describes the mapping of FDL terms to communicating frame processes.

Message	Meaning
`SetSlot(s,v)`	Set receiver frame's slot s to value v
`Lookup(s,o)`	Lookup of slot s in receiver frame. If s is found, its value is returned to the originator frame o. Otherwise, the receiver frame forwards the message to its prototype
`Call(m,v)`	Remote invocation of method m with parameter p in receiver frame
`CreateFrame(state)`	Requests creation of a new frame with initial state state.
`Go(p)`	Tells the receiver frame to move to place p. (The receiver frame then requests migration by sending the message `Migrate(p,state)` to the underlying process manager)

Table 1 Fundamental messages of the MFM

- A recursive "let" construct within the FDL that supports the simple definition of mutually recursive frame structures.

The first two points provide a range of tradeoffs with respect to data dynamics and message trafffic. See (Kirste, 1996b) for detailed discussion of the MFM.

5 CONCLUSION

In this paper, a pragmatic approach to automatic dynamic application partitioning has been outlined. It is based on the introduction of an end-user data model supporting the creation of frame-based structured application definitions, which are used to guide partitioning. Furthermore, the basic frame model has been enhanced to support parallel execution and migration of active frames.

The mobile frame model is currently in the state of a theoretical concept with a precise mathematical specification (including the frame definition language). The fundamental viability of mobile frames has been verified by an abstract implementation of the model and its language using the functional language Haskell (Hudak et al., 1992). Here, frame processes are represented by Haskell functions of type `PROC = [Rep] -> [Req]`. I.e., a process is a function mapping a sequence of replies to sequence of requests. (See (Kirste, 1996b) for the details.)

The complete model also provides extended support for the flexible visualization of arbitrary frame structures, based on the concepts of *Facets*, *Display Methods*, and *Viewers*, see (Kirste, 1996a).

A major drawback of the work presented in this paper is the lack of actual experiments with the behavior of frame-based application partitioning. These experiments are required for answering, *e.g.*, the following questions:

- Which strategies should be used for optimizing an implementation of the mobile frame model (*e.g.*, caching of inherited slots).
- How to design frame structures such that frame borders coincide with low traffic interfaces.
- How to describe a frame's resource requirements to guide clustering.

Future research will address these questions.

ACKNOWLEDGMENTS

This work has been supported by the German Science Foundation (DFG) grant En 123/20-1 as part of the researcher group "MoVi" (Mobile Visualization).

REFERENCES

Cardelli, L. (1994). Obliq – A language with distributed scope. White Paper, Digital, Systems Research Center.

Davies, N., Blair, G., Cheverst, K., and Friday, A. (1994). Supporting adaptive services in a heterogeneous mobile environment. In (MCSA'94, 1994).

Hudak, P., Peyton-Jones, S., Wadler, P., Boutel, B., Fairbairn, J., Fasel, J., Guzman, M., Hammond, K., Hughes, J., Johnsson, T., Kieburtz, D., Nikhil, R., Partain, W., and Peterson, J. (1992). *Report on the functional programming language Haskell, Version 1.2.* SIGPLAN Notices 27.

Joseph, A., de Lespinasse, A., Tauber, J., Gifford, D., and Kaashoek, M. (1995). Rover: A Toolkit for Mobile Information Access. In *Proc. Fifteenth Symposium on Operating System Principles*.

Kirste, T. (1995). An infrastructure for mobile information systems based on a fragmented object model. *Distributed Systems Engineering Journal*, 2(3):161–170.

Kirste, T. (1996a). A flexible presentation model for distributed information systems. In *Proc. 4th Eurographics Workshop on Multimedia (EGMM'96)*, Rostock, Germany.

Kirste, T. (1996b). The mobile frame model. Technical report, Darmstadt Technical University, Department of Computer Science, Interactive Graphics Systems Group, Germany. Current

version available at
`ftp://dakeeper.igd.fhg.de/outgoing/kirste/mfm.ps.gz`.

Malone, T., Lai, K.-Y., and Frey, C. (1995). Experiments with Oval: A radically tailorable tool for cooperative work. *ACM TOIS*, 13(2):177–205.

McKeehan, J. and Rhodes, N. (1994). *Programming for the Newton*. AP Professional.

MCSA'94 (1994). *Proc. Workshop on Mobile Computing Systems and Applications (MCSA'94)*, Santa Cruz, CA. IEEE Computer Society.

Minsky, M. (1975). A Framework for Representing Knowledge. In Winston, P., editor, *The Psychology of Computer Vision*. McGraw-Hill, New-York.

Roberts, R. and Goldstein, I. (1977). The FRL Primer. AI Memo No. 408, Artificial Intelligence Laboratory, MIT, Cambridge, Mass.

Satyanarayanan, M., Noble, B., Kumar, P., and Price, M. (1995). Application-aware adaption for mobile computing. *Operating Systems Review*, 29(1).

Shibata, Y. and Katsumoto, M. (1993). Dynamic Hypertext and Knowledga Agent Systems for Multimedia Information Networks. In *Proc. Hypertext'93 (November 14–18 1993, Seattle, Washington)*, pages 83–93. The Association for Computing Machinery.

Smith, W. (1994). The Newton Application Architecture. In *Proc. IEEE Computer Conference*, San Francisco.

SUN Microsystems (1995). *The Java Language Specification Release 1.0 Alpha 2*. Mountain View, CA.

Ungar, D. and Randall, B. (1987). Self: the power of simplicity. In *Proc. OOPSLA'87 Conference*, pages 227–241, Orlando, Florida. Published as *SIGPLAN Notices 22*, Dec. 1987.

Watson, T. (1994). Application design for wireless computing. In (MCSA'94, 1994).

6 BIOGRAPHY

Thomas Kirste has received his MSc and PhD in Computer Science from the Darmstadt Technical University. From 1989 to 1993, he worked as research assistant for the Computer Graphics Center (ZGDV), focussing on extensible Hypermedia systems.

From 1993 to 1995, he was a research assistant at the Darmstadt Technical University, working on the topic of mobile computing. Since 1995, he is responsible for the technical coordination of the MoVi project.

His research interests include mobile computing, mobile visualization, and hypermedia. His focus is the development of object models and language concepts. Also, he feels a strong inclination towards functional languages.

7
Generic Personal Communications Support for Open Service Environments

T. Eckardt[*], *T. Magedanz*[*], *C. Ulbricht*[**], *R. Popescu-Zeletin*[**]
[*]*Technical University of Berlin*, [**]*DeTeBerkom GmbH*
[**]*Voltastraße 5, 13355 Berlin, Germany, phone +49-30-4 67 01-0, fax +49-30-467455, [ulbricht]@deteberkom.de*

Abstract

The vision for future telecommunications is often described by the slogan „*information at any time, at any place, in any form*" driven by both the society's increasing demand for „universal connectivity" and the technological progress in the area of mobile computing and personal communications. In order to realize this vision a *Personal Communications Support System (PCSS)* has been developed within a project of the R&D-programme from the DeTeBerkom GmbH, a subsidiary of the Deutsche Telekom AG, by Technical University of Berlin. The PCSS is designed to support advanced personal communications capabilities, i.e. *personal mobility, service personalization, and advanced service interoperability*, in a uniform way to numerous communication applications in distributed multimedia inhouse environments. From a functional perspective, the PCSS provides enhanced Intelligent Network (IN) and Universal Personal Telecommunications (UPT) capabilities with respect to customer addressing (based on logical names instead of numbers) and advanced customer control capabilities. From a design perspective, the centralistic IN/UPT approach to the realization of service logic has been replaced by a highly distributable, object-oriented approach based on X.500/X.700/Telecommunication Management Network (TMN) concepts. This paper presents the basic aspects of the PCSS, including design criteria, system functionality, platform design and evolution issues.

Keywords

Customer Control, Format Conversion, IN, Media Conversion, PCS, PCSS, Service Personalization, Service Interworking, TMN, UPT, User Mobility

1 INTRODUCTION

The vision for future communications is *„information any time, any place, in any form"*, based on the idea of an open „electronic" market of services, where an unlimited spectrum of communication and information services will be offered at different qualities and costs by different service providers. This service spectrum ranges from simple communication services up to complex distributed applications. In particular, new intelligent information services, such as hotel and flight reservation services are common place.

The prerequisite for this vision is global connectivity, based on a fast developing web of interconnected communication networks, comprising both fixed and wireless networks (Eckardt 1995). In addition, the provision of a global service infrastructure, based on network-independent open service platforms is the other fundamental prerequisite, hiding the complexity of network diversity and allowing the fast and efficient creation, provision and management of future services.

This vision requires appropriate communication services and information systems, which are compatible with human communication processes. Users have to be supported in order to cope with the increasing complexity of global information availability and the increasing number of communication services which are based on different network technologies. Also the increasing reachability of people due to advanced mobile communication services requires adequate means for information filtering and communications control.

One important aspect of this vision is that the communication between people regardless of the communication service used should be based on *person-oriented identifiers*, such as the personal name or an individual personal number, rather than on service-specific terminal numbers, which becomes awkward in an environment of increasing communication service options. For example, there should be only one universal personal number for communicating with this user via phone, fax, paging, e-mail, video conferencing, etc. instead of having to use five quite different numbers and addresses.

Due to society's increasing demand for „universal connectivity" and technological progress, mobile and personal communications are becoming fundamental attributes of future telecommunication systems. The trends toward ***personal communications*** can be viewed in terms of three areas:

Mobility in fixed and wireless networks

Within the future communications environment users will have global access to an individual set of services, any time, any place and in any form. Hence the mobility of users will be enabled in the future environment by means of both terminal mobility and personal mobility. *Terminal mobility* and the corresponding wireless network interfaces and protocols (i.e. cordless, cellular and satellite) are fundamental for the provision of „real" global connectivity. It has to be stressed that besides voice the transfer of multimedia data becomes the ultimate target in wireless environments. This has severe impacts on the design of mobile multimedia terminals and the wireless networks capabilities. In addition, *personal mobility* will also enable global service access, allowing people to make use of any kind of terminal located at their whereabouts for obtaining access to their services. Since there exists a broad spectrum of terminals, e.g. POTS terminal versus multimedia workstation, services have to be designed to be accessed by terminals and networks with different capabilities. For example, users could generate by speech an e-mail from a public phone booth or from their PDA.

Also *session mobility* will be supported in the future vision, allowing users to suspend a specific service session and later resume that very session from the same mobile or a different terminal. This means that the network will keep the context of a service session for a particular customer in order to allow moving people not only to have access to their own set of services but also to their last working session. In the context of personal and session mobility the notion of ubiquitous computing is important. In a target scenario it may be possible to allow mobile users for example within an organization, to use some „blanc" PDAs, which can be personalized after authentication in the user's working environment.

Personalization of communication services access and delivery
Personalization describes the customer's ability to define his own working environment and service working conditions stored in a „Personal Service Profile". This profile defines all services to which the user has access, the way in which service features are used, and all other configurable communication aspects, in accordance with the user's needs and preferences, with respect to parameters, such as time, space, medium, cost, integrity, security, quality, accessibility and privacy.

Interoperability of interfaces and services
Interoperability is one step beyond personalization and describes the capacity of a communications system to support effective interworking between different services, supported by and offered on heterogeneous networks, with the long-term aim of achieving fully interworking applications.

The aspects of *service personalization*, *service integration and interworking*, enabling people to define *if, when, where, for whom* and in *what form* they will be reachable, are still in their infancy. Today no adequate standards exist for allowing people to define their own individual communications and services environment, referred to as „*personal services communications space*" (Guntermann 1993). However, advanced customer control capabilities are required for the realization of real personalized communications.

In this context standards in the field of Intelligent Networks (IN) (ITU-T Q.1200) and Universal Personal Telecommunications (UPT) (ITU-T F.851) provide only limited capabilities without reflecting the emergence of Telecommunications Management Network (TMN) standards (ITU-T M.3010) in the context of service management. This problem has led to many research activities in the field of IN/TMN integration (Magedanz 1995, 1995a), where one thesis promoting the use of TMN concepts for providing IN service capabilities is gaining increasing attention.

Based on these considerations, in 1993 the DeTeBerkom GmbH, a subsidiary of the Deutsche Telekom AG, has initiated a project within its R&D-programme in the field of IN/TMN integration with the focus on personal communications support and evolution towards future object-oriented service platforms, such as the Telecommunications Information Networking Architecture (TINA) (TINA 1995). In the course of this project, a TMN-based *Personal Communications Support System (PCSS)* (Eckardt 1994, Eckardt 1995a, Magedanz 1995) has been developed, which provides advanced personal communications capabilities within distributed multimedia inhouse environments.

2 OVERVIEW OF THE PERSONAL COMMUNICATIONS SUPPORT SYSTEM

The *Personal Communications Support System (PCSS)* has been developed to provide generic capabilities for personal mobility, service personalization, and advanced service interoperability in a uniform way to numerous communication applications, ranging from simple telephony up to multimedia conferencing services. From a functional perspective, the PCSS provides enhanced UPT capabilities with respect to customer addressing (based on logical names instead of numbers) and advanced customer control capabilities. From a design perspective, the centralistic Intelligent Network approach to the realization of service logic has been replaced by a highly distributable, object-oriented approach.

2.1 The PCSS Approach

The PCSS approach differs from the conventional Intelligent Network (ITU-T Q.1200, ITU-T F.851] approach in a number of elementary points:

The PCSS generally uses personal references as communication addresses instead of physical addresses such as terminal IDs or network access points. The PCSS user is relieved from the need to configure a personalized call management in terms of originating and destination terminal addresses presuming an exact and fixed user-terminal relationship. In fact, in scenarios of nomadic personal communications, the fixed user-terminal relationship is replaced by a a flexible, dynamically changing association between nomadic users and various terminals in the current proximity of the user. The general goal of the PCSS approach is to bring a uniform, user-centric, i.e. non-technical approach to communications. It replaces terminal-oriented by person-oriented communications.

The PCSS provides capabilities for a uniform, *service-neutral* configuration and personalization of all types of teleservices subscribed to or used by a user. The advantage of this approach is a more focused, integrated view of the control, configuration and management of the total communications environment offered to the customer.

Similar to UPT, the users have to register at each new location in order to indicate to the network where to route a call directed to their personal number. But the PCSS user *registers at locations*, not at terminals. In this regard, we defined several registration procedures to be supported by the PCSS: automatic registration - that is an automatic user tracking applying Active Badges (Want 1992, Harter 1994) developed at Olivetti Research Ltd. in Cambridge UK, manual registration, and scheduled registration using a kind of electronic personal diary stored in the network.

2.2 PCSS Functional Model

The main dynamic mapping function within the PCSS is a multi-stage mapping of a PCSS address or personal identifier of the called party to an appropriate physical device address of the Customer Premises Equipment (CPE) at the user's temporary location. It is characterized by a distinctive model for the dynamic CPE selection defining a selection process in 4 stages (see Figure 1):

- The evaluation of a user's call logic constitutes the 1st *stage* of the multi-stage dynamic CPE selection process. It provides the *management of the user's reachability* . In case of call forwarding it represents a person to person mapping, otherwise one of the following features will be chosen: call accepting, call blocking, announcement, voice box, etc..

- At the 2nd stage *a mapping of the user's identifier to the location i*s made based on user registration data if the exact recipient of the communications invitation has been settled and no further call management will be performed.

- The 3rd stage performs a mapping from a *location to a virtual communication endpoint* corresponding to a group of terminals, i.e. Virtual Access Point (VAP). A VAP encompasses knowledge on terminal capabilities, supported services, etc. and a selection mechanism.

- The 4th *stage selects an appropriate terminal ID* from the group of devices contained by the VAP. This function will be performed by the Virtual Access Point selected in the previous stage. The processing at this stage is parameterized by a service type, used media, and optionally by user preferences.

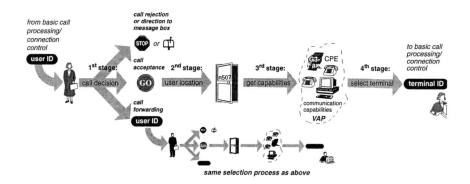

Figure 1   PCSS mapping of user ID to terminal ID

2.3 PCSS Object Model

The PCSS component model has been defined to structure the functionality required for the multi-stage PCSS call processing into distinctive functional components. This component model defines a number of functional components and elementary objects that are required to provide the PCSS call processing. The term *component* is used to denote a distributable server, a management subsystem or a X.500 DSA which provide the functionality of a „host" for PCSS objects and profiles. The units of individual state, interaction and addressing are termed *objects*, e.g. the *Call Session, Call Logic Agent, Registration Agent, Virtual Access Point, Service Access Point*. Two of these objects are called „agent[1]" since they represent and act on behalf of an individual user. The components may be understood as a kind of container or process for multiple objects of a specific type. The hosting components for „agents" are called „server" and all other hosting components are called „manager" or „MIB".

---

1. Please note, that the term 'agent' used in this context is not to be dismissed as an agent in the X.700 CMIP context.

Figure 2 gives an overview of the basic components of the PCSS which cooperate in order to provide the PCSS call processing. We applied an object-based modelling that defines the individual elementary functional components of the PCSS component model in terms of the operational interfaces they provide.

The interface descriptions of the functional components *Call Logic Server* and *Registration Server* are specified according to the Guidelines for the Definition of Managed Objects (ITU-T X.722). These functional components are modelled and implemented as management systems. Each of the components contains a number of dedicated *Managed Objects* (MOs) in specific *PCSS Management Information Bases* (MIB). Besides these X.700 components, a number of profile stores are indicated, which have been realized as X.500 DSAs, e.g. user profile store, zone profile store, virtual access point store, and service access point store. These different profile stores may be arbitrarily distributed over one or multiple DSAs.

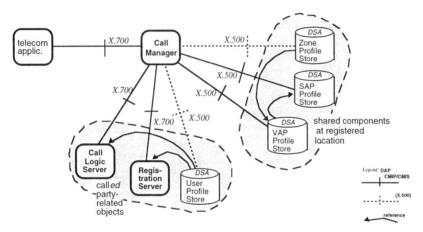

Figure 2   Basic components of the PCSS

The *Call Manager* contains and maintains a set of *Call Sessions*. A Call Session is an object representing the state of a particular PCSS call. The *Call Logic Server* incorporates multiple *Call Logic Agents* with each of the agents representing the personalized call logic of a single user. The third component, the *Registration Server* contains multiple *Registration Agents* that represent the different registered users within that location area which is handled by the Registration Server. Currently, the *Virtual Access Points* (VAPs) profiles and *Service Access Points* (SAPs) profiles are passive objects, i.e. X.500 directory entries. Two of the active components, the *Call Logic Server* and the *Registration Server,* have been realized as dedicated management processes comprising a CMIS agent and multiple MOs representing the Call Logic Agents and Registration Agents.

Figure 3 outlines the basic interactions of the objects above described, required to provide the *PCSS Call Processing*. The activity of calling a party (0) within a teleservice results in the request for the creation of a *Call Session* object dedicated to that particular call (1). It *coordinates* and provides the total multi-stage *PCSS Call Processing*. The called party addresses the Call Session Object with the personal reference of the called party in order to receive a physical address where the called party can be reached. At the lower left, the *Call Logic Agent* of the called party is shown. The Call Session object requests the evaluation of the Call Logic Agent (2). In this scenario the evaluation results in a „Call Forwarding" to a different person. This may happen numbers of times: multiple Call Logic Agents are addressed during a chained, multiple *Call Forwarding* (2a-n). Finally, the Call Logic Agent of the last addressed party is assumed to accept the call. In the next step the corresponding *Registration Agent* will be approached by the Call Session object to retrieve that last party's current location (3). The Registration Agent contains the information on the user's current location in terms of a Zone ID. This is a link to the Zone Profile that holds any required information on the location, especially the reference to the corresponding *Virtual Access Point* (VAP) (4). Each VAP represents a number of *Service Access Points* (SAPs) that contain information on specific communication endpoints and terminals at a certain location. The VAP will dynamically select an appropriate SAP (5a) thereby considering the state of the terminal (i.e. busy or idle) (5n). The final SAP including the proper communication address will be replied (4) to the Call Session object and finally to the Basic Call Process in the requesting teleservice (4,6). Subsequently, the Basic Call Process can establish a connection to the resulting Terminal ID given by the SAP_x.

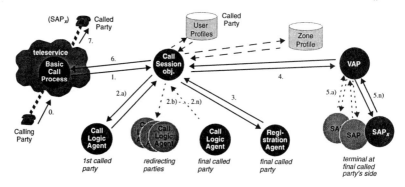

Figure 3 Object interactions within the PCSS Call Processing

2.4 PCSS Platform

When designing the PCSS, we concentrated on a uniform modelling of PCSS functionality completely based on current X.500 Directory, X.700 Management, and Telecommunications Management Networks (TMN) standards. The underlying idea has been: *to use concepts of OSI/TMN management for both service management and service control to enhance basic telecommunications services to the level of IN services or Service Features.*

The uniform application of object orientation and management concepts throughout the total design allowed us to develop a systematic and regular overall model or „umbrella" for personal communications resulting in a PCSS Platform supporting a wide range of applications. This PCSS Platform (shown in the centre of Figure 4) comprises a *PCSS Applications Framework* built on top of an advanced management platform (RACE H.430, 1993) integrating X.500 and X.700. The management system accesses specifically defined PCSS User Profiles via X.500 *Directory User Agents* (DUAs) applying the *Directory Access Protocol* (DAP).

Additionally, X.700 *CMIS/CMIP* managers are used for a location transparent access to dynamic PCSS Managed Objects (e.g. Registration, SAPs, VAPs) using a naming space established by the X.500 Directory Service (Tschichholz, 1993).

The PCSS Applications Framework on top of that management platform is the basic interface to the PCSS data. It consists of an extensive set of objects which provide access to all information and data stored within the scope of the PCSS profiles and services. The main purpose of this completely object-oriented interface is to serve as a common foundation for applications of any complexity. The application framework and its various component objects support various telecommunications services. At the same time, it enables a rapid development of PCSS-specific management applications, e.g. for user registration or user profile management, and various kinds of personal communications-related multimedia applications.

Every PCSS application is based on the retrieval and modification of data stored in the profile databases or managed by management services within the scope of the PCSS. The main purpose of the PCSS application framework is to provide a uniform and intuitive but nevertheless powerful interface to PCSS related information for all existing and forthcoming front-end and helper applications. The most important of these front-end applications are the *User Profile Management Service (UPMS)* and the *User Information Service (IVIS)*, which are described below.

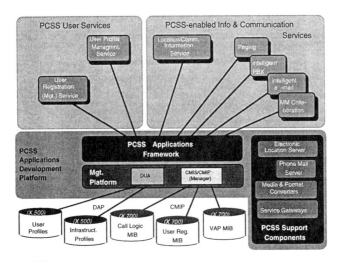

Figure 4   PCSS prototype platform

3 USER PROFILE MANAGEMENT SERVICE

While the PCSS application framework supports the low level data access, it does not provide a graphical user interface to personal data tailoring and management. Therefore, a graphical user interface based front-end application is needed for the PCSS users to view and edit the contents of their own, personal user profiles. It must enable the users to easily adapt the user specific information stored within their personal Generic Service User Profile (GSUP). By this way, this PCSS front-end application empowers the users to customize the rich set of user related services provided by the PCSS and to suit the PCSS communication environment and the provided PCSS service functionalities to individual preferences.

Due to the highly complex PCSS information model and the various service features provided, it is essential for this application to support the user with an intuitive and easy to use graphical user interface. The front-end application has to abstract the several complex information structures and to simplify their visualization, and it has to support the users in their interaction with the application. It must protect the users from making mistakes, and it must offer support if the users nevertheless encounter errors.

The *User Profile Management Service* (UPMS) offers all the capabilities required to enable the PCSS users to adapt their personal data and to customize their individual PCSS communication environment. To fulfil these needs, the UPMS provides the following functionalities for the manipulation of the contents of the personal GSUP to the PCSS users:

- *user authentication* to ensure secure access to user profiles,
- *user profile configuration* for the modification of personal user data,
- *user registration* for the manual registration of current location information,
- *personal schedule configuration* for automated registration of location data by schedule, and
- *communication configuration* for customization of the personal call logic.

These several features of the UPMS front-end application are realized by a set of basic application components, which can be considered as application building blocks for the realization of the UPMS. They are described below in more detail.

3.1 User Identification and Authentication check

The connection and access to the PCSS database must be strictly secured to protect the confidential personal information contained within the Generic Service User Profile (GSUP) and the other related infrastructure and organizational profiles as well as to the registration information against access by unauthorized users. Therefore, the PCSS users have to pass a PCSS identification and authentication procedure successfully before using the UPMS.

The UPMS users do not only have to give their unique identification for access to the PCSS, but they also have to enter an individual password matching the given identifier. The users are granted access to the PCSS and the UPMS only if they enter the correct password. Unlike the personal identifier, this individual password can be defined by the users, of course. It is stored in encrypted form as part of the general user data within the scope of the personal GSUP.

3.2 Personal User Data Editor

The personal Generic Service User Profile of all PCSS users contains basic, service independent information. This information, especially the general user data, are not directly related to the several communication services provided by the PCSS, but nevertheless they are part of the individual PCSS user environment. Therefore, the *personal user data editor* enables the users to customize these data.

The *general user data* like common name, surname, and private address, the personal title and organizational status can easily be viewed and modified by the users (see figure 5).

The PCSS system supports the modelling of complex organizational structures within companies. It enables the management of service access points, zones, events, and individuals, and also the definition of groups containing references to these objects. So departments or projects as well as any other assembly of PCSS objects can easily be represented.

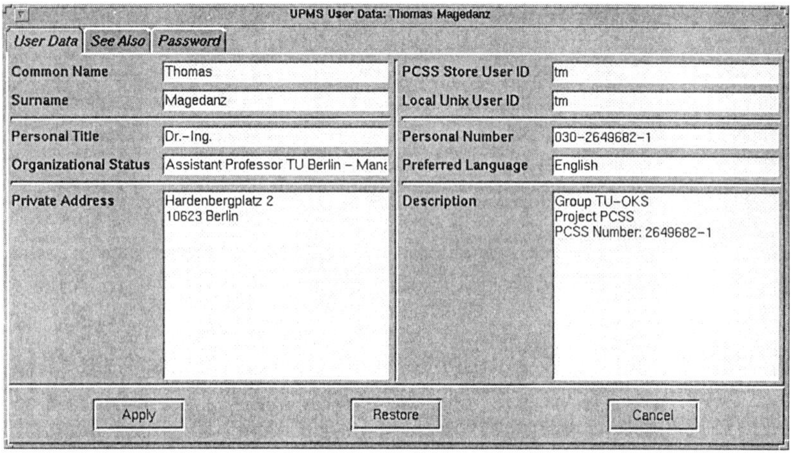

Figure 5 Personal User Data Editor Module of the UPMS

The UPMS users can use the personal user data editor to manage other available objects or groups of objects within the scope of their personal GSUP, hidden below the menu button *see also*. The users can tailor their personal list of references by selection from lists of available individuals, groups and events. The editor provides simple functionalities to rearrange the list as well as to add references to or remove them from the list. Within the scope of the personal user data editor, the users are also able to replace their password needed for secure authentication to the PCSS.

3.3 User Registration Manager

According to the user oriented approach to communication by the PCSS, abstract user identifiers must be mapped to physical communication addresses. Therefore, the PCSS has to gather needed information about the current location of the participating users. The location data is managed by the PCSS registration server of the PCSS X.700 management services. The personal GSUP contains references to these location information.

The PCSS registration server can retrieve current location information from different sources. Unlike the location information to be gained by automatic registration, the user can individually define location information via the *user registration manager*.

The manual registration procedure enables the user to provide the PCSS with his current location data manually. The registration manager provides the users with a list of all available zones from which they can select their current location. The user registration manager selects a location for the users. This default choice for the manual registration can be, in this precedence, the location the users are currently registered to, the users specified default location or the location of the terminal actually used for registration via the UPMS. In addition, the users are offered a list of recently registered locations for easy retrieval of location information. For convenience, the manual registration procedure allows the users not only to specify their real current location for registration, but also each other location within the scope of the PCSS.

3.4 Personal Schedule Editor

The registration server manages current location information registered manually by the PCSS users via the user registration manager. Unlike the current location information depending on the presence of the users and the use of the UPMS, the *personal schedule* enables the users to specify their future locations during absolute or even periodically recurring time intervals independently of the registration server.

The personal schedule is described by a scheme of rules. Each of these personal schedule rules describes a time interval and an associated future location. During their evaluation by the PCSS, the schedule entries are assigned priorities according to their order in the personal schedule. When looking for the location associated with a certain time interval, it is matched against the time frames contained in each schedule one after another. The first schedule entry matching the given time interval then determines the current location.

The PCSS users must be enabled to setup and to configure the contents of the personal schedule to meet their individual requirements. Therefore, the UPMS supports the users in the customization of their personal schedule. Using the graphical user interface provided by the *personal schedule editor*, the users can define and modify the entries of the personal schedule. Assigning locations to time intervals, the users can describe events or appointments taking place once or regularly as well as future locations assigned to certain complex time intervals.

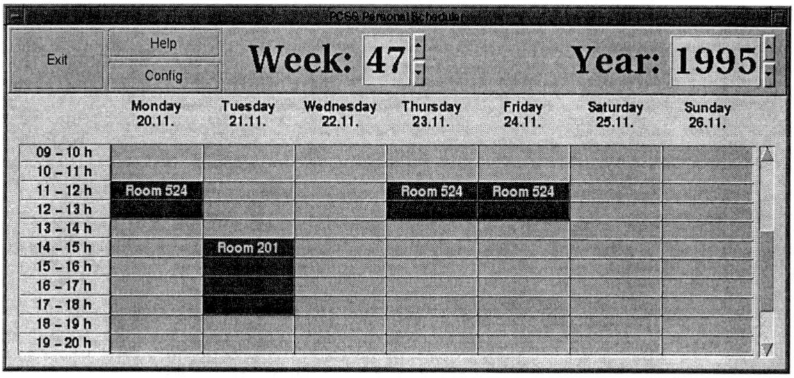

Figure 6 Weekly Schedule Plan

By dragging the mouse within the actual week plan, the users can highlight a number of adjacent schedule fields to create a new schedule entry. Following a dialogue allows to create a new schedule entry by specification of a time interval and the assignment of a location.

For periodically recurring time frames, an optional type of *repetition* as well as the end of its iteration can be given. There are several different types of repetitions selectable by the user. By using these repetitions, frequently visited locations can be indicated. *No Repetition* is the default, but repetitions for every week, every month, every year, week in month or day in week are possible to choose.

3.5 Call Logic Editor

The most important capabilities provided by the PCSS are the mapping from abstract personal identifiers to physical communication addresses and the advanced incoming call management features. These high level PCSS functionalities are performed by the PCSS call manager component. Concerning the address resolution and the incoming call processing performed by the PCSS, the most important information are the data stored within the *personal call logic*.

Similar to the personal schedule, the personal call logic is described by a scheme of user defined rules. These rules determine the personal *call logic matrix*. The rules contain instructions for the handling and processing of calls incoming to the PCSS. Each call logic rule contains actions to be executed if an associated set of conditions is matched by the actual parameters of incoming calls. Using the various different supported condition types, the evaluation of incoming call handling actions can depend on the time of the incoming call, on the calling user or device, and on the requested communication medium. Furthermore, all conditions can be combined to even more complex compound conditions.

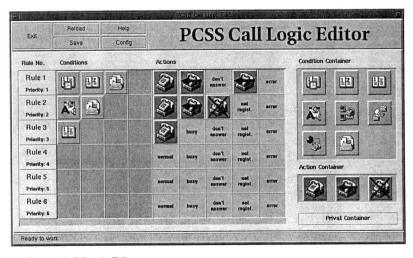

Figure 7 Call Logic Editor

During the processing of incoming calls by the PCSS, the personal call logic of the recipient is evaluated by the PCSS call manager. The priorities are assigned to call logic rules according to their position in the call logic matrix. When searching for the actions to be performed on the incoming call, the conditions specified in each rule are matched against the actual call parameters. The first rule matching all given conditions determines the set of actions to be executed.

For the setup of an individual communication environment within the scope of the PCSS, the UPMS must enable all PCSS users to customize the service specific information contained in their personal GSUP. Because its significant effect on the call processing by the PCSS, one of the main aims of the UPMS is to support the users in the setup and configuration of their personal call logic. The *call logic editor* provides the PCSS user with a powerful and easy to use graphical interface for the setup and customization of the personal call logic matrix.

4 USER INFORMATION SERVICE

The idea behind the realisation of an „User Information Service" application is to use the services and information managed inside the PCSS not only for the purposes of the PCSS and its participants themselves, but also to make the core functionality and data visible to any participating user or external users outside the PCSS. A first approach to this idea is the realisation of an **Interactive Visitor Information Service (IVIS)**, which responds to the needs of a visitor of getting information and contacting a company or organisation and their individuals in a fast and efficient way.

In the given scenario of an organisation which uses the PCSS environment for its inhouse communication, the implementation of IVIS will realise a kind of „electronic receptionist", which provides a visitor with a set of „functionalities" as in real life, combining it with the information managed inside the PCSS. The task of connecting a visitor and a person which corresponds to the needs of the visitor is performed interactively with the visitor using the application IVIS. Therefore one major task of the application IVIS is the search for a person represented in the databases of the PCSS. During the process of searching the user will be supported by a highly intuitive and simple user interface through the application IVIS.

The first prototype of the application IVIS will only act as a „passive" member of the PCSS architecture and interacts with the user like a common information system. Therefore IVIS uses the contents of the different PCSS Profiles, e.g. the User Profiles, as a smart database and presents this information to the user in a easy understandable way.

The application IVIS will lead a visitor through the „universe" of the organisation by representing their people, their locations, current events, etc. Any visitor who wants to meet one specific individual inside the company will be supported by the application IVIS and the PCSS in presenting information about a person, who matches exactly the visitors needs and requirements. Using the up-to-date information stored inside the PCSS database, a visitor will be led by the application IVIS in a simple and intuitive way without involving a third person in this process. Her interactions will be supported with the presentation functionalities of IVIS described above by showing photos, displaying textual descriptions, etc. Additionally, the visitor will be provided with actual location information of the desired people and with information how to reach this location through a map or floorplan. By viewing the information about the employees of an organisation, the visitor will also be provided with a couple of alternate persons and related topics about any specified individual. This will prevent him from additional complex and time consuming searches inside the PCSS, if the desired partner is not available or does not want be disturbed by a visitor.

In the following example there was be selected the button „PCSS" inside the search screen containing the „projects". The list of buttons now represents all individuals related to the project PCSS, including a small photo of each person. Additional textual information about the currently selected project is presented to the user in the upper part of the screen.

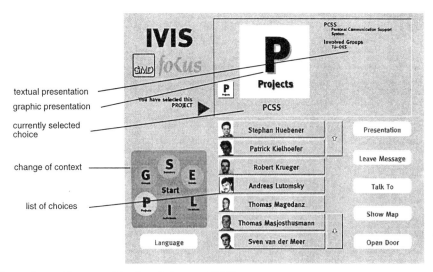

Figure 8   Contents of the Search Screen of IVIS

Beneath the area of the search additional interaction-components will allow the user to invoke other and future components of the PCSS like „intelligent map", communication, or more generic functionality like printing, language-selecting, etc. Another group of interaction elements besides the list of selectable choices of the search will show a small representation of the start screen with the same functionality. The characteristic look of this group makes it easy for a user to associate its purpose. So a visitor may freely change between the different main contexts without returning to the start screen. If the user changes his current context inside the search, e. g. from groups to project the application IVIS will save the contents of the current context for a later use. So a user does not need to start the search new from the beginning, if he is temporarily changing the context.

5   PCSS EVOLUTION

Current enhancements of the outlined PCSS concentrate on dynamic media type and format conversion required for entirely supporting the above outlined vision of „information at any time, at any place, **in any form**". Therefore, a powerful set of information converters, including both format and media converters, which can be dynamically configured according to a given personal communications scenario are under development (Magedanz 1996).

Furthermore, another point of PCSS evolution is the migration from a TMN architecture towards TINA architecture [TINA-95]. In this context the PCSS has become recently an auxiliary project of the TINA consortium work, labelled „PCS in TINA". The basic objective of this TINA auxiliary project is the incorporation of the developed PCSS concepts into the TINA service architecture in order to support advanced personal communications within TINA (Eckardt 1996).

6 ACKNOWLEDGEMENTS

The following persons have contributed to the development of the presented PCSS: S. Arbanowski, F. Gadegast, L. Hagen, S. Huebener, S. van de Meer, U. Scholz, M. Vetter, H. Wang.

7 ACRONYMS

CMIP	Common Management Information Protocol
CPE	Customer Premises Equipment
DAP	Directory Access Protocol
GSUP	Generic Service User Profile
ID	Identifier
IN	Intelligent Network
IVIS	Interactive Visitor Information Service
MIB	Management Information Base
MO	Managed Object
PCS	Personal Communications Support
PDA	Personal Digital Assistent
PCSS	Personal Communications Support System
SAP	Service Access Point
TINA	Telecommunication Information Networking Architecture
TMN	Telecommunications Management Network
UPMS	User Profile Management Service
UPT	Universal Personal Telecommunications
VAP	Virtual Access Point
X.500	Directory
X.700	TMN / OSI Management

8   REFERENCES

Eckardt, T. and Magedanz, T. (1994) On the Personal Communication Impacts on Multimedia Teleservices. International Workshop on Advanced Teleservices and High-Speed Communication Architectures (IWACA), Heidelberg, Germany, September 26-28, 1994.

Eckardt, T. and Magedanz, T. (1995) On the Convergence of Distributed Computing and Telecommunications in the Field of Personal Communications. Kommunikation in Verteilten Systemen (KiVS), Zwickau, Germany, February 22-24, 1995.

Eckardt, T. and Magedanz, T. (1995a) The Role of Personal Communications in Distributed Computing Environments. 2nd Interantional Symposium on Autonomous Decentralized Systems (ISADS), Phoenix, Arizona, USA, April 25-26, 1995.

Eckardt, T. and Magedanz, T. and Stapf, M. (1996) Personal Communications Support within the TINA Service Architecture - A new TINA-C Auxiliary Project. Proc. of 6th TINA Conference, Heidelberg, Germany, September, 1996.

Guntermann, M. et.al. (1994) Integration of Advanced Communication Services in the Personal Services Communication Space - A Realisation Study. Proceedings of the RACE International Conference on Intelligence in Broadband Service and Networks (IS&N), Paris, France, November 1994.

Harter, A. and Hopper, A. (1994) A Distributed Location System for the Active Office. IEEE Network, Special Issue on Distributed Applications for Telecommunications, January 1994.

ITU-T Draft Recommendation F.851 (1991) Universal Personal Telecommunications - Service Principles and Operational Provision. November 1991.

ITU-T Recommendation M.3010 (1992) Principles of a Telecommunications Management Network. Geneva, 1992.

ITU-T Recommendation X.722 (1992) Information Processing - Open Systems Interconnection - Structure of Management Information - Part 2: Guidelines for the Definition of Manged Objects. Geneva, 1992.

ITU-T Recommendations Q.1200 series (1992) Intelligent Network. Geneva, March 1992.

Magedanz, T. and Popescu-Zeletin, R. and Eckardt, R. (1995) A (R)evolutionary Approach for Modeling Service Control in Future Telecommunications - Using TMN for the Realization of IN Capabilities. International TINA Conference, Melbourne, Australia, February 13 - 16, 1995.

Magedanz, T. (1995) On the Integration IN and TMN - Modeling IN-based Service Control Capabilites as Part of TMN-based Service Management. 6th IFIP/IEEE International Symposium on Integrated Network Management (ISINM), Santa Barbara, California, USA, May 1-5, 1995.

Magedanz, T. and Pfeifer, T. (1996) An Intelligent Personal Communications Support System enabling Information any time, any place, in any form. Submitted to IEEE Global Telecommunications Conference, London, United Kingdom, November 18-22, 1996.

RACE Common Functional Specifications (CFS) H.430 (1993) The Inter-Domain Management Information Service (IDMIS), Issue D. December 1993.

TINA-C Doc. No. TB_MDC.018_1.0_94 (1995) Overall Concepts and Principles of TINA. February 1995.

Tschichholz, M. and Donnely, W. (1993) The PREPARE Inter-Domain Management Information Service. RACE IS&N Conference, Paris, France, 1993.

Want, R. and Hopper, V. and Falcao, V. and Gibbons, J. (1992) The Active Badge Location System. ACM Transactions on Information Systems, Vol. 10, No. 1, 1992.

9 BIOGRAPHY

Cordula Ulbricht finished her studies in computer sciences at the Technical University of Berlin in 1994. Since then she has been working as a project manager at the DeTeBerkom GmbH, a subsidiary of the Deutsche Telekom AG, in the field of Open Communications Systems, Intelligent Networks and TINA. Presently her work focuses mainly on the development of future telecommunication services considering the aspects of mobility.

PART FOUR

Mobile Agents and Multimedia Applications

8
Supporting User Mobility

Martin G. Brown
Olivetti Research Ltd.
24a Trumpington St., Cambridge, CBI 1QA, UK
phone: +44 (1223) 343373
fax: +44 (1223) 313542
email: mbrown@cam-orl.co.uk

Abstract
The availability of wireless network connections to laptop computers and PDA's has created interest in the issues surrounding mobile computing. However, enabling users to be genuinely mobile in their work requires more than a wireless connection. Distributed system services are needed to support the locating of people, equipment and software objects, and, especially for mobile multimedia applications, network transport protocols which can adapt to a wide range of networking conditions must be developed. This paper reviews some important mobility issues, looks at some of the systems requirements raised by user mobility, and describes some practical experiences with mobile applications at Olivetti Research Ltd. (ORL).

Keywords
Mobility, mobile multimedia, personalisation

1 INTRODUCTION

User mobility is a more general concept than wireless terminal mobility or mobility enabled by portable computers. Instead of requiring the user to carry around a laptop or PDA it is in many instances more convenient to simply use any networked computer or information appliance which is to hand, including those which may have a wireless connection to the network.

Current computing systems have the idea of user ownership deeply ingrained in them, even diskless workstations become private property once someone has logged into them. This runs counter to trends in the workplace where, increasingly, people's working patterns are more mobile, and rooms, desks and equipment may be used only temporarily - they are not "owned" by anyone in particular.

Desktop and portable computers are usually highly personalised, both in their hardware and software configuration and because they contain unique copies of files. It is not easy to borrow another computer. These problems would be avoided if, instead of private, personalised hardware, we had public, *depersonalised hardware* which would be equally accessible to everyone. To achieve this kind of mobility the hardware on which user interaction occurs must be stateless, or perhaps only have state during the course of a user interaction. The computing environment would still be personalised, of course, but now that would be taken care of elsewhere in the network. In this view the only "private" property is in persistent software objects and associated files residing on networked servers. By removing the close binding between individuals and their personal computers we give ourselves the freedom to allow computer interactions to persist and *follow* us as we move between tethered, or untethered computers, both in our work and leisure.

Recently several companies, including Oracle Corporation and Sun Microsystems, have announced the intention to market products, known as *network computers* which contain elements of these ideas. Only a limited amount of software need be stored locally and applications would be downloaded from network servers, freeing users of systems administration issues, and all user status would be stored at networked locations. These developments are linked to widespread interest in highly portable architecture-independent languages such as Java[*] and, if successful, may lead to the widespread adoption of computers which are well suited to support user mobility.

It is natural to also expect multimedia applications to be always available to us by following our movements. What is required is an architecture providing an infrastructure which makes user mobility transparent, in the sense that adaptation to varying resources is automatic, and which includes integrated support for the special problems of multimedia data transport.

Section 2 reviews the benefits that mobile multimedia could bring. Section 3 looks at some important systems which are required to support user mobility and reports on some practical experience at ORL with various aspects of mobility.

2   THE PROMISE OF MOBILE MULTIMEDIA

User mobility will make possible many new and exciting applications, as well as making life much easier for people who are mobile in their work. It should be possible to bring a personal computing environment to not only alternative work locations, and to the home, but also to endpoints sited in public places, such as restaurants, banks and airports. See Weiser (1993) for generalisations of these concepts.

The development of mobile multimedia applications will create demand for activities such as the browsing of archives of audio, video and images; mobile video-conferencing; mobile video-on-demand; and persistent multimedia connections for monitoring remote activity. Monitoring applications could include site security and industrial safety. Monitoring officers could carry a portfolio of audio-visual inspections with them during the course of their work,

---

[*] Java is a trademark of Sun Microsystems

and could also continue the activity off-site. Some of these potential mobile multimedia applications, in the context of wireless networking, have been discussed by Lee (1995).

There is strong evidence of an increasing commercial interest in networked multimedia applications involving mobility. For example, in the UK, BBC Television has embarked on a long-term plan to create a "newsroom of the future" which must take into account the work habits of roving journalists and programme makers who wish to access multimedia information. They would like journalists remote from the studios to not only be able to easily contribute new multimedia material, but also to take part in editorial decisions and browse multimedia library material. Other commercial applications include live feeds from news agencies; the rapid dissemination of graphics and video to mobile workers in public services; and the transport of images to organisations like insurance companies. On a more personal level, some people may like to have live family photos on their desk, or answer their front door remotely.

3 SYSTEMS REQUIREMENTS FOR TRANSPARENT MOBILITY

There are some key technologies and indispensable system services needed to support user mobility. Designing the required infrastructure involves an interesting combination of computer science issues. I will focus on some areas of interest at ORL: distributed software support for multimedia; location; intelligent adaptation to changing network and endpoint characteristics; and wireless mobility appropriate for multimedia. The security issues related to mobility will not be considered here.

3.1 Distributed Multimedia Applications

The ability to distribute applications across networked computers in order to take advantage of intelligence elsewhere on the network is an important ingredient of a mobile architecture. The standardisation of interfaces and protocols for handling distributed software objects achieved by the Object Management Group (OMG)* has been an important milestone in software engineering. Use of the OMG Interface Definition Language (IDL) enforces modularity and facilitates code re-use. CORBA (OMG, 1995) is a language-independent standard to which multiple vendors conform, allowing interoperability between software objects on different computer architectures. The standard provides a "software bus" for monitoring and controlling software objects which are distributed across networks. Microsoft's Active X✣ architecture will offer an alternative and widespread option for creating distributed systems.

However, support for multimedia in conventional distributed systems is inadequate. The methods available to move data between software objects do not match the special transport requirements of multimedia. Live multimedia streams must have a minimal end-to-end latency and low jitter, and some stream types may require a different quality of service from others.

---

* OMG , OMG IDL and CORBA are trademarks of the Object Management Group Inc.
✣ Active X is a trademark of the Microsoft Corporation

Standard remote method invocations can be used to control the software objects through which multimedia streams flow. But for movement of data a transport protocol specialised to support multimedia is necessary. Until these ideas are absorbed into the distributed system standards a "back door" for live multimedia data transport will still be needed.

The Streams project at ORL and Cambridge University Computer Laboratory is using CORBA both to distribute multimedia sources, sinks and pipeline elements across networked computing devices and to access the distributed system services required to support mobility. Of particular interest is the ability to monitor and control directly networked sources and sinks of multimedia data such as video cameras, display devices, audio input/output devices and multimedia file stores.

3.2 Intelligent Adaptation

Users may interact with a wide variety of stationary and mobile systems in the course of their day so it will be necessary for mobile applications to intelligently adapt to different environments and to different users.

Network Adaptability

Moving between a tethered workstation and a computer with a wireless network connection may cause a change in the characteristics of the network connection, such as a change in the bandwidth, and a change in latency and error rate. A different change in connectivity will happen when moving to the home or to a remote site which may not have high-speed network connections. If information on network characteristics were stored in a network resource database then some high level adaptation decisions could be made by the application. But to handle dynamic changes in the network characteristics the transport protocol must be aware of the way conditions are varying so that it can notify the application of changes in the quality of service.

Endpoint Adaptability

Applications will also have to adapt to the capabilities of different endpoints. Kantarjiev (1993) and Richardson (1994) have noted the problems of migrating X Window* displays between workstations with different display characteristics. Displays can differ both in their resolution and colour-handling, and unless applications are written with user mobility in mind ad hoc solutions must be adopted. The user-interface may also differ from the standard keyboard and mouse, PDA's often only have a pen interface, and in the future may have a speech interface. It is unusual to design user interfaces which can dynamically adapt to alternative input and output devices. The Model-View-Controller architecture for applications may offer some solutions (Gray, 1990 and Stajano, 1995). Here, the Model is the core of the application, moving between different application states. It presents an interface to which multiple Controllers (input devices) and Views (output devices) can be attached.

---

* The X Window System is a trademark of The X Consortium.

Another interesting adaptation issue is how intelligence should be partitioned between networked servers and endpoints. Bartlett (1994) noticed that he obtained much better performance by locating the intelligence for his wireless Web browser in a powerful networked workstation, while only requiring the endpoint, a PDA, to behave as a dumb alphanumeric display. Another example of partitioning of intelligence and display at ORL is described by Wood (1996). Here a dumb ATM-networked LCD screen called a VideoTile shows an X Window display by being refreshed, when necessary, with blocks of screen pixels from a process on a networked workstation. Kantarjiev (1993) examined various points along the line from dumb X display to full X server in an effort to optimise a wireless X display. It may be that the positioning of the intelligence within a mobile application will have to be automatically reconsidered as the mobile user moves between endpoints. Some endpoints with poor network connectivity would be able to sink more data if a better compression scheme were used, but the endpoint may then not have the resources to decompress that data and deal with the resulting increased data rate.

3.3 Location

All models of mobility require a mechanism for identifying the location of both users and computing equipment. One might simply assume that the user is adjacent to the machine he is logged into, but this is open to ambiguity. It is possible to imagine cumbersome methods of indicating location information to the computer system. The user could log in to a computer and signal his or her presence, but this makes assumptions about the location of the computer and ancillary hardware which may not be valid - computer hardware and peripherals are frequently repositioned for a variety of reasons. Wireless mobile computers can be located to some volume served by their base station but the wireless cell sizes may range from several rooms within a building up to many miles in an unpredictable fashion, so this is unsuitable for reliably collocating users with other users or equipment.

Active Badges

ORL has developed a method of automatically locating users and equipment called the Active Badge* Location System (Want, 1992 and Harter, 1994). Considerable experience has convinced us that such a location system is an important enabling technology for automating features of mobile applications, allowing user interactions to become simple and transparent. Schilit (1993) has also experimented with the use of infra-red passive location systems to permit customisation of mobile applications.

The installed Active Badge system at ORL maintains a dynamically updated record of the location of both people and hardware and has made possible interesting experiments involving mobility. If one is to make progress in a field there is no substitute to building systems and gaining experience and insight by using them in day-to-day work.

Medusa The Medusa multimedia installation at ORL (Wray, 1994) makes use of an extensive switched ATM network to connect together various kinds of directly network connected

---

* Active Badge is a registered trademark of Ing. C. Olivetti & C., S.p.A.

endpoints. These endpoints include video cameras, high quality audio input and output, VideoTiles (networked colour LCD displays), storage devices and workstations. Applications use badge location information on both users and equipment to make video and audio connections to people who are referred to by name, rather than network address. If you are near enough to suitable multimedia hardware then the multimedia connection will be automatically routed to you, without any login requirements.

Teleporting The ORL teleporting system (Richardson, 1994) is an example of a user-mobile application where the interface is made transparent by the use of Active Badge location information. Teleporting offers a means of redirecting the user interface of ordinary applications which run under the X Window System. Instead of being fixed to one display as part of a standard "X" session, applications can be part of a "Teleport session" which can be dynamically called to any suitable display on the network. When called up on a new display, the applications within the Teleport session appear in exactly the state the user left them on a previous display. This allows users the freedom to move around whilst retaining immediate access to their computing environment wherever there is a suitable display. This is remarkably useful, and has convinced us that user mobility is a goal worth pursuing. Apart from the obvious benefit of being able to carry one's work when moving between rooms, it is very useful to be able to show complete multi-window screens to others, whilst retaining all of one's own environment, and not even requiring the other party to log out. Because VideoTiles display the X Window world, it is also possible to teleport to any VideTile on the ATM network. It has a pen interface, and no keyboard, but this is very appropriate and comfortable for using a Web browser teleported to a VideoTile. VideoTiles are stateless networked computers and if rebooted or power cycled immediately recover their display from the remote X server.

Follow-Me Video It is of interest to extend the concept of teleporting to include other data types such as multimedia. In order to become familiar with the issues involved we have experimented with both Follow-Me audio and Follow-Me video. It is possible to take part in a multi-party audio conference using Medusa software. Each end of the conference can be called to another location using buttons on your Active Badge. The Follow-Me video prototype makes it possible to materialise live video connections on a nearby workstation or VideoTile. The video images could be views of your office, simulating a security application, or views of the car park, simulating the ability to see what commuter traffic conditions are like. This is only an experimental prototype, but experience with it will help us to develop our distributed multimedia and adaptive transport protocol work.

JavaTel extends some of these ideas to areas beyond the range of the Active Badge system. Web browsers present a platform-independent view of the World Wide Web which has some of the characteristics of user mobility, despite its bandwidth limitations. It is possible to automatically download and execute Java applets within some Web browsers. This programmable intelligence at the users endpoint can be used to make a Web page behave like a dumb graphics display. At ORL, Wood (1996) has written a Java applet which simulates the behaviour of an ORL VideoTile, creating a globally mobile X session. This is an interesting foretaste of what true user mobility might hold.

A Dynamic Location and Resource Database

Mobile applications will need to be able to discover the current locations of not only users, but also computers, peripheral equipment and resident software objects. It will also be necessary to know what software objects could be created, if required, on a particular platform. The characteristics of each endpoint that the user interacts with must be stored, and, for any particular user, the system will need to be able to find out what facilities are available in their immediate vicinity. A dynamically updated object naming, location and resource database is required to maintain all this information. It must contain resource descriptions of hardware and software objects allowing mobile applications to make queries and receive callbacks when hardware and software objects change position or state. A constant stream of location updates should arrive at the database from a wide variety of location and movement detection sensors. Clifford Neuman (1993) has identified some of these system support issues, and at ORL the SPIRIT (Spatially Indexed Resource Identification and Tracking) project will provide a comprehensive solution.

The availability of such a service will make possible the automation of many user interface issues and will make user mobility a transparent process. For example, information on a machine's power and the software base available to it will allow automatic resource allocation of particular tasks to processing engines specialised to carry them out. A networked speech recognition service might need to be run on a powerful machine equipped with a large amount of memory and local storage (Brown, 1995) and it could be optimised by adapting to different users by accessing speaker-dependent speech data.

The appropriate information would now be available to automatically adapt user interfaces to match the capabilities of different platforms. Qualities like screen resolution, colour capability and input/output peripherals would be known to the system. If the database contained details of network connection characteristics for different machines then automatic intelligent adaptation of multimedia streams could be done as the user moves to parts of the network which have different performance characteristics. As an example, an alternative video compression scheme might be chosen for a low bandwidth connection to a home compared to that for a high bandwidth ATM connection in the workplace. The local processing capabilities of the endpoints would also influence such a decision.

3.4 Wireless Mobility

Imielinski (1994) has presented an overview of wireless mobility, but wireless network technologies are not generally suitable for transporting good quality live multimedia. Handling live multimedia brings special problems, particularly of latency, synchronisation and continuity. For example, in a live videophone call the audio and video streams must be delivered with sufficient synchronisation to avoid obvious lack of lip-sync, the latency must not be intrusive, and there should not be any breaks in connection when the mobile does handoffs to another base station. To save bandwidth media streams over wireless connections may be compressed in some way, this places constraints on the tolerable error rate for these connections.

So there are a number of important considerations for good quality live wireless multimedia: the wireless connection needs to have sufficient bandwidth for transporting multimedia; it must offer separate qualities of service for each media stream, so that compressed streams are well handled and streams can be prioritised; and handoffs between different base stations must not interrupt the streams or add latency to them.

The ability to offer different qualities of service to different media streams is usually associated with connection-oriented networks. Wireless ATM can have the property of maintaining end-to-end quality of service, just as in wired ATM, and so is particularly interesting for mobile multimedia. ORL has been working for some time on a wireless ATM network designed with the characteristics of multimedia in mind (Porter, 1996). It provides the equivalent of a wireless ATM drop cable from the base station (which is connected to the wired ATM network) to the wireless endpoint. Since it supports standard ATM adaptation and signaling protocols it is possible to treat the wireless section like a normal connection to the wired ATM network. Wireless ATM VideoTiles behave just like wired ones, with the exception that the wireless bandwidth is limited to 10 Mbps, soon to be extended to 25 Mbps.

4 CONCLUSIONS

A multimedia system supporting user mobility needs a highly distributed software system which must be extended to handle the specialised transport requirements of multimedia streams. By building and maintaining an object naming and location database which stores system resource information, it becomes possible to automate many aspects of mobility and so make user mobility transparent.

5 ACKNOWLEDGMENTS

This paper includes many ideas frequently discussed among research staff ORL. In particular I would like to thank Andy Hopper, Andy Harter, Frazer Bennett, Glenford Mapp, Sai Lai Lo, Tristan Richardson, Ken Wood, Alan Jones, John Porter, Damian Gilmurray, Oliver Mason, Harold Syfrig, Mike Addlesee, Dave Clarke, Gray Girling, Frank Stajano, and also Brendan Murphy from Cambridge University Computer Lab., for their ideas and criticism.

6 REFERENCES

Bartlett, J.F. (1995) The Wireless World Wide Web, *Workshop on Mobile Computing Systems and Applications, Dec 8-9 1994, Santa Cruz, IEEE Computer Society Press*, 176-178.

Brown, M.G., Foote, J.T., Jones, G.J.F., Sparck Jones, K. and Young, S.J. (1995) Automatic Content-Based Retrieval of Broadcast News, Proceedings of ACM Multimedia '95, San Francisco CA. 35-43

Clifford Neuman, B., Augart, S.S. and Upasani, S. Using Prospero to Support Integrated Location-Independent Computing , *Proceedings of the USENIX Mobile and Location-Independent Computing Symposium*, Aug 1993, Cambridge MA. 29-34

Gray, P.D. and Mohamed, R. (1990) *Smalltalk-80: A Practical Introduction*, Pitman.

Harter, A. and Hopper, A. (1994) A Distributed Location System for the Active Office, *Network*, **8(1)**.

Imielinski, T. and Badrinath, B.R. (1994) Mobile Wireless Computing, *Communications of the ACM*, **37(10)**, 19-28

Lee, K. (1995) Adaptive Network Support for Mobile Multimedia, *ACM MobiCom 95, The First Annual International Conference on Mobile Computing and Networking*, Nov 1995, Berkeley, CA.

Kantarjiev, C.K., Demers, A., Frederick, R., Krivacic, R.T. and Weiser, M. (1993) Experiences with X in a Wireless Environment, *Proceedings of the USENIX Mobile and Location-Independent Computing Symposium*, Aug 1993, Cambridge MA, 117-128.

OMG (1995) *The Common Object Request Broker: Architecture and Specification*, Object Management Group.

Porter, J., Hopper, A., Gilmurray, D., Mason, O., Naylon, J. and Jones, A. (1996) The ORL Radio ATM System, Architecture and Information. ORL technical report, available through the ORL Web server at http://www.cam-orl.co.uk/

Richardson, T., Bennett, F., Mapp, G., and Hopper, A. (1994) Teleporting in an X Window System Environment, *IEEE Personal Communications Magazine*, **1(3)**, 6-12.

Schilit, B.N., Theimer, M.M. and Welch, B.B. (1993) Customizing Mobile Applications, *Proceedings of the USENIX Mobile and Location-Independent Computing Symposium*, Aug 1993, Cambridge MA, 129-138.

Stajano, F. and Walker, R. (1995) Taming the Complexity of Distributed Multimedia Applications, *Proceedings of the 1995 USENIX Tcl/Tk Workshop*, Toronto, July 1995.

Want, R., Hopper, A., Falcao, V. and Gibbons, J. (1992) The Active Badge Location System, *ACM Transactions on Information Systems,* **10(1)**, 91-102.

Weiser, M. (1993) Some computer science issues in ubiquitous computing. *CACM*, **36(7)**, 137-143.

Wray, S., Glauert, T. and Hopper, A. (1994) The Medusa Applications Environment, IEEE Multimedia, **1(4)**.

7 BIOGRAPHY

Martin Brown obtained a B.Sc. in physics in 1974 and a physics Ph.D. specialising in computer simulation, both at Imperial College, London. After spending some time in the broadcast audio and videographics industry he joined Olivetti Research in 1991. Here he has worked on many aspects of the Medusa distributed multimedia system, with a particular interest in the audio. He manages a joint research project with Cambridge University Engineering Dept. called VMR which uses speech recognition and information retrieval techniques to perform keyword indexing into video mail archives.

9
Designing secure agents with O.O. technologies for user's mobility

David Carlier
RD2P - Recherche & Développement Dossier Portable
CHR Calmette, 59037 Lille Cédex - France
tel: +33 20 44 60 46, fax: +33 20 44 60 45, email: david@rd2p.lifl.fr

Patrick Trane
TIT - Tokyo Institute of Technology
Ookayama 2-12-1 Meguro-ku Tokyo 152 - Japan
tel: +81 35499 7001, fax: +81 35734 2817, email: patrick@cs.titech.ac.jp

Abstract
Using different kinds of computers from different locations has become a classical phenomenon. A user is said to be mobile when he does not always communicate with the outside from the same location. More and more people are being included in this category, such as people using a workstation linked to the Internet at their office, a microcomputer equipped with a telephone modem at home or a portable computer communicating via a wireless link. Features of these stations can be drastically different [GC94] especially for mobile computers due to numerous constraints such as weight, size, low communication flow and energy consumption [FZ94], [IV94]. Communication with a mobile user depends on the terminal used. This paper proposes an agent-based system in which one agent is associated to one user in order to simplify his tasks. Object-oriented technologies will be used as a way to guarantee agent's data privacy, protect agent acceptor's sites from intrusions and, in case of loss, make agent's recovery easier to perform.

Keywords
Mobile Computing, Oriented-Object Technology, Fault Tolerance

1 DESCRIPTION OF THE PROPOSED SYSTEM

This proposed system aims at enabling anybody to use the same services and data whatever the kind of terminal. To do so, a *representation agent* is widely used. A representation agent is a piece of software able to travel throughout the wired network and excecute itself on acceptor sites [CT95].

A unique representation agent is associated with a given user. An agent takes in charge three main tasks :

- The agent is the user's representation, able to act on his behalf even when he is logged out.
- The representation agent manages data sent from/to a mobile user. This data must be specific to the user's current environment in order to allow him to be able to use his applications from different terminals. For instance, if the communication flow and the resolution of the current user's computer screen are weak, a large image may be reduced.
- The representation agents are located at acceptor sites providing an execution environment for agents. An agent is a mandatory intermediary between a user and all his interlocutors. The agent remains fixed during the whole communication even when the user is moving.

The last property allows an easier integration of mobile computing into current systems. However, when there is no external communication and when the agent is too far from its associated user, the agent is allowed to *migrate* to get closer to the user. For example, if an European person travels to Australia, his associated agent must follow him so as to avoid extra use of the network and to reduce message latency; if the user is communicating with an Australian station, it would be very inappropriate to have messages travelling from Australia to Europe and vice-versa. This step is called the agent migration. Using an agent is particularly suited to address the following problems for mobile computing described in [MDC93] :

- *Task delegation* : It allows the user to save energy and get more processing power. The delegation is performed by sending a small program or script to the user's agent. The computation results are returned to the user.
- *Asynchronism management* : As the user is often disconnected, the representation agent can be viewed as an ever active dedicated component on the wired network. Its behavior can be defined with the help of scripts.
- *Communication strategies* : All data sent to the user are first filtered by the agent with respect to both the communication flow and the device features. For example, it is no use transferring a 16-million-color picture to the user if his mobile station is only equipped with a black-and-white color screen. On the other hand, emitting data requires more energy than receiving data. Instead of sending a request each time the user wants to access a document, it is more adapted to send a short script describing data he wants to receive regularly. For example, if he is used to consulting news relative to a specific newsgroup, each time the user is connected, new items are automatically sent.

2 AN OBJECT ORIENTED APPROACH

The use of representation agents on acceptor sites involves several security problems. Some requirements must be taken into account so that, on the one hand, an agent cannot

80   *Part Four   Mobile Agents and Multimedia Applications*

voluntarily or otherwise damage an acceptor site and, on the other hand, an agent is not able to read or modify another agent's data located on the same site.

Some constraints are therefore applied on the agent processing domain. Two methods are proposed :

- The first one is to interpret the agent code. The user defines his own agent and sends it to an acceptor site. An agent interpreter on this acceptor site verifies that all instructions are correctly processed. The main drawback is the slow execution speed : running interpreted code is much slower than running native code.
- The other possibility is to predefine the code of the agent. If a user wants to get an agent on an acceptor site, he requests it to create one from a predefined code. This code can be written in native code, so the execution speed is improved in comparison with the first method. The features and the behavior are, however, predefined, that is, the agent cannot damage an acceptor site and cannot use another agent's content, as will be indicated hereafter.

The first method allows a user to define his agent the way he wants, according to the interpreted language. In the second one, the code is predefined. This code mainly enables the agent to communicate with the user and the outside world. It is also used to manage data and scripts sent by the user. The agent consists of (1) the code part *ie* the agent code and (2) its individual part, *ie* personal data and scripts. Thanks to the data and scripts, the user can define its own behavior. Moreover, as the agent code is the same for all agents, only the agent data and scripts are affected by the migration phase.

The use of object-oriented concepts enable an easy design of the system. An object is composed of two sets of elements : the interface (code), and the structure (data). Creating an object is achieved by sending a *create* message to an object server.

A representation agent can then be viewed as an object. On each acceptor site, an agent objet server is responsible for the agent creation. To create an agent within an acceptor site, a *create* message must be sent to the agent server on the site. Once a blank agent is created, the user personalizes it by sending a *personalize* message with a set of data related to the user and a set of scripts defined by him.

To process an agent migration, a *create* message is sent to the target acceptor site by the agent that wants to migrate. It then personalizes it and transfers its own personal data and scripts. When these operations are correctly performed, it destroys the copy remaining on the user site.

The agent code itself is not important. It may be different according to acceptor sites. However, the interface and the behavior of an agent must be independant of the site.

3   REPRESENTATION AGENT OBJECT STRUCTURE AND SCRIPT OBJECT

3.1   Agent structure

The user can add a script into his associated representation agent by sending it the message *AddScript*. When receiving such a message, the agent creates a new instance

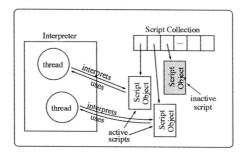

Figure 1 Representation agent object structure

of the class *ScriptObject* and adds it to its *ScriptCollection* as shown in Figure 1. One schedule is associated to each script sent to the agent to define the time and the frequency of its execution. The *AddScript* message is followed by two arguments : the script itself and the associated execution schedule. Once this method is executed, a script identifier is returned to the user. This identifier allows him to modify or remove the script from the agent. When the execution of a script is no longer planned in the schedule, the script can automatically be removed from the *ScriptCollection* of the agent object.

When a script is activated according to the schedule, a thread of the agent's script interpreter is created. Its role is to manage the execution of a given script during its running time. After the end of the execution, this thread is removed and the script object returns to an inactive mode.

3.2 Script structure and script execution

As mentioned in the previous part, a script can be in two different states : inactive or active. In the former case, the script object can be just considered as data. It is only made of two elements : one script text and one schedule. On the other hand, the script object structure is extended by an execution block. This one is used as a memory dedicated to the execution data of the script. This memory is divided into three elements : a data stack, a set of registers (such as an ordinal counter) and a data space for the variables storage as represented in Figure 2.

A method *Start* run on a script object enables to create an interpreter thread associated to this object and to create an execution data space. The thread is then activated by the script object. The method *Abort* kills the thread associated to the script and frees the execution data space.

3.3 Agent object migration

A representation agent can migrate only when no communication is in progress. All its active scripts must not be communicating. The migration must allow the representation agent to suspend both its execution and those of its scripts, to travel accross a network

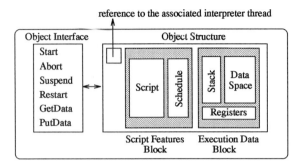

Figure 2 Active script object structure

from the current agent's acceptor site to another one and finally to end up with the same execution environment as before the migration.

It is necessary to identify all data to be transferred from one site to another. Minimizing the quantity of transferred data is an important issue. As it is entirely re-created according to the new site hardware and software features, the agent interface does not need to be sent to the target site. Even if the agent code may differ, both its interface and its behavior do not change according to their location on various acceptor sites. However, the data contained in an agent object must migrate to the target site. The scripts can be considered as data. Since the running scripts contain data required by an interpreter to continue their execution, the formerly defined script structure allows them to keep their execution context even after a migration. For all scripts for which the execution has been suspended, a new interpreter thread will be created and will use the execution data contained by the script object. Four methods must be added to the script object interface :

- *Suspend* which stops the execution of a script before a migration.
- *GetData* which allows the migrating agent to get data of a script object in order to send them to the new script object on the target acceptor site.
- *PutData* which is used to put script data from the old agent script object into the corresponding new agent's script object.
- *Restart* which enables the script object to continue its execution from the instruction where the script was suspended.

4 LOSS AND RECOVERY OF AGENTS

4.1 Identification of problems

The system presented here entirely relies upon the presence of the agents. As the agents are located on acceptor sites and may migrate from one site to another, if an acceptor site fails, two main problems may be identified : (1) all agents located on this acceptor site are lost, and (2) all agents migrating towards this faulty are will not reach their destination. Although the problem of acceptor sites diagnosis is beyond the scope of this paper [TC96], it is important to mention that it provides a useful tool to quickly warns the system to forbid any further migration towards faulty sites and initiate a recovery process for lost agents.

Recovery of agents must be done with care. Two main problems are identified. Reconstruction of an agent after its acceptor site has failed and restarting from a consistent global state without any message losses. The solution proposed also focuses on low-overhead dynamic reconfiguration strategies. Problems are as follow :

Checkpointing [AB94] must be performed with care so that the saved states form a consistent global state. Agent A sends a message M_1 to agent B. A has taken a checkpoint C_a before sending this message. B takes a checkpoint C_b right after having received this message as shown in Figure 3. Subsequently, A fails and restarts from its last checkpoint C_a. At this stage, the system global state is inconsistent for A's local state shows no message sent to B while B's local state indicates that a message has been received from A. This remains true even if B restarts from C_b.

Rollback-recovery [KT87] from consistent checkpoints may also cause message losses. In Figure 3 again, B sends a message M_2 to A. A receives it and subsequently fails. Both A and B rollbacks to their respective last checkpoints. B's local state shows that it has already sent M_2. A's local state indicates that it has not been received. The sytem recovers from a consistent state. However, the channel from B to A is empty. Consequently, M_2 is lost.

4.2 Proposed solution

The O.O. agent approach, that is, code and data (refer to part 2), is particularly suited to provide fault-tolerance taking into account the problems mentioned this above. However, to ensure consistency and integrity, an agent has to be modelled as follow :
code: a block of code
data: (a) a data block, (b) a queue of incoming messages and in case the agent was active when the failure occured, (c) its history since last message reception

As the agent classes located on acceptor sites make the agent code always available the code reconstruction problem is directly eliminated. To be able to restart properly, a counter of outgoing messages is also required. A change in an agent's local state occurs if either (1) the agent sends a message *i.e.* the outgoing message counter increases, or (2) the agent receives a message, *i.e.* its incoming queue changes, or (3) the agent completes a task, *i.e.* a new checkpoint is performed.

To be able to recover properly from a sudden failure, two messages are sent together with the original one. A *shadow* of the original message is sent to the shadow of the recipient.

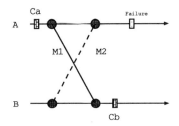

Figure 3 Identification of problems

A message informing the shadow of the sender that the sender has sent a message is also sent so as to increase the message counter. To ensure fault-tolerance, four actors are required. The sender (Agent A), its shadow (Agent A'), the receiver (Agent B) and its shadow (Agent B'). The process is ordered as follow : *(first)* Agent A sends a copy of the message he wants to send to Agent B', *(second)* Agent A sends a message to its shadow A' to increment its message counter and finally *(third)* Agent A sends the message to the receiver B. All transactions are simultaneous and atomic. When these operations are completed, Agent A checkpoints and sends a copy of its state to its shadow. Agent A' then replaces its data with the copy it has received. Agent A also keeps a copy in case it has to regenerate the shadow. In case the shadow has to be generated, it just takes both the checkpointed data block and the queue of incoming messages of the agent and transforms itself into a real agent.

4.3 Implementation

The agent is a mandatory intermediary between the user and the external world. The agent is responsible to update data within its shadow when necessary, that is, after a transaction or after a long enough computation modifying the agent data. As an agent script is processed by a interpreter thread associated to it, the shadow data management can be delegated to the interpreter and to the script object itself.

Two kinds of script object modifications implying a change in the shadow state are identified. The first one is induced by a call of certain script object methods. For instance, the call of the *Start* method involves the creation of the execution data block within the script object. In the same way, it must create a similar block in the script shadow of the shadow of the agent. The *Abort* method, which ends the script running, frees this part both in the agent script and in the agent shadow script. The *PutData* method must set the execution data block of both the agent script and its shadow. The above shadow changes can be taken into account by the code of the method itself. The second kind of modification is due to the interpreter running. It changes the script execution data block values consequently changes the agent state. It must periodically and just after a transaction update the shadow.

5 CONCLUSION

This paper defines a system enabling a mobile user to obtain facilities from an associated assistant : the representation agent. Its structure takes advantage of the object-oriented technology. The object instanciation and behavioral notions enable the creation of secure agents respecting rules such as privacy of the agents located on the same site and acceptor site integrity. This object structure also allows to split agents into two kinds of data : static data and dynamic data. The last one are very sensitive and each update must be taken into account so that even if an agent is lost, it is possible to restore this agent with the same state as before the failure.

Future works include the implementation of a prototype allowing the management of small representation agents responding to the requirements described in this paper.

ACKNOWLEDGEMENTS

The authors wish to express their sincere appreciation to members of their respective laboratory and more especially to Professor Vincent Cordonnier, Pierre Paradinas from RD2P - University of Lille 1, and Professor Nanya from Tokyo Institute of Technology who have made this collaboration possible. The authors also would like to thank Sylvain Lecomte for lively discussions regarding this research.

REFERENCES

[AB94] A. Acharya, B. Badrinath, *"Checkpointing Distributed Applications on Mobile Computers"*, in proceedings of the 3rd International Conference on Parallel and Distributed Information Systems, September 94

[CT95] D. Carlier, P. Trane, *"Security Requirements for Mobile Computing Systems"*, Technical Report of IEICE, FTS 95-70, pp 57-65, Tokyo, December 1995

[FZ94] G.H. Forman, J. Zahorjan, *"The Challenges of Mobile Computing"*, in IEEE Computer, pp 38-47, April 1994

[GC94] S. Gadol, M. Clary, *"Nomadics Tenets - A User's Perspective"*, Sun Microsystems Laboratories Inc. Technical Report, SMLI-TR-94-24, June 1994

[MDC93] B. Marsh, F. Douglis, R. Cáceres, *"System Issues in Mobile Computing"*, Technical Report TR94-020, Matsushita Information Technology Laboratory (Princeton), February 1993

[IV94] T. Imielinski, S. Viswanathan, *"Adaptative Wireless Information Systems"*, in proceedings of SIGDBS Conference, Tokyo, October 1994

[KT87] R. Koo, S. Toueg, *"Checkpointing and Rollback-Recovery for Distributed Systems"*, in proceedings of IEEE Transactions on Software Engineering, Vol. SE-13, No 1, pp 23-31, January 1987

[TC96] P. Trane, D. Carlier, *"Diagnosis Algorithm for Mobility Oriented System"*, in proceedings of the 2nd International Conference on Application-specific Systems, Architectures and Processors, IEEE, Chicago, USA, August 1996 (to appear)

10
The Idea: Integrating Authoring Concepts for Mobile Agents into an Authoring Tool for Active Multimedia Mail

Schirmer Jürgen, Kirste Thomas
Darmstadt Technical University
Interactive Graphics Systems Group. Telephone: +49 6151 155 241.
Fax: +49 6151 155 480. email: `schirmer@zgdv.de`
URL: `http://zgdv.igd.fhg.de/`

Abstract

Mobile agents are a competitive concept of client-server computing and are especially suitable in mobile environments, that are characterized by low bandwidth communication facilities and ad hoc connection/disconnection to stationary systems. They can be used for information retrieval and information filtering, in which case they evaluate replies and return only the relevant data. *Mobile agents* as a metaphor of active objects are created on a mobile device such as a Personal Digital Assistant (PDA), will be launched into the information galaxy and are fulfilling the mobile user´s task on the services available on networked stationary systems. One transmission channel for these itinerant agents is email. This paper introduces $\text{ACTIVE}M^3$ as an example of an active mail framework, which can be regarded as a first approach in authoring mobile agents in a graphical interactive manner. $\text{ACTIVE}M^3$ integrates two known concepts: *active mail* and *multimedia mail*.

The idea of *active* or *computational mail* has existed since 1976 (3); transfer programs embedded in standard mail which automatically execute at the receivers site. This *hiker* –the active mail message– can also carry multimedia elements (audio, animation, image etc. and indeed sub-agents) in his *backpack*, which it can also render and orchestrate at the receivers site in a sophisticated manner. Therefore, synchronization concepts are introduced. To summarize, the objective of $\text{ACTIVE}M^3$ is the development and implementation of a system for the realization of *active multimedia mail* and a composing tool as an easy to use interactive authoring environment. An extension towards a more generalized authoring environment for this *remote programming* concept enables authoring of mobile agents.

This paper introduces the basic ideas. The results of the proposed concept for a composer for *active multimedia mail* are presented through a realized prototype.

Keywords

Active Mail, Computational Mail, Multimedia Mail, MIME, Remote Programming, Mobile Agents, Itinerant Agents, Synchronization

1 INTRODUCTION

Essentially, *active mail* (6) simply means to send executable programs within a mail message, instead of passive data. On a somewhat more abstract level, *active mail* messages can be regarded as *active objects* traveling through the network searching for, distributing, and collecting information; interacting with receivers; making decisions based on context knowledge; replicating themselves, etc. *Active mail* is used for remote program installation *, for initiating synchronous connections between several remote users without being intrusive (*e.g.*, a collective writing facility)–so *active mail* could be a platform for Computer Supported Cooperative Work (CSCW)–, for survey generation and meeting scheduling etc.

Today *active mail* is also a suitable application in mobile environments characterized by low bandwidth communication facilities and small display devices. *Active mails* are launched every-time the user opens a wireless link to a stationary system. Afterwards these messages can hike between several heterogeneous locations and through different networks, collecting information, evaluating information, making transactions, etc. So indeed *active mails* are *itinerant agents* (9).

In conjunction with support for structured multimedia content data and user-interface toolkits, *active mail* messages could for instance build up a form on the receivers site, consisting of text fields (for editing purposes), multiple choice elements, anchor buttons for linking multimedia elements to the form, etc. accept and evaluate the answers of the receiver (*e.g.*, check the consistency) and automatically send back a reply to the originator of the mail. The active part –the "hiker"– of the mail can also structure the contents of the *backpack*, the part of the mail containing multimedia elements, in a more sophisticated manner. Synchronization in time and space will be possible. We will call this enhanced version of *active mail* –*active multimedia mail* (7).

---

*E.g. updating the latest version of an application.

To emphasize, elements of the backpack could be also sub-agents, which describe tasks that have to be synchronized in a certain manner. For instance the installation of a program as one task and the establishment of a persistent connection to a shared application as the task following at the end of the first task. An active mail message is basically a program written in a suitable interpreter language. At the receivers site, the interpreter is started automatically upon message access, executing the program contained in the message.

This of course implies, that *creating* an *active mail* message is equivalent to writing a program. The average user can, however, not be expected to do a programming job, when he simply wants to send a message to a friend or colleague. One of the challenges of *active multimedia mail* is therefore to create an *authoring tool* that makes the writing of *active multimedia messages* as simple as writing a standard text message.

2 THE IDEA

One step further towards a generalized authoring environment for asynchronous communication is the integration of visual specification technic for *mobile agents*. A novice user without programming knowledge wants to have a possibility to define his goals and needs in a graphical interactive manner.

A main feature of an agent is to help the user to accomplish his task. In a mobile scenario, in which the end-user has access to stationary systems over a wireless link via a portable computer, this task could for instance be, to find the cheapest offer for a hotel, to make a reservation, or to collect some technical product information, etc.

Today *agent technologies* are heavily discussed in the research community as well as in the commercial field. The most important instance of this new technology is characterized by the metaphor of a *mobile agent*. Basically *mobile agents* are programs, typically written in a script language, which may be dispatched from a client computer and transported to a remote server computer for execution. CORBA, HTTP and SMTP are possible transmission channels for *mobile agents*. SMTP or email is an adequate transmission channel concerning a mobile scenario. Every time the user connects to a stationary system, all outgoing messages inside the *spool area* will be delivered. Besides, all notifications and alert mechanisms to inform the end-user about any change in resource are based upon email exchange.

The advantage of this approach to integrate *mobile agents* into email is, that it is embedded into a communication environment, which is broadly available [†]. Finally, the application domain for *mobile agents* like information finding/retrieval, information gath-

---

[†] Email is ubiquitous.

ering/filtering, the update of programs, meeting scheduling, etc. is quite similar to the one suggested earlier for the concept of *active mails*.

3 STATE OF THE ART: MOBILE AGENT TECHNOLOGY

Agents have their origin in the research area of *artificial intelligence* (2). As mobile agents, they are suggested as a suitable enhancement of *client server computing*. A main feature as opposed to traditionally *client server computing* is, that the processing of queries (scripts) is asynchronous and does not depend upon a continuous link between two parties. A more precise definition of a *mobile agent* is given as follows: *"A mobile agent is an encapsulation of program-code, data and execution state, which is able to migrate between networked computers autonomously and goal-oriented while executing."*

There exists many variations in the literature concerning this definition. Migration are often reduced to transmission of program code (script) without execution state and includes *unidirectional, bidirectional* and *multi-directional* mobility. Autonomy and goal-orientation are often related to the fact, that the agent acts on behalf of the author, that he has a user defined intention and strategy. Moreover *mobile agents* are suggested to be able to cooperate with their peers, that means that they are able to communicate and to negotiate. Intelligence skills of agents, the ability to learn, to plan and to inference, are also desirable for *mobile agents*. This implies the presence of a knowledge base and rule pattern, as well as an inference machine.

3.1 Research activities

Research activities in *mobile agent* technology focus on suitable infrastructures which support for instance the migration of agents, security aspects and the dynamical access of unclassified services (*electronic commerce*), etc. (see for instance (18) (1) (4)). There exists also a several suggestions to enrich general purpose languages towards mobility and security (AgentTCL (13), Java-To-Go (16), SafeTCL (7), etc.) . Research activities concerning communication skills of agents and knowledge sharing are done in the artificial intelligence domain.

The Knowledge-Sharing Effort of DARPA [‡] is developing an Agent Communication Language (ACL), which is known as Knowledge Query Manipulation Language (KQML) (12) and which constitutes a language based protocol for knowledge sharing for distributed inter-operation among knowledge-based systems. This effort is also developing the

---

[‡]Defense Advanced Research Projects Agency

Knowledge Interchange Format (KIF) (19), which means that knowledge of one agent can be represented in any language and can be transformed into KIF for exchange.

Research activities in the AI domain focus also on Agent Based System/Software Engineering (ABSE, ASE) §.

It is a higher aggregated approach of object-technology with standardized message semantics, which was invented to facilitate the creation of software able to inter-operate, that is exchange information and services with other programs (in an expressive ACL) and thereby solve problems that cannot be solved alone. In this approach to software development, application programs are written as software agents, i.e. software components that communicate with their peers by exchanging messages in an expressive agent communication language.

Another research activity in the AI field are *user interface agents*. These are intelligent agents, which build an extension of the traditionally user-interface metaphor of *direct manipulation* towards the metaphor of *indirect management*. This *indirect management* is established by personal assistants, who are collaborating with the user and help to automate repetitive user tasks. The agents have the ability to observe the user behavior, to learn and to inference (e.g. information filtering, information retrieval, mail management, meeting scheduling etc. based on user preferences; intelligent help agents, that correct user mistakes; see also WebHunter at MIT (20)).

3.2 The work carried out in the commercial field and the Internet

Information Retrieval and Filtering

Due to rapid growing of the Internet and world wide web, technologies such as search engines are becoming well known and are often referred too as agents (spiders, robots, web-crawler, infobot etc.). These are search engines, which establish an index database after contacting web-servers in a synchronous fashion. A migratable executable script and real autonomy with inferencing capabilities does not exist.

Tacoma (24) is a research activity which extend WWW-servers to host itinerant agents, which perform information retrieval. As mentioned before WebHunter is an approach for information retrieval based on learning capabilities about user preferences.

---

§Besides the term Agent Oriented Programming (AOP) is also in widespread use. AOP is a specialization of Object-Oriented Programming (OOP). Agents are objects that have mental states (such as beliefs, desires and intentions) and a notion of time. They are programmed using commitment rules. These are simple forward chaining rules which connect messages, the agent's internal state and its actions.

Databases

Oracle Mobile Agent (21) is an approach for client-agent-server communication. Mobility in regard to real transmission of scripts is not available; here, the only object which is mobile is the application domain and we can rather speak of *static* agents for mobile scenarios. The infrastructure consists of a message manager at the client side (a mobile device) and the *agent event manager* on the server side (or an the corporate LAN), which collects messages of the clients and passes them to application specific *message handlers* resp. *transaction handler*, which constitute the agent.

Languages

Java (23) is an example of embedding behavior into the World Wide Web, it brings HTML documents to life and visualizes information at the client site. Java do not support real mobility. Mobility is given through the capability to down-load programs (applets) from a server to a client (unidirectional). These applets can neither go any further nor can they go the other way round. Java addresses mainly security issues concerning the exchange of programs between heterogeneous networked systems. Recently there exists some research activities which enrich Java with communication and itinerant skills (see JavaToGo (16) and JavaAgentTemplate (15)).

A more sophisticated approach for a mobile agent language (and an *electronic commerce* scenario) is provided by General Magic's **Magic Cap** and **Telescript** (26) system. **Telescript** is an object-oriented language, which supports migration (statements: go, places, ticket, passport). This is an example of a well-designed electronic shopping system, which makes use of migratable mobile scripts, but suffers from the small amount of platforms be supported (MagicCap, Personal Link,UNIX) and the specific application domain (electronic shopping).

3.3 Application Domain

To conclude this section, a short impression of example application domains for mobile agents is given. *Knowbots* are stationary agents (e.g. launched from a mobile device, unidirectional), which implements a special kind of user profile (mail enabled application, dissemination of information, watchers, alert mechanisms for notification of any change in resource e.g. file size; personification of server behavior, etc.). The modeling of business processes as a collections of autonomous, problem solving agents which interact when they have interdependencies is an application domain for *workflow-management* systems (e.g. meeting scheduling). *Mobile* agents are suggested for distributed *information retrieval* resp. information filtering and data-retrieval. *Mobile agent* technology are also suitable for *information dispersal*, for instance to update the latest version of a program, and in *computer supported cooperative work* scenarios to establish a persistent connection to a groupware application. They are introduced to perform *transactions* in the database world and begin to gain attention in the financial world. A challenging domain in regard to the complexity of interaction, security aspects, trustworthiness, etc. is the *electronic commerce* field (electronic markets, markets for buying and selling goods, e.g. job, accommodation, car, flea-market, etc.; electronic shopping, open-bidding auction).

So far, the state of the art in regard to *mobile agent* technology has been introduced. Challenges have been identified. The introduction of a visual specification technic for *mobile agents*, which assists the novice user without programming knowledge in the definition of his goals and needs, is one challenging approach. The following chapter focuses on the development of a generalized authoring environment for *mobile agents*, that is currently being established. First of all, the already existing specialized authoring environment ACTIVEM^3 is introduced.

4 THE ACTIVEM^3 COMPOSING TOOL

The main goal of ACTIVEM^3 was to develop an easy-to use authoring system for *active multimedia mail*, based on the metaphor of *direct manipulation*.

The overall look-and-feel of the composing tool is derived from Graphical User Interface Builders (GUIB) like iX-Build: the most important aspect of an *active mail* message is the way it presents itself to the receiver – *i.e.*, its user interface. Figure 1 gives an impression of what ACTIVEM^3 looks like.

The user creates an active message by interactively assembling message elements onto a *mail form*. This form may contain presentation/interaction elements such as buttons, text-areas or in-line images. In addition, it may contain *anchor*-buttons, leading to multimedia content objects which are presented in external presentation tools, such as image, audio or video. Besides assembling active mails it is also possible to send normal email, linear multimedia mail, or active mails without a form-metaphor, etc. The construction and visualization of alternative mails is supported through the concept of different views.

After a message has been completed, ACTIVEM^3 converts the final form into a MIME(Multipurpose Internet Mail Extension) (5)-conform message. The active component of the message is represented as a LISP dialect (HyperPicture Command Language) (17) script contained in a special MIME-content part; multimedia objects are contained in another set of MIME-parts, the *backpack* (for in depth technical discussion of ACTIVEM^3 see (11)).

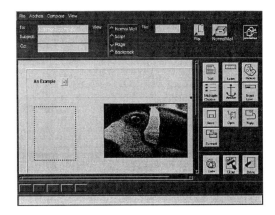

Figure 1 The user interface of ACTIVEM^3.

4.1 Authoring the temporal structure of multimedia presentations

As mentioned before, ACTIVEM^3 can describe interactive multimedia applications. These applications are characterized by the presentation of static and dynamic media which may be correlated in time and space. One possibility for instance is to use timed petri-nets (Petri-Net Based Hypertext (22) and Object Composition Petri-Net (25)), directed weighted graphs, as the model for building synchronization interrelations among initially unrelated backpack objects. In our scenario the functionality of petri-nets to describe interrelations seems to be too powerful for the needs of a novice user, who may only want to send a little slide show about his recent holiday. Therefore, we suggest to establish basic synchronization descriptions with a subset of synchronization operators as proposed by the synchronization model of *path-expressions* (14). The selection of these operators and the selection of visible multimedia objects inside the *backpack* enables the interactive description of interrelationships among the data. A visualization is given by a directed graph (see Figure 2).

The reason for choosing *path-expressions* is that we need a user-level abstraction of the underlying concept of a *finite-state-machine* (used to implement synchronization), which

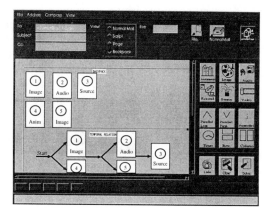

Figure 2 Multimedia inclusion and interactive description of synchronization with basic operators.

enables the adding of hand-crafted functionality in a very easy and intuitive manner in case the description power of the authoring mechanism is not sufficient.

4.2 A Generalization towards Authoring of Mobile Agents

We propose, that the construction of *mobile agents* is quite similar to the composition of *active mails* and the synchronization interrelation.

Fundamental components ¶, which may be offered from third parties, will be available on a special panel and a workbench will be inserted into an additional view, in which the user can construct his own special agent and where he can get direct feedback of his construction.

An interesting approach in creating agents in a graphical interactive manner is given in the KidSim system (10). It is based on the concept of *graphical rewrite rules* and *programming by demonstration*. KidSim is an environment, which allows children to construct a

---

¶For instance results of the componentware research. Componentware is an approach to create custom agents by combining components using scripts, which would contain domain specific expertise (e.g. financial planning) and would use the services of the components (text formatting, equation solving) to perform some of the actions.

computer game. It has to be evaluated if the concept of *programming by demonstration* should be changed to *programming by example* in the context of *mobile agents*, because this is the way, people learn to program. Its additionally worthwhile to evaluate the concepts of *visual programming* (8) in this context. Proposed systems that are based on *dataflow-diagramms* seems to be not very intuitive for the needs of a mobile user (e.g. display-size of a PDA).

An agent is sometimes more than only an executable script. It has a mental state (beliefs, intention, goal), can communicate with other agents, and can inference to fulfill a defined task. So, agents are suitable for an open system, in which services available do not have to be known in advance. A novice user may visually program agents for his needs and goals, without concrete knowledge of existing services.

Consequently, the agent consists of a knowledge-base part and a behavior part. For the behavior part we may introduce a *visual programming* concept, but the user needs some additional mechanisms to construct the knowledge-base part. He (or the system) has to specify the knowledge of an agent in an appropriate knowledge representation language with a vocabulary (ontolingua) suitable for a special domain. This knowledge may also be derived and extended from a history of successfully performed tasks and user preferences.

The user may also wish to specify whether the agent should share his knowledge with other agents, add security properties, add electronic cash, add a list of locations to be visited or to be excluded. In addition he wants to specify the priority and the expiration time of the agent. To summarize, the user may wish to specify the degree of autonomy, intelligence, social ability, and mobility. Trustworthiness for such a system and the constructed agent is a very important attribute, and must be addressed as well as security aspects.

We conclude, that constructing a mobile agent for general purpose is a large design issue, for which influences of AI has to be regarded as well. ACTIVEM^3 is a first step and an appropriate basis towards such an effort.

5 CONCLUSION AND FUTURE WORK

In this paper, we have outlined the general concept of *active multimedia mail*. We then introduced a prototype composing tool for *active messages*, the ACTIVEM^3 tool. A user can develop "programs" without programming knowledge and it would be worthwhile to evaluate the concepts of *visual programming* and if they are well suited as the underlying metaphor of such a composing tool. As part of a Global Information System, ACTIVEM^3 can be regarded as an approach in generating *mobile agents* acting on behalf of the author in a graphical interactive manner.

6 ACKNOWLEDGMENT

The synchronization concept and the work regarding *mobile agents* are performed in the MOVI project funded by the German Science Foundation (DFG).

REFERENCES

Intelligent Agents. *Communications of the ACM*, 37(7), July 1994.

R.H. Anderson and J.J. Gillogly. Rand intelligent terminal agent (rita): Design philosophy. Technical Memorandum R-1809-ARPA, Rand Corporation, February 1976.

Das Ara Projekt. Available via WWW
  (URL: http://www.uni-kl.de/AG-Nehmer/Ara/ara_D.html).

Freed N. Borenstein, N. MIME (Multipurpose Internet Mail Extension). RFC 1521, IETF, 1993.

N. Borenstein. Computational mail as network infrastructure for computer supported cooperative work. In *CSCW '92*, Toronto, 1992.

N.S. Borenstein. Email With A Mind of Its Own: The Safe-Tcl Language for Enabled Mail. In *ULPAA '94*, 1994. (also available as Internet Draft).

M.M. et al. Burnett. *Visual Object-Oriented Programming*. Manning Publications Co, Greenwich, 1995.

Colin Harrison David Levine Colin Parris Gene Tsudik David Chess, Benjamin Grosof. Itinerant Agents for Mobile Computing. *IEEE Personal Communications, The Magazine of Nomadic Communication and Computing*, 2(5), October 1995.

David Canfield Smith et al. KIDSIM: Programming Agents Without a Programming Language. *Communications of the ACM*, 37(7):55–67, July 1994.

J. Schirmer et al. ActiveM3 -An Authoring System for Active Multimedia Mail-. In *Multimedia Computing and Networking 1996*, volume 2667, January 1996.

Tim Finin et al. DRAFT: Specification of the KQML Agent-Communication Language. Technical report, DARPA Knowledge Sharing Initiative, June 1993.

Robert S. Gray. AgentTcl: A transportable agent system. In *CIKM Workshop on Intelligent Information Agent, 4th International Conference on Information and Knowledge Management*, Baltimore, Maryland, December 1995.

Petra Hoepner. Synchronizing the presentation of multimedia objects - oda extensions. In L.Kjelldahl, editor, *Multimedia, Systems, Interaction and Applications (1st Eurographics Workshop, Stockholm, Sweden)*, pages 87–100. Springer-Verlag, 1991.

Java Agent Template, Version 2.0. Available via WWW
  (URL: http://cdr.stanford.edu/ABE/JavaAgent.html).

Java-To-Go: Itinerative Computing Using Java. Available via WWW
(URL: http://ptolemy.eecs.berkeley.edu/ wli/group/java2go/java-to-go.html).

T. Kirste. HCL Language Reference Manual, Version 1.0. ZGDV-Report 68/93, 1993.

W. Lamersdorf M. Merz. Agents, Services and Electronic Markets: How do they Integrate? In *to appear in: IFIP/IEEE International Conference on Distributed Platforms*, Dresden, 1996.

Richard E. Fikes Michael R. Genesereth. Knowledge Interchange Format, Version 3.0, Reference Manual. Technical report, Computer Science Department, Stanford University, Stanford, California, June 1992.

WebHunter. Available via WWW
(URL: http://webhound.www.media.mit.edu/projects/webhound/doc/Webhound.html).

White Paper Oracle. Oracle Mobile Agents, Part A22547. Technical report, Oracle, March 1995.

P. David Stotts and Richard Furuta. Temporal Hyperprogramming. *Journal of Visual Languages and Computing*, pages 237–253, 1990.

SUN Microsystems, Mountain View, CA. *The Java Language Specification Release 1.0 Alpha 2*, March 1995.

TACOMA. Available via WWW
(URL: http://www.cs.uit.no/DOS/Tacoma/Overview.html).

Arif Ghafoor Thomas D.C. Little. Synchronization and Storage Models for Multimedia Objects. *IEEE Journal on Selected Areas in Communications*, 8(3):413–427, 1990.

J.E. White. Telescript Technology: The Foundation for the Electronic Marketplace. Technical report, General Magic Inc, White Paper, 1994.

11

Animation within Mobile Multimedia On-line Services

C. Belz, M. Bergold, H. Häckelmann, R. Strack
Computer Graphics Center (ZGDV)
Wilhelminenstraße 7, D-64283 Darmstadt, Germany
Phone: +49 6151 155-254, Fax: +49 6151 155-480
Email: belz@zgdv.de, URL: http://www.zgdv.de

Abstract

Suitable concepts have to be identified or developed that guarantee the effective usability of mobile multimedia information and services. A platform dedicated to the mobile infrastructure that bundles such concepts is currently being developed within the European ACTS project *Mobile Media and ENTertainment Services* (MOMENTS). The platform will offer multimedia products – information and services – to end users within service trials in a GSM-/DCS-1800-based mobile infrastructure. Further advanced developments, like wide band user channels, will be shown in a Technological Demonstrator. This paper concentrates on the handling of animation within wireless multimedia services.

Technical requirements on the usage of animation that result from the mobile infrastructure and multimedia on-line services are outlined. The potential of animation within mobile services is illustrated via selected application scenarios. Standardization activities for the modeling and encoding of animation are described. A feasible architecture derived for the handling of animation within wireless multimedia on-line services is outlined.

Keywords

Mobile Multimedia, On-line Services, Animation, VRML, Modeling, Encoding, GSM, DCS1800, Wireless Services, World Wide Web

2.4 Application Communication Manager

The *Application Communication Manager* is an application specific interface to the components of our architecture. It can be realized as a built-in interface in especially designed mobile applications or as a separate component that uses a defined interface of an existing application (see Figure 2). Tasks of the ACoM include the management of a list of all requests and related replies of an application together with applied contexts and the decomposition/composition of complex compound replies. The sending ACoM divides compound replies into several subreplies of certain defined types and hands them over to MHs for type specific exchange. These MHs have to be started by the ACoM. They transfer the subreplies to their counterpart at the receiving site (started by its NS) and hand them over to the receiving ACoM that will reassemble the reply object.

2.5 Message Exchange

The protocol that we use for the communication over the Object Bus is a combination of an efficient protocol for coding header information and the possibility to express the contents of messages in an application dependent format, like SQL, HTTP, MIME, or KQML. Since the content of a message should not be evaluated by the Network Scheduler all data that are necessary for planning the transfer, e.g. priority, size, and QoS demands, have to be part of the header. Our protocol allows splitting of larger messages into smaller pieces and sending them separately. The headers of these pieces are provided with offset values so that a transfer can be continued at this offset without a full retransmission after a disconnection has occurred.

3 USE OF CORBA

The acceptance of a new approach is often measured by its integration in the "real world". Due to this, the use of widely known and accepted standards and the integration of legacy systems into the new environment is crucial. We propose the Common Object Request Broker Architecture (CORBA) of the Object Management Group (OMG, 1991) as a platform for the design of object oriented, distributed systems – even in the context of mobile visualization.

We show, how this standard can be used to support resource poor mobile hosts and minimize the amount of data by optimization of the data flow. To do this we map the abstract architecture of the OBus to CORBA and point out which work is necessary to gain interoperability between standard CORBA and the modified broker architecture implementing the OBus.

3.1 CORBA Motivation

The design of software systems in general and of software for special circumstances like mobile environments in particular is a non trivial and expensive challenge. The problems of today's software engineering can only be faced by the use of modern paradigms and powerful tools that support the user by hiding great parts of the complexity one has to cover. The role of CORBA in this context is a platform for integration and distribution: objects can be used in a comfortable way independent of their location in the network, their implementation language

1 INTRODUCTION

Global information management systems – such as the World-Wide Web (WWW) – show that the on-line access to vast amounts of distributed multimedia information is possible not only for the expert, but also for the "naive" end user. At the same time relatively cheap and widely available wireless data communication services are available to the mass-market.

Within the coverage of the cellular network, the vision of information access for "everyone, anytime, anywhere" has become reality. However, because of the low bandwidth of wireless narrowband wide-area networks (such as the 9.6 Kbit/s of GSM, DCS-1800) and the limited resources of mobile hardware, the handling of distributed multimedia applications and services faces severe problems.

Users of mobile systems will expect to get access to all of the multimedia data making up modern information applications including time-dependent data like speech, sound, animation, or video – at least within the limits defined by the mobile system's input and output capabilities. They have become accustomed to comfortable, easy-to-use interactive multimedia systems based on the concepts of direct manipulation. A step backwards in interactivity will cause a serious acceptance problem. So in order to build "everyone, anytime, anywhere" information applications that will be *used*, the problem of how to make these services interactive across a slow data link has to be solved. Suitable concepts have to be identified or developed that guarantee the effective usability of mobile multimedia information and services.

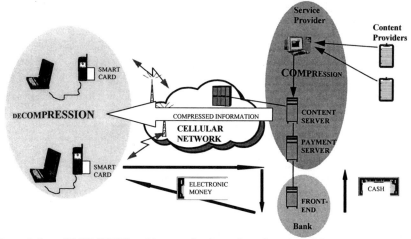

Figure 1 Overall MOMENTS architecture for the projected service trials.

The overall objective of the ACTS* project *MObile Media and ENTertainment Services* (MOMENTS) is to demonstrate the technical feasibility and business viability of a wireless media highway for the distribution of advanced multimedia products. This covers two

---

* Advanced Communications Technologies and Services.

application categories, on-line services and entertainment. The aim of the project is to contribute significantly to the understanding of the users' perception of the values of wireless multimedia services, identify how commercial exploitation of the services using third generation systems can be accelerated, create new enabling technologies, in particular for the presentation of visual material, and make valuable contribution to standardization.

Figure 1 illustrates the overall MOMENTS architecture for the projected service trials that will be conducted in three different European countries.

MOMENTS will allow a realistic assessment of wireless multimedia services and verify the identified business opportunities. System enhancements will be developed and demonstrated in the Technological Demonstrator to access the benefits of wider band user channels and determine how they can be incorporated into UMTS. New techniques that will be evaluated and applied include presentation technologies, optimized for the inherently narrow cellular user channels; wide band user channels allowing transmission rates of the third generation mobile systems (n x 9.6 Kbit/s, n >= 2). This paper focuses on the handling of animation within wireless multimedia services.

Within Chapter 2, the term animation is briefly defined as used in this paper. Subsequently, a state-of-the-art in regard to existing standards and standardization activities for animation is given. Technical requirements to use animation within wireless multimedia on-line services are outlined in Chapter 3. Furthermore, typical mobile application scenarios are presented. A feasible approach for the efficient handling of animation is outlined in Chapter 4.

2 ANIMATION

As the term animation can be used in various ways it is necessary to briefly describe these definitions first and to point out which of them is applied here. Subsequently, standards and standardization activities in the area of animation are outlined.

2.1 Definition

The definitions of the term animation mainly differ in the kind of data that is transmitted to the user's mobile data terminal (MDT), and the handling of the data on the MDT.

On the one hand, the term animation can refer to the playback of animated sequences on the MDT. Changes within an original graphical scene are rendered and recorded as video images. The resulting sequences are transmitted to and displayed on the MDT.

On the other hand, animation can be regarded as changing the appearance of a graphical scene where the graphical description of the scene and the behavior of the objects of the scene (i.e., non-rendered images) are transmitted to the MDT. The rendering of the data is performed on the MDT.

The use of the term animation within this paper refers to the latter definition. The first definition is rather a transmission of video images. The problem arising here is the size of the rendered images. General video is far too large for transmission over mobile narrowband WANs. However, substantial work is being performed in this area to apply video transmission to mobile networks (Nieweglowski, 1996).

Furthermore, user interactivity is an important feature for multimedia on-line services and therefore is also applied to animation. This interactivity comprises user-oriented navigation through the scene as well as the possibility to interactively change the appearance of a graphical scene.

2.2 State-of-the-Art

The description of any animation, independent of the dimension (2D or 3D), consists of two conceptual functional units:
- *Static scene description:* describing the geometric objects of the scene (static over time). and
- *Behavior description*: describing the behavior of the objects within the scene over time (dynamic over time).

In regard to the static scene descriptions several (de-facto) standards are already available (e.g., CGM (Henderson, 1993), VRML 1.0/1.1, DXF (Autodesk, 1992), etc.). Current standardization activities in the context of VRML focus on extending VRML 1.0 (Bell, 1996) functionality in regard to the dynamic aspects of objects.

Currently, the specification of VRML 2.0 (Mitra, 1996) is still in process and will not be finalized before August 1996. VRML 2.0 enables features like interaction and animation. Interactivity is realized by applying sensors to the scene which possess the ability to set off events. These sensors are either sensitive to user interaction (e.g., moving the mouse pointer in a distinct area of the world or by clicking an object of the world) or to time passing by. Furthermore, VRML 2.0 offers basic animation features such as rotation, translation, etc. and additional features such as interpolating colors, positions etc. by introducing interpolators to VRML. Besides, it provides the possibility of including animation scripts or references to theses scripts (URL) in the scene description. These scripts can be written in a variety of languages including Java and C. The animation effects are achieved by means of events, i.e., a script takes input from sensors (receives an event) and generates events based on that input which then can change other nodes in the world. Passing events to the corresponding nodes within a scene graph is provided by the routing mechanism. Figure 2 illustrates the event routing mechanism in VRML 2.0.

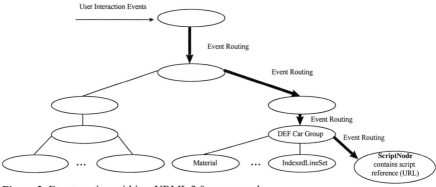

Figure 2: Event routing within a VRML 2.0 scene graph.

In order to reduce the size of VRML 2.0 files, the VRML Architecture Group (VAG) agreed on specifying a binary format for VRML 2.0. Currently, the specification of the binary format based on a proposal by Apple (Pettinati, 1996) is still under development. The specification builds upon the architecture of Apple's 3DMF binary metafile format (Apple, 1996), a file format for 3D applications.

3 ANIMATION WITHIN MOBILE SERVICES

For the usage of animation within mobile multimedia on-line services different technical requirements hold. They are derived from the mobile GSM-/DCS1800-based infrastructure and the MDTs that will be used for the service trials. Furthermore, distinct technical requirements result from the need to use animation with other media simultaneously. Typical application scenarios illustrate the added value that could be achieved with animation.

3.1 Technical requirements

The technical requirements in regard to animation in mobile on-line services are as follows:
- The hardware platform of the MDTs consists of laptops with restricted resources.
- The animation facilities should be oriented on existing (de-facto) standards, if applicable.
- The format for the description of animated graphics should provide sufficient compression to allow "real time operation" at transmission rates of 4.0 Kbit/s and 8 Kbit/s. Then, the transmission rates of GSM or DCS-1800 data channels will allow to transfer other media like speech simultaneously[†].
- The integration of the animation facilities into the browser at the MDT is recommended from the user's point of view.
- The system shall be able to synchronize animation with other media, especially speech.

3.2 Application scenarios

Within mobile on-line services animation can be applied to a variety of services. Sample scenarios illustrating the usability of animation are listed below:
- **Entertainment**: In the area of entertainment animation plays the role of an appetizer in order to increase the attractiveness of presented information for the user. Animation can be applied to leisure information services which inform the user about music, movies, TV programs, etc. Furthermore, animation as an appetizer can entertain the user during waiting periods (e.g., downloading data from the server to the user terminal). Animation during such downloading processes does not only entertain the user but additionally gives feedback to the user that the downloading is still in process. An example for such an appetizer presented to the user during a downloading process is illustrated in Figure 3.
- **Location dependent information**: Information provided on city maps, restaurant guides, traffic jams, etc. belong to this area. As an example, animation in the field of city

---

[†] Simultaneous voice and data calls are not possible.

104 Part Four Mobile Agents and Multimedia Applications

maps could be performed by highlighting in a certain color the route a user has to follow in order to reach a certain target or by letting a car or small figure drive/walk along the streets to a target.
- **Financial information**: Financial information comprises economic indicators, interest rates, stock exchange rates, etc. Within such a service the user would be able to receive on demand a 3D presentation of distinct stock exchange rates over a certain period. It might be interesting for the user to compare the presented rates with those over the same period last year or watch the profits/losses he/she would gain with these exchange rates. This additional information could be presented by including it in the former 3D presentation.

Figure 3 Example of an appetizer presented to the user during a downloading process.

4 HANDLING OF ANIMATION WITHIN MOBILE ENVIRONMENTS

A feasible approach in regard to the handling of multimedia information in mobile services should use existing (de-facto) standards wherever possible (see Chapter 3.1).

The geometric objects of a scene are defined by a standardized graphical modeling language. The behavior of the scene's objects is realized by a script that references the objects of the scene. This script can be either directly hooked into the scene description itself or can be stored separately. The advantage of this concept is the potential to define several object behaviors or animations in the context of one scene description. Furthermore, the animation script may encompass additional static components that update or extend the current scene description. Figure 4 illustrates this principle.

On the MDT the following steps are performed to present the animation to the user:
1. The scene description (those parts needed) is transmitted and successively parsed.
2. The scene is rendered (at least partially) as a starting point of the animation.
3. The animation script is executed step by step, whereby the changes of the scene are continuously re-rendered. Those objects of the scene that are needed and that were not transferred so far to the MDT are transmitted and parsed.

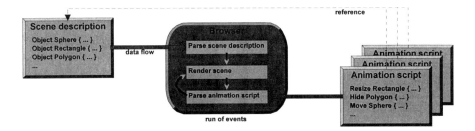

Figure 4 Principle of animation presentation on the MDT (within a browser); several animation scripts can refer to one scene description.

The overall scene is managed on the MDT via a scene graph that can be updated and extended according to events triggered by the user or the system (e.g., predefined updates at distinct time slots that constitute parts of the animation). The following example illustrates this mechanism.

Let us suppose an initial scene description contains elements like a globe (see Figure 5). For every service that the user is able to select an external script is available. These scripts contain information about the elements which represent a 3D presentation of the service selected and their behaviors (e.g., rotation commands). If the user selects a service (e.g., financial information) an event is triggered that initiates the respective external script. The script is responsible for including the elements – describing the 3D presentation of the service selected (e.g., financial information) – in the scene graph (see Figure 5). The scene graph contains additionally now a subtree of nodes that starts for example with *DEF FinancialInformation* – representing the 3D presentation of the financial information service selected by the user. The newly included elements will then be animated, e.g., letting the 3D presentation of the financial information service orbiting the globe (see Figure 6).

Figure 5 Example of a scene graph that is updated and extended via event triggering with an external script. The script encompasses both additional static scene components and new behavior descriptions.

The previously described approach could be based on VRML 2.0 principles applied to the mobile environment. The separation of the static scene description from its behavior description and the separation of one overall scene into several components illustrate the objective on the modeling level. To optimize the transfer of the animation to the mobile terminal encoding issues have to be considered. Comparable to vector graphics formats like CGM – CGM specifies three different encodings: binary, clear text and character encoding – one encoding that is optimized for the mobile environment may have to be specified. When specifying a new encoding, the impacts on lower layers (supporting e.g., TCP header optimization, LZW packet coding, etc.) have to be considered.

Figure 6 Example for an animated presentation of the service currently selected by the user.

5 SUMMARY

In this paper, an approach to handle animation as an integral part of mobile multimedia services has been presented.

Relevant standardization activities in regard to the usage of animation within on-line services have been outlined. Requirements derived from the mobile GSM-/DCS1800-based infrastructure and the MDTs used for the projected service trials have been depicted. Typical application scenarios illustrate the added value that could be achieved with animation.

The approach defined for animation within multimedia services could be based on VRML 2.0 principles applied to the mobile environment. The separation of the static scene description from its behavior description and the separation of one overall scene into several components illustrate the objective on the modeling level. Examples illustrate the approach described.

To optimize the transfer of the animation to the mobile terminal encoding issues have to be considered.

6 ACKNOWLEDGMENTS

This work has been performed in the framework of the project ACTS AC002 MOMENTS, which is partly funded by the European Community. The authors would like to acknowledge the contributions of their colleagues from: Nokia Telecommunications Oy, Bertelsmann AG, Citicorp Kartenservice GmbH, DataNord Multimedia SrL, E-Plus Mobilfunk GmbH, Gemplus, Omnitel Pronto Italia S.p.A, Orange PCS Ltd., Reuters AG, and Zentrum für Graphische Datenverarbeitung e.V.

The authors would also like to thank Hewlett-Packard, the information technologies provider of the project.

7 REFERENCES

Apple (1996) 3D Metafile Reference. URL: http://product.info.apple.com/qd3d/QD3DMetaFile/3DMetafileRefTOC.html.
Autodesk Inc. (1992) Drawing Interchange and File Format - Release 12.
Bell, G., Parisi A. and Pesce, M. (1996) VRML 1.0C. URL: http://vag.vrml.org/vrml10c.html.
Henderson, L.R: and Mumford, A.M. (1993) The CGM Handbook. Academic Press Limited.
Mitra, Honda Y., Matsuda, K., Bell, G., Yu, C. and Bell, G. (1996) The Moving Worlds VRML 2.0 Specification Draft #1. URL: http://vag.vrml.org/VRML2.0/DRAFT1/spec.main.html.
Nieweglowski, J. and Leskinen, T. (1996) Video in Mobile Networks. Accepted by the European Conference on Multimedia Applications, Services and Techniques (ECMAST'96) in Louvain-la Neuve, Belgium.
Pettinati, F., Hecker, R. and Davidson, K. (1996) Apple VRML Binary Proposal. Draft 0.5 URL: http://product.info.apple.com/qd3d/QD3D.HTML.

8 BIOGRAPHY

Constance Belz received a diploma in computer science in 1995 from Darmstadt Technical University (THD). Since November 1995 she is working as a research assistant at the Computer Graphics Center (ZGDV) in Darmstadt. Her work as a staff member of the R&D Dept. "Mobile Information Visualization" concentrates on mobile multimedia applications in general and the modeling and handling of animation for a mobile environment in particular.

Michael Bergold received a diploma in computer science in 1996 from Darmstadt Technical University. Since March 1996 he is working for the Dept. "Mobile Information Visualization" of the Computer Graphics Center (ZGDV) in Darmstadt. His work concentrates on the applicability of Java technologies for efficient content presentation in a mobile environment.

Heiko Häckelmann received a diploma in 1995 from Darmstadt Technical University. Since January 1996, he is working part time for the Dept. "Mobile Information Visualization" at the Computer Graphics Center (ZGDV) in Darmstadt. His work at ZGDV concentrates on the handling of animation within mobile data terminals.

Rüdiger Strack received a diploma in computer science in 1990 from Darmstadt Technical University (THD). He gained his PhD (Dr.-Ing.) at THD in 1995.

From 1990 to March 1995 he worked as a research assistant at the Fraunhofer Institute for Computer Graphics (Fh-IGD) at Darmstadt. His work concentrated on the areas imaging and distributed multimedia environments. Since 1993 he was acting as a project manager at Fh-IGD.

Since April 1995 he is head of the R&D Dept. "Mobile Information Visualization" at the Computer Graphics Center (ZGDV) in Darmstadt. At ZGDV he is responsible for research and development in the areas of mobile computing and distributed multimedia information systems. This includes the coordination of several industrial and scientific projects.

PART FIVE

Networking and Protocols for Mobile Communication

12
Theoretical Analyses of Data Communications Integrated into Cordless Voice Channels

Radhakrishna, CANCHI. and Yoshihiko, AKAIWA.
Graduate School of Information Science & Electrical Eng.,
Kyushu University
6-10-1, Hakozaki, Fukuoka 812-81, JAPAN
Phone : +81-92-642-3890   Fax : +81-92-632-5204.
E-mail : radha@mobcom01.is.kyushu-u.ac.jp

Abstract
Dynamic integration of data into voice channels of 2nd generation cordless systems provides an effective channel utilization. This paper proposes and theoretically examines an Inhibit and Random Multi Access (IRMA) protocol for data terminals in the integrated voice and data system. Analytical expressions are derived to quantify the effect of data inhibition on its performances, i.e. throughput and delay for both infinite and finite population model. We investigate the data performance in two extreme situations: 1. No voice load and 2. Full voice load. The numerical results indicate that a comparable performance can be achieved.

Keywords
TDMA, IRMA, Throughput, Delay

1 INTRODUCTION

The last decade has witnessed an unprecedented development in the arena of wireless data communication systems (Kaveh Pahlavan,1994). The advent of digital cellular and cordless systems primarily for voice applications and local and wide area data networks for data oriented services, has stimulated the ideas and efforts for integration of services in order to evolve Personal Communication Networks. At the other end, the advances in computer technology and signal processing, spurred by the miniaturization, have shown the feasibility of mobile computing as well as personal voice and data communication using a portable device.

Various schemes have been proposed in literature for voice and data integration in wireless networks ((Akaiwa,1994), (GangWu,1994), (Jeffrey,1995), (Sanjiv,1994)). This paper examines an Inhibit and Random Multi Access (IRMA) protocol (Akaiwa,1994) for data transfers to be integrated into voice oriented digital channels of cordless systems such as PHS during the silence periods of speech activity. The data transmission performance degrades due to the constraint, i.e. data inhibition, provided in the integrated voice and data system. We derived analytical expressions to quantify the effect of data inhibition on the performances, i.e., throughput and delay, of the data communications. In our analyses, we consider both infinite and finite number of data terminals. Finally, a finite population case of integration is modeled and analyzed by assuming an approximate Markovian process. The expressions for average throughput and average delay are obtained for a simple two state speech model.

2 SYSTEM DESCRIPTION

The integrated system consists of a base station and a number of terminals, separate for voice and data, that share a common radio channel with a pair of frequencies. The basic access scheme assumed for voice terminals is TDMA with an assignment of channel on demand and an IRMA for data terminals. The base stations broadcast in the format of frames of fixed length that is further divided into equal time slots. The length of each slot is identical to that of a packet. Each packet is an integral of actual information bits and some overheads required for synchronization and signaling. Our area of investigation is limited to the communication between cordless terminals and their base stations.

3 IRMA PROTOCOL FOR DATA COMMUNICATIONS

The data communications are controlled by the respective base stations by broadcasting appropriate control signals at the beginning of each slot. By assuming a voice activated transmission, we dynamically integrate the data into the channel during silence periods of speech activity. Whenever a silence period is detected in a channel occupied with voice communications, the base would indicate this event by broadcasting the corresponding signal in the down link. The data terminal transmits in the form of bursts of duration equal to one time slot. Since the system is voice oriented, the crucial issue is how to avoid collision between voice and data packets. The first packet of the resumed talkspurt indicates the restart of voice activity with a view to avoiding any loss of voice packets. In order to resolve any collisions between voice and data whenever speech transmission is resumed, data terminals are inhibited from transmissions in the following slot and then allowed to access the channel randomly. In other words data terminals follow an IRMA protocol for channel access.

4 ANALYSES

4.1 Case 1: Infinite data terminals and no voice load

The system is assumed to have infinite data terminals and the propagation delays over radio channel is quite small compared to the packet transmission time. Each user have almost a packet ready for transmission. We assume no voice traffic, i.e., all slots are available for data transfers. However, data transmission is inhibited after every collision. The average offered load, G, follows the Poisson distribution. The activity on the channel can be divided into cycles where each cycle contains zero or one or more idle events, an active event and an inhibit event in case of a collision and busy events during reserved slots. The throughput of the channel, S, is the average successful duration in a given cycle of operation. The normalized throughput without any reserved slots is derived as

$$S = \frac{G e^{-G}}{2 - e^{-G} - G e^{-G}} \tag{1}$$

This is compared with the throughput of Slotted ALOHA $S = G e^{-G}$ as in the Figure 1. The maximum throughput achievable is less than that of S-ALOHA and is because a slot is unutilized following every conflict. The throughput obtained for the infinite model reaches a maximum value of 30.2% at an offered load of 76.8%. This throughput value is comparable to that of Slotted-ALOHA where the maximum throughput is 36.8% at an offered load of unity. Moreover, at lighter loads, the effect of inhibition on data transfers is insignificant and throughput is identical to that of S-ALOHA. Furthermore, in typical ALOHA networks the channel traffic is usually in the range of 0.10 or less(Norman Abramson,1994), in which case our system offers the effective usage of the channel.

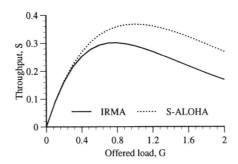

Figure 1 Throughput characteristics of infinite data terminals.

The delay analysis accounts newly arrived and retransmitted packets as separate variables(Kleinrock,1984). The retransmissions take place randomly in one of the following k slots ($1 \leq k \leq K$) after a collision is confirmed and each slot being equally likely with probability $1/K$, where K is a backoff parameter. The more accurate expressions of normalized throughput S and delay D_n, by considering the delay experienced by collided packets, are obtained as

$$S = G \frac{q_r}{q_r + 1 - q_n} \tag{2}$$

where

$$q_n = \left[\frac{G}{K} e^{-G} + e^{-\frac{G}{K}} \right]^K \left(\frac{1}{2 - e^{-G} - G e^{-G}} \right) e^{-S},$$

$$q_r = \left[\frac{e^{-\frac{G}{K}} - e^{-G}}{1 - e^{-G}} \right] \left[\frac{G}{K} e^{-G} + e^{-\frac{G}{K}} \right]^{K-1} \left[\frac{1}{2 - e^{-G} - G e^{-G}} \right] e^{-S} \quad \text{and}$$

$$D_n = 1 + H \left(1 + \frac{K+1}{2} \right) \tag{3}$$

where $H = (G/S - 1)$ = average retransmissions given in terms of S, G and K.

Figure 2 depicts the delay performance given by the equations (2) and (3) for the infinite terminals model. The delay that a data packet experiences, in terms of slots, is obviously higher than that of S-ALOHA since a slot is unutilized by data terminals for every collision. Nevertheless, the delay values are comparable to that of S-ALOHA.

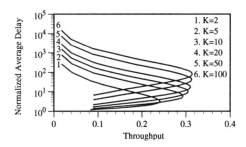

Figure 2 Delay characteristics of infinite data terminals.

4.2 Case 2: Finite data terminals and heavy voice load

A finite population case of integration is modeled and analyzed by assuming an approximate Markovian process. The expressions for average throughput and average delay are obtained for a simple two state speech model. In this section, we derive the expressions for throughput and delay with the help of Steady (Equilibrium) State Analysis.

Since the system is voice oriented, the blocking probability of the integrated system is same as that of voice only system. In this section we donot consider the signaling traffic of voice users and call set up procedures but we assume fixed number of connections by voice users all the time. Though this fixed number of connections can be evaluated by considering some performance measures, such as blocking probability, we assume a typical heavy (full) loading conditions during busy hour where all channels (slots) are occupied voice. In other words, the fixed number of connections equal to the number of slots in a TDMA frame. Let each TDMA frame be divided into N equal time slots, of which N_v slots, where $0 \le N_v \le N$, are always occupied by voice calls. In the present case $N_v = N$.

Voice Traffic Modeling
For voice traffic, the talkspurts and silence periods within a voice call are also exponentially distributed with means $T_t = 1/\alpha$ seconds and $T_s = 1/\beta$ seconds, respectively. The steady state (equilibrium) for Talkspurt and Silence can be obtained as

$$P_T = \frac{P_{S2T}}{P_{T2S} + P_{S2T}} \quad \text{and} \quad P_S = \frac{P_{T2S}}{P_{T2S} + P_{S2T}} \tag{4}$$

where P_{T2S} and P_{S2T} are Talkspurt to Silence and Silence to Talkspurt transition probabilities in a slot duration T sec, respectively.

Data Traffic Modeling
The data traffic is assumed to be an approximate Poisson process with arrival rates λ_d messages per second. The data message length is assumed to have a geometric distribution with mean $1/\delta_d$. Consider a finite number of data terminals M. The behavior of each terminal can be described by a model with three states: data generating(DG), contending for a channel (DC), and transmitting data packets (DT) successfully.

By assuming the discrete random process points at the beginning of frames, consider a point of time t at the beginning of a frame "x" (x is an integer) as shown in Figure 3. If l_x, m_x, k_x denotes the number of data terminals generating, contending and transmitting respectively, then the frame x can be described by state vector $\{l_x, m_x, k_x\}$. Since the total number of terminals is finite and constant, M, the state vector can be simplified to $\{m_x, k_x\}$ where l_x is given as $\{M - m_x - k_x\}$. Since the transition to the state $\{m_{x+1}, k_{x+1}\}$, at $t+1$, depends only on the events at t, we can follow a Markovian analysis. A steady state distribution exists for this process since $0 \leq m_x \leq M$ and $0 \leq k_x \leq N$.

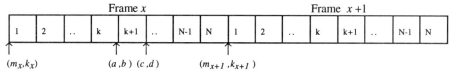

Figure 3 TDMA frame and instants of state transitions.

If π_{ij} denotes the steady state probability of the system in state (i,j) and p_{ijrs} denotes the one step state transition probability from state (i,j) to (r,s), the steady state probabilities given by solving a set of linear equations

$$\Pi = \Pi P \text{ and } \Sigma \pi_{ij} = 1, \tag{5}$$

where $\Pi = \{\pi_{ij}\}$ = steady state vector, and $P = \{p_{ijrs}\}$ = State Transition Matrix.

Since each frame has N slots and the state at the end of the N th slot gives the state at $t+1$, i.e., at the beginning of frame $x+1$. It is enough, if we evaluate the state at the end of N th slot for a given state at t. To arrive at this, we consider all possible state transitions within the frame of x by adopting an approximate Markovian analysis again.
While (a,b) being the state vector at the end of k th slot, $k+1$ th time slot (channel) of frame x exists in one of the following states as far as the data terminals are considered :

       1. Talkspurt state
       2. Occupied by a data terminal
       3. Inhibit state
and   4. Idle state

Let P_{TS}, P_{DT}, P_{IN} and P_{IDLE} denotes the probabilities of a channel being in the above states, respectively and can be obtained as

$$P_{TS} = P_T \frac{N_v}{N} = \frac{P_{S2T}}{P_{T2S} + P_{S2T}} \frac{N_v}{N} \tag{6}$$

$$P_{DT} = \left(1 - P_{TS}\right) \frac{b}{N} \tag{7}$$

$$P_{IN} = 1 - P_{TS} - \left(1 - P_{TS}\right)(1-p)^a + \left((2 P_{TS})-1\right) P_{ST} \tag{8}$$

Then the probability of an idle state is given by

$$P_{IDLE} = 1 - P_{TS} - P_{DT} - P_{IN} \tag{9}$$

The probabilty that a voice slot is in silence state is given as $P_{SIL} = 1 - P_{TS}$.
The elements of P are obtained by an iterative equation given as

$$p_{ij}(c,d) = \sum_{b=0}^{N} \sum_{a=0}^{M-b} q_{abcd} \, p_{ij}(a,b) \tag{10}$$

where $a = m_x(k) = mx$ @ the end of k th slot.
$b = k_x(k) = kx$ @ the end of k th slot.
$c = m_x(k) = mx$ @ the end of $k+1$ th slot.
$d = k_x(k) = kx$ @ the end of $k+1$ th slot.
$p_{ij}(a,b)$ = The probability of state $\{a,b\}$ given the state at the start of frame, i.e.(i,j).
q_{abcd} = one step(slot) transition probability from state (a,b) to (c,d) taking into account of the probabilities of silence state, inhibit state of the channel etc.

The matrix P is constructed by using the iterative equation (10) for all possible values of i,j, where $0 \leq i \leq M - i$ and $0 \leq j \leq N$ and then steady state vector $\{\pi_{ij}\}$ is calculated.

The total average data throughput for the finite population model during the silence periods of channel is given by

$$S = S_r + S_c \tag{11}$$

where S_r, the ratio of the average number of data packets successfully transmitted in a frame to the number of slots(N) in a frame by taking average silence and inhibit probabilities, P_{SIL} and P_{IN} into account, and S_c, the average number of data terminal that are successful in a slot in an equilibrium state given a data generation rate σ, can be evaluated as

$$S_r = \frac{P_{SIL}}{N} \sum_{k=0}^{N} \sum_{m=0}^{M-k} k \, \pi_{mk} \quad \text{and} \quad S_c = M\sigma - \sum_{k=0}^{N} \sum_{m=0}^{M-k} (m+k) \, \pi_{mk} \, \sigma \tag{12}$$

The total average delay experienced by data packets for this case is given as

$$D = D_c + D_r (\text{Packet Length}) = D_c + D_r ((1/\delta_d) - 1) \tag{13}$$

where, By Little's results,

$$D_c = \frac{\sum_{k=0}^{N} \sum_{m=0}^{M-k} m \, \pi_{mk}}{S_c} \quad \text{and} \quad D_r = \frac{\sum_{k=0}^{N} \sum_{m=0}^{M-k} k \, \pi_{mk}}{S_r} \tag{14}$$

5 NUMERICAL RESULTS OF THE PERFORMANCES

In order to obtain the numerical solutions, the iterative procedure of relaxation is applied to the equations (5), i.e.,

$$\Pi = \lim_{n \to \infty} \Pi_o P^n \tag{15}$$

where n is chosen sufficiently large so that truncation error is negligible and the calculation is terminated when iterative values of Π converges. For numerical results the system with the parameters listed in Table 1 is considered.

Table 1 System Parameters

Parameter	Symbol	Nominal value	
TDMA frame length (*m.sec*)	T	5	
Slots per TDMA frame	N	4	
Average talkspurt duration (*sec*)	T_t	1	
Average silence duration (*sec*)	T_s	1.35	
Voice source coding(*Kbps*)	R_s	32	
Data terminals	M	25	(variable)
Data generation rate (*packets / slot*)	σ	0.004	
Mean data length rate	δ_d	0.1	
Probability of data transmission	p	0.1	(variable)

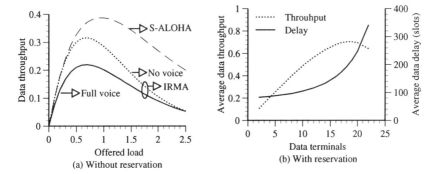

Figure 4 Performance characteristics of finite population model.

The performance achieved with the IRMA protocol with the transmit probability 0.1 is shown in Figure 4. Figure 4(a) shows the throughput curves for No voice load and Full voice load for the finite data terminals without any slot reservation facility. The maximum throughput per silence duration during heavy voice load conditions found to be 22%. Figure 4(b) represents the average throughput and delay performances as the number of terminals with slot reservation facility per base station is varied. The maximum number of data terminals per base station for stable operation is 19 and the corresponding throughput and delay are 70.44% and 216.79 slots respectively. Figure 5 shows the performance for 16 data terminals as the transmission probability is varied. The optimum range of transmit probability over which high throughput and low delay can be achieved is found to be 0.05 - 0.15.

6 CONCLUSIONS

An IRMA (Inhibition based Random Multiple Access) protocol suitable for data integration into the voice channels of digital cordless system has been theoretically analyzed. The analytical results indicate that a comparable performance of data communications can be achieved by integrating data during the silence duration of speech activity despite the constraint,i.e. inhibition, on the random access of data terminals.

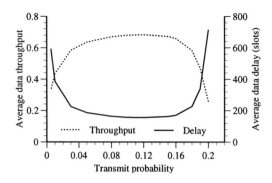

Figure 5 Performance curves versus transmit probability with M = 16.

7 REFERENCES

Akaiwa, Y. and Nakajima, A.(1994), Integration of wireless local area data communication to digital cordless telephone channel, *Proceedings of IEEE International Conference on Personal Wireless Communications*, 190-3.

Gang Wu, K. Mukumoto and A. Fukuda (1994), Integration of Voice and Data in Wireless Information Networks with Data Steal into Voice Multiple Access, *IEICE Transactions on Communications*, Vol.E77-B, 939-47.

Jeffrey E. Wieselthier and Anthony Ephremides(1995), Fixed and Movable-Boundary Channel-Access Schemes for Integrated Voice/Data Wireless Networks, *IEEE Transactions on Communications*, VOL.43, 64-74

Kleinrock, L. (1976) *Queuing Systems*, Vol.2, Computer Applications, New York:

Kaveh Pahlavan et al.(1994) Wireless Data Communications, *Proceedings of the IEEE*, Vol. 82, 1398-430.

Sanjiv Nanda (1994), Stability Evaluation and Design of the PRMA Joint Voice Data Syatem, *IEEE Transactions on Communications*, VOL.42, 2094-104:

Norman Abramson , Multiple Access in Wireless Digital Networks, *Proceedings of the IEEE*, Vol. 82, 1360-70.

8 BIOGRAPHIES

RADHAKRISHNA CANCHI received an M.Tech degree in electronics engineering from Regional Engineering College, Warangal, India. He has been a research engineer in C-DOT, India since 1989. He has been involved in the development of TDMA Point to Multi Point radio communications at C-DOT. He was awarded Hindu-Hitachi Scholarship and undergone a training at Hitachi Telecommunication Division, Yokohama for six months in 1992. Now, he is a student of doctorate course in Intelligent Systems at Kyushu University on MONBUSHO (Japanese Government) 1994 Scholarships. He is a student member of IEEE.

YOSHIHIKO AKAIWA received his doctorate degree in electronics engineering from Kyushu University in 1979. He joined Central Research Laboratories, NEC in 1968. Later he moved to Kyushu Institute of Technology where he was a professor from 1988 to 1996. Now he is a professor at Kyushu University. He has been involved in the research and development of mobile radio communication techniques for 20 years. He received the paper award and the distinguished achievement award from IEICE, Japan and the Avant Garde award from IEEE Vehicular Technology Society. He is a member of IEICE, Japan and a senior member of IEEE.

13
Random Access, Reservation and Polling Multiaccess Protocol for Wireless Data Systems

Theodore V. Buot
Centre for Telecommunications Information Networking,
University of Adelaide, 33 Queen St, Thebarton, SA 5031 Australia
tel: 61-8-3033233   fax: 61-8-3034405
email:tbuot@ctin.adelaide.edu.au

Abstract
This paper propose a Medium Access Control (MAC) protocol for wireless data networks. The main feature of this protocol is the combination of Polling and Slotted ALOHA contention in the multiaccess environment to insure stability and robustness. The frame structure in Advanced TDMA is used to guarantee that polling or random access are allocated with timeslots in every frame. The performance of such protocol is evaluated using simulation and Transient Fluid Approximation (TFA). It is shown that the protocol is adaptive to varying access rates when the system is near or in the overload region.

Keywords
wireless data, reservation, polling, random access, ALOHA, Transient Fluid Approximation

1 INTRODUCTION

Reservation protocols are found to be attractive in wireless data systems where the traffic is characterized as non-continuous. Under this assumption, the terminals alternate between silence and active modes depending on the users' activity. Protocols that exploit this feature combine random access and reservation in order for the busy terminals to transmit longer messages thereby increasing the efficiency of the radio resource. For simplicity, the random access mechanism commonly used is Slotted ALOHA. Examples of these protocols are Reservation ALOHA (Lam,1980), Packet Reservation Multiple Access (Goodman et al, 1989) and Reserved- Idle Signal Multiple Access (Wu et.al., 1994). However, reservation protocols suffer from decreased throughput if the average message length is relatively short. As described by Tobagi (1980) adaptive protocols are required to maintain the performance.

In the wireless environment, a substitute for random access is **polling**. Polling ensures stability but it enduces large access delay even if the system is under light load conditions. Conversely, random access algorithms are prawn to instability or unfairness problems but attains short access delay in the low load region. To mention, Slotted is unstable in some region but it attains a certain level of fairness. In contrast, stack algorithm is very stable but it exhibit unfairness due to its LCFS counter discipline. In this paper a marriage between random access and polling is investigated. The random access will provide short access delay under light loads while polling will enhance the access performance at high load. The use of combined polling and random access was proposed in (Lu and Chen, 1994), and later in (Li and Merakos, 1994) all for integrated voice/data wireless system. In those protocols, the use of polling is mainly for high priority users. In this paper, we combine polling, random access and reservation to develop an **adaptive reservation protocol**.

2 PROTOCOL DESCRIPTION

In this paper, we named the proposed protocol as **SCARP** which stands for **Silence-Contention-Acknowledgment-Reservation-with-Polling**. This protocol is a variant of Advanced TDMA in which the addition of a polling state has a threefold advantage. First is its stability which provide fast recovery of the Slotted ALOHA when it operates in the high delay stable operating point or in the unstable region. In this case, a higher retransmission probability parameter can be used in the Slotted ALOHA thereby improving its performance in the underload region. Secondly, the frame structure does not have to be optimized according to the traffic statistics since the polling mechanism enhances its adaptability to various traffic types as it provides extra capacity to the access rate. So the SCARP protocol can manage to maintain higher throughput even if the average message length is relatively short.. And thirdly is the ability for the base station (central control) to poll immediately users that has just finished transmitting if in case some packets are erroneous. If this function is incorporated in the random access, this will cause system overload if the probability of packet error is high.

2.1 Channel Structure

The channel structure of Advanced TDMA is used with the exception of the Fast Paging Acknowledgment slot. In an n-slot TDMA frame, random access (r) slots are allocated for random access in the uplink. These r slots are distributed in the frame to minimize the latency and to provide enough time for acknowledgment. As explained in Dunlop et al (1994), every r slot is paired with an a slot for acknowledgment and slot allocation. The random access feedback requires only a small portion of the information field. Therefore most of the information field in the a slots are mainly used for slot allocation purposes. Thus, it is possible to provide multiple acknowledgment in one a slot. Also, a time shift is provided for the uplink and downlink greater than the round trip propagation delay + processing time for the contention process. The slot size is uniform with a duration of τ.

2.2 State Transition Cycle

In the protocol, we assume that data information are generated by multiple users registered to a single base station. During the initial transmission, the terminal listens to one of the radio channels in the nearest base station and synchronize in order to contend together with the existing users. In the random access, the terminal identify itself as a new user and is subject for authentication. Upon authentication, a terminal identity, *TI* is allocated to every terminal in the base station. The *TI* must include a base station identifier in order to avoid confusion with the neighboring cells. With the limited length of the information field, only one access is allowed in every *r* slot.

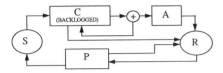 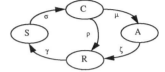

Figure 1 State Transition Cycle of SCARP   **Figure 2** Simplified Markov Model

The terminal activity is assumed to be alternating between *idle* and *active* modes. A terminal is said to be idle if it has no packet in its buffer.. We describe the protocol cycle by starting with a user in the idle mode. An idle user is identified as in the silent state (S). When it becomes active, it goes to the contention state (C) and attempts for contention in the first incoming *r* slot in the TDMA frame. The contention process is Slotted ALOHA with no capture. Unsuccessful users retransmit in the next *r* slot with a probability *p*.

After a successful contention, the terminal goes to the acknowledgment state (A) and waits for a slot to be allocated. When a slot is available during the contention, the terminal is assigned immediately with a slot in the first *a* slot. Otherwise it waits on a first come first serve basis. A priority assignment is also possible as well as multislot allocation can be implemented. When a slot is allocated, the terminal moves to the reservation state (R) and start transmitting its packets. After all the packets in its buffer are transmitted, the terminal loose its reservation and goes to the polling state (P). At the polling state a terminal waits for the acknowledgment if all the packets are transmitted correctly. If all packets are succesfuly transmitted, the terminal goes back to the *S* state. Terminals in the silent and contention states are polled with lower priorities than the terminals in the polling state. This allows terminals in the contention states have two access mechanisms. If the packet error probability is low, the rate of polling for contending terminals is high thereby reducing the average access delay.

3 PERFORMANCE APPROXIMATION

Here we present an approximation based on TFA as in (Wu et al,1995). In TFA the transition intensities are similar to the Semi-Markov Flow Graph therefore its accuracy is subject to the distribution of the time a user spend in a state. Another thing to consider in flow graphs is that every process must be unique and exclusive to a state resulting to an imbedded Markov chain

(Clymer,1990). Since the polling mechanism involves a common process to both the A and P states. The Markov model of the state transitions are simplified in the Figure 2. The exclusion of the polling state is due to two reasons. One is due to the assumption of a noiseless channel. The other is the small probability that a user in the polling state will become active and attempt a random access then become successful. From the model in Figure 2 the following stationary transition probabilities are assumed to be exponentially distributed:

$$\sigma = 1 - e^{(-\tau/Ts)} \tag{1}$$

$$\gamma = 1 - e^{(-\tau/Tr)} \tag{2}$$

In TFA, we are interested in the system state in the *k*th iteration. Therefore the nonstationary transition probabilities must be solved in each iteration as follows:

$$\mu_{(k)} = C_{(k-1)} p (1-p)^{C_{(k-1)}-1} (1-\sigma)^{S_{(k-1)}} + S_{(k-1)} \sigma (1-\sigma)^{S_{(k-1)}-1} (1-p)^{C_{(k-1)}} \tag{3}$$

$$\zeta_{(k)} = \min\left\{ \left[\frac{Ni - R_{(k-1)}}{Na} + R_{(k-1)} \gamma \right], A_{(k-1)} \right\} \tag{4}$$

The calculation of p assumed a *last out first serve* polling sequence. Therefore we are interested in calculating the probability that a polled user is busy and has not succesfully contended in the r slots it has passed. First, the polling cycle has to be calculated as

$$t_{p(k)} = \left\lfloor \frac{S_{(k-1)} + C_{(k-1)}}{\frac{(Ni - R_{(k-1)})}{Na} + R_{(k-1)} - \mu_{(k)}} \right\rfloor . \tag{5}$$

The probability that a particular user will be successful in the access slots is

$$\rho_{s(k)} = \frac{\mu_{(k)}}{C_{(k-1)} + S_{(k-1)} \sigma} . \tag{6}$$

Thus the probability that a user becomes successful in the *t*th access slot after it becomes active is geometrically distributed expressed in (7).

$$P_s(t)_k = \rho_{s(k)} (1 - \rho_{s(k)})^{t-1} \tag{7}$$

where $t = \{1,2,3....t_p\}$. Then the probability that the user is successful before it is polled U_p, is expressed in (8) and (9) and the probability of succesful polling is in (10).

$$F(t)_k = 1 - \exp(-t/T_s) \qquad (8)$$

$$U_p(t_p)_k = \frac{\sum_{z=1}^{t_p} \sum_{m=1}^{t_p-z} P_s(m)_k F(z)_k}{\sum_x F(x)_k} \qquad (9)$$

$$p_{f(k)} = F(t_p)_k \left(1 - U_p(t_p)_k\right) \left(1 - \left[\frac{\mu_{(k)}}{C_{(k-1)} + S_{(k-1)}\sigma}\right]\right) \qquad (10)$$

From (10) we can calculate the rate of successful polling as the product of the probability of successful polling and the rate of polling in (11). The successful polling rate is subject to (12)

$$\tilde{\rho}_{(k)} = p_{f(k)} \left[\left[\frac{N_i - R_{(k-1)}}{N_a}\right] - \varsigma_{(k)} + R_{(k-1)}\gamma\right] \qquad (11)$$

$$\rho_{(k)} = \min\{\tilde{\rho}_{(k)}, (C_{(k-1)} + S_{(k-1)}\sigma - \mu)\} \qquad (12)$$

From the above transition probabilities, the number of users in each state for the *k*th iteration is the sum of the flow minus the sum of the outflow added to the value in the previous iteration conditioned that the total number of users in all states, M is fixed. Then we have

$$S_k = S_{(k-1)} - S_{(k-1)}\sigma + R_{(k-1)}\gamma \qquad (13)$$

$$C_k = C_{(k-1)} - \mu_{(k)} + S_{(k-1)}\sigma - \rho_{(k)} \qquad (14)$$

$$A_k = A_{(k-1)} - \varsigma_{(k)} + \mu_{(k)} \qquad (15)$$

$$R_k = R_{(k-1)} - R_{(k-1)}\gamma + \varsigma_{(k)} + \rho_{(k)}. \qquad (16)$$

After calculating the mean values of the state population, the access delay is calculated as the averaging of polling and of contention and acknowledgment delays expresses in (17).

$$\overline{D_a} = \left[\frac{C_\infty + S_\infty \sigma}{C_\infty} + \frac{\left(\frac{\mu_\infty}{\mu_\infty + \rho_\infty}\right)}{\left(1 - \frac{R_\infty}{N_i}\right)}\right] N_f \qquad (17)$$

124   *Part Five   Networking and Protocols for Mobile Communication*

The parameter $\frac{\mu_\infty}{\mu_\infty + \rho_\infty}$ is also a measure of the proportion of random access against polling. And lastly, the throughput is calculated as the mean number of users transmitting divided by the number of slots per frame, ($T = R_\infty/N$).

4 SIMULATION MODEL

In the simulation, a noiseless channel is used with 16 slots per TDMA frame (see Figure 3). Two access slots were allocated sufficient to provided fast access with minimum overhead with the used traffic statistics. The process starts with all users in the silent state. Since a single arrival model is used for message generation, a user is allowed to wait indefinitely if it is either in the contention and acknowledgment states. In the case of polling, a Last Out First Served (LOFS) discipline is used. Users just being polled and users just finished transmitting are held at the end of the polling sequence. Since a noiseless channel is assumed, only users in the silent and contention states are polled.

5 OBSERVATIONS

The advantage of using combined polling and random access is clearly shown in the comparison with SCARP and ATDMA. From the results, the ATDMA is unstable when the access rate is high caused by the reduction of the average message length. Also, when the operating point of the Slotted ALOHA is already close to the maximum throughput (0.36), the tendency that the contention process to flip-flop between the two stable regions cannot be avoided causing a large access delay. Even when the access rate is relatively low, the SCARP protocol still exhibit superior delay performance. Polling compensates the contention process when the S-ALOHA operated in the high delay region as it pulls back the operating point to the low delay region. Under the SCARP protocol, it is also possible to use higher retransmission probabilities since the contention process can already avoid the unstable state. This enable the system to obtain shorter access delays. The results are plotted in Figures 4-8.

6 CONCLUSION AND RECOMMENDATION

In this paper, combined polling, random access and reservation shows a significant advantage for access protocols where the traffic statistics is unpredictable. The effectiveness of polling in the overload region prevents the system from operating in the unstable region thus allowing the operating point to return to the low delay region during time of low access rates. The simulation shows that the polling shares more than 75 percent of the successful access in the overload region. Also, the bistable characteristics disappear in the combined access scheme where the polling mechanism exploits the high delay region. In this paper, a noiseless channel is considered. However, it is necessary to evaluate the performance of this protocol under a noisy channel. It is also recommended to test the performance under bursty traffic.

7 REFERENCES

Lam, S.S. (1980) Packet Broadcast Networks - A Performance Analysis of the R-Aloha Protocol, in *IEEE Transactions on Computers*, Vol.C-29, No.7, Jul. 1980, pp 596-602.

Goodman, D.J., Valenzuela, R.A., Gayliard, K.T. and Ramamurthi, B. (1989) Packet Reservation Multiple Access for Local Wireless Communications, in *IEEE Transactions on Communications*, Vol.37, No.8. Aug.1989, pp 885-890.

Wu, G., Mukumoto, K. and Fukuda, A (1994) Performance Evaluation of Reserved Idle Signal Multiple-Access Scheme for Wireless Communications Networks, in *IEEE Transactions on Vehicular Technology*, Vol.43, No.3, Aug. 1994, pp 653-658.

Tobagi, F. (1980) "Multiaccess Protocols in Packet Communication Systems," *IEEE Transactions on Communications*, Vol. COM-28, No.4, Apr 1980, pp. 468-488.

Lu, C.C. and Chen, K.C. (1994) A combined polling and random access protocol for integrated voice and data networks, in *Proceedings of Personal, Indoor and Mobile Radio Communications Conference (PIMRC'94)*, Sep. 18-23, The Hague, Netherlands.

Li, F., and Merakos, L. (1994) Voice/Data Channel Access Integration in TDMA Digital Cellular Networks, in *IEEE Transactions on Vehicular Technology*, Vol.43, No.4, pp.986-996, Nov. 1994.

Dunlop, J., Cosimini, P. and Robertson, D. (1994) Optimisation of Packet Access Mechanism for Advanced Time Division Multiple Access (ATDMA), in *Proceedings of 44th IEEE Vehicular Technology Conference*, Stockholm, June 1994, pp. 1040-1044.

Wu, G., Mukumoto, K., Fukuda, A., Mizuno, M. and Taira, K. (1995) A Dynamic TDMA Wireless Integrated Voice/Data System with Data Steal into Voice (DSV) Technique, in *IEICE of Japan Transactions on Communications*, Vol.E78-B, No.8, August 1995, pp 1125-1135.

Clymer, J. (1990) *Systems Analysis Using Simulation and Markov Models*, Prentice-Hall, Inc, pp. 169-179.

8 FIGURES

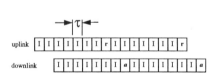

Figure 3 Frame Structure of 16-slot TDMA

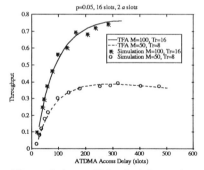

Figure 4 Access Delay in Advanced TDMA Protocol

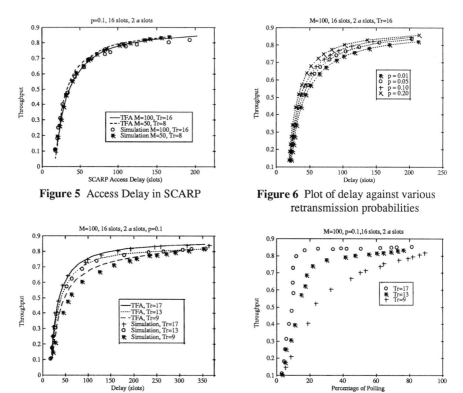

Figure 5 Access Delay in SCARP

Figure 6 Plot of delay against various retransmission probabilities

Figure 7 Access delay for Different Message Length

Figure 8 Plot of the Percentage of Polling in Making Reservation

9 BIOGRAPHY

Theodore Buot received the B.S.(Eng.) in Electronics and Communications from the Cebu Institute of Technology, Cebu, Philippines, in 1989 and the M.Eng. in Telecommunications from the Asian Institute of Technology, Bangkok, Thailand, in 1992 under the grant of the Government of The Netherlands. In 1993, he worked as a research engineer in TMX International. In 1994 he received a grant for postgraduate study at The University of Adelaide. Currently, he is enrolled as a PhD student at the university's Centre for Telecommunications Information Networking (**CTin**). His interests include integrated systems, access protocols and teletraffic engineering. He is a Student Member of IEEE.

PART SIX

Methods and Algorithms for Mobile Information Access

14
Challenges in Mobile Information Systems and Services
-Extended Abstract -

R. Strack
Computer Graphics Center (ZGDV)
Wilhelminenstraße 7, D-64283 Darmstadt, Germany
Phone: +49 6151 155-231, Fax: +49 6151 155-480
Email: strack@zgdv.de, URL: http://www.zgdv.de

1 INTRODUCTION

Mobile Computing introduced mobility as a new dimension to computer applications. Users of mobile devices are freed from location constraints and do not longer need to stick to their cumbersome desktop systems. Mobile devices such as laptops, notebooks, or PDAs – providing the necessary communication facilities and basic services already by themselves or connected to a data capable (cellular) phone – can be used to access remote services and resources. In order to realize the vision of "All information at your fingertip" for both experts and end users, applications must be provided that are capable of interactively accessing, manipulating, and visualizing distributed multimedia information. Users expect to handle all types of multimedia data (including time-dependent data) with mobile devices as with desktop systems at least within the limits of those systems.

However, due to the low bandwidth of wireless narrowband wide-area networks (such as the 9.6 Kbit/s of GSM, DCS-1800) and the limited resources of mobile devices, the handling of distributed multimedia applications and services faces severe problems. Thus, effective solutions have to be provided as an underlying basic technology for mobile information systems and services. We call this research field mobile information visualization.

Besides the fundamental problems raised in regard to response times and transfer rates required for mobile multimedia applications and services, mobile information visualization addresses numerous practical usability issues. This includes:

- visualization and navigation tools for complex/huge information spaces (e.g., the WWW);
- downloading mechanisms for visualization purposes;
- adaptable and configurable user interfaces; usage of real-world metaphors; and
- support of new input devices (pen, touch screen) and interaction techniques (handwriting, speech recognition).

Below, selected research projects and prototypes of ZGDV are outlined that contribute to that field.

2 MOBILE MULTIMEDIA DATA HANDLING

The objective is to develop a mobile multimedia middleware (MoWare) that provides a layered architecture for developers of mobile multimedia applications with a set of APIs, tools, and services at different levels of abstractions. MoWare, that is currently being established, addresses the provision of enhanced services for multimedia data handling and interfaces for the integration of communication mechanisms and/or services into applications.

One basic concept being established for the MoWare is the internal representation of a multimedia task defined by the user as a pipeline. Each component of the pipeline – a pipeline is either pre-composed or being established dynamically – represents an atomic process. A framework is being established that enables the dynamic utilization of available resources via resource dependent delegation of the processes. This framework encompasses a resource information base as well as a resource and a task manager.

3 SUPPORTED SEARCH BY CONTEXT VISUALIZATION

The objective of SuSe (Supported Search Service) was the establishment of a tool that supports the location and discovery of information within complex information structures on stationary and mobile devices. The prototype[*] of an interactive resource discovery system visualizes the current context to improve the exploration of the organization and the content of a resource space.

The SuSe server is a meta-information system containing indexes of summaries and a classification repository. The summaries are reasonably small sized descriptor objects, which contain meta-information to describe a document and to point anyone to the original data. The classification repository describes the relationship between these characteristics. This enables a structured view onto the available documents.

The user interface of SuSe combines the advantages of browsing and searching. It uses the information categories of the server to guide the user, who is searching for information in its resource space. A request is visualized via a search tree. At each level of the search tree the user can refine his request by defining a further descriptive search or by selecting the sub-context, that best matches his information needs. Several user interface metaphors and the 3D-perspective are used for the support of intuitive interaction even for the first-time, naive users.

4 HYPERFUNK – A PERSONAL MOBILE INFORMATION SYSTEM

HyperFunk (Hypermedia in Mobilfunk) was a project for the development of innovative information services with wireless data transmission[†]. The HyperFunk prototype represents a

---

[*] The SuSe prototype was developed within a DeTeBerkom project, tailored for a typical "real-world" application scenario of a town council.

[†] It was carried out by ZGDV in cooperation with DeTeMobil GmbH.

dynamic extensible mobile information system that provides access to both public and private data. The modules and concepts of HyperFunk can be used as a basis for a wide variety of mobile applications due to their configurability and extensibility. Examples can be found in all fields where access to remote documents is necessary, for example in medicine, maintenance, and sales.

On the architectural level HyperFunk consists of two main components: Mobile data clients (Windows95-based) of the user and the stationary data server (UNIX based). The server offers various information services that can be accessed by the mobile user through the data services of the GSM-based D1 network.

In regard to information visualization and interaction HyperFunk offers a dynamic object system based on HCL, a LISP-like dialect. Specialized object behavior can be migrated transparently from the server to the client. The object system enables the introduction of new information types as well as their interaction and presentation methods: The system provides an easy modeling of the information by information providers. The device of the mobile user recognizes new classes of information and loads automatically respective method definitions from the information provider.

The user interface of HyperFunk represents the information space through real world metaphors: The user is able to interact with the system based on familiar analogies to concepts they already know. The main user interface metaphor, a service center, visualizes a building that provides services and personal information in different rooms of the building. HyperFunk offers a wide variety of interaction methods which help users to navigate and find information in an intuitive way.

15
A Buffer Overhead Minimization Method for Multicast-based Handoff in Picocellular Networks

Eunyong Ha, Yanghee Choi, Chongsang Kim
Seoul National University
Department of Computer Engineering, Seoul National University,
Silim-Dong, Kwanak-Ku, Seoul 151-742, Korea
Telephone : +82-2-887-8992
email : `eyha@twins.snu.ac.kr, yhchoi@smart.snu.ac.kr,`
`cskim@sparc.snu.ac.kr`

Abstract
A picocellular network consists of picocells whose radius is of the order of ten meters. The smaller cell size causes frequent handoffs of mobile hosts. In order to support seamless communication service to mobile users, we need a new handoff scheme. In this paper, we propose a multicast-based pre-handoff scheme , called SGMH (SubGroup Multicast-based Handoff), which minimizes buffer overhead for seamless comnunication service as well as handoff processing delay. We show the performance in terms of blocking probability through analysis and simulation.

Keywords
Handoff, Buffer Overhead, SGMH, Picocellular Network, Multicast

1 INTRODUCTION

For the next generation of mobile communication systems an ubiquitous coverage is required. It is widely accepted that the coverage of places with high user densities can only be achieved with small cells(microcell and picocell). Picocells permit higher reuse of channels resulting in an increase in system capacity by several orders of magnitude. The smaller cell size in microcellular or picocellular systems, however, increases the number of handoffs (Nanda,1993). This implies that the time between handoffs and the time to complete handoffs becomes shorter. Thus in order to provide good quality of connection to mobile user, a faster handoff or pre-handoff method is needed.

In a two-tier system of microcells overlaid with macrocells, by assigning low mobility users to microcells and high mobility users to macrocells, it decreases the total number of handoffs (Yeung,1995)(Chih,1993).

In (Acampora,1994),(Yu,1995) and (Ghai,1994), they construct a multicast tree for handoffs in the near future and when a handoff occurs, the reserved connections are used. They drop the handoff processing delay but consume lots of resources.

In this paper, we propose a multicast-based pre-handoff scheme, called SGMH (Sub-

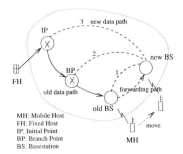

Figure 1 Point-to-point handoffs

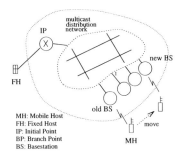

Figure 2 Multicast-based handoffs

Group Multicast-based Handoff), for providing mobile user with seamless connection service. SGMH can not only minimize the buffer overhead of multicast-based handoff methods but also provides seamless connection service to mobile user.

This paper is organized as follows. In section 2, we will survey about various handoff methods in the literature. In section 3, we propose SGMH method and explain about SGMH in detail: the concept of SGMH, an architecture of picocellular networks, method of subgroup selection, the traffic model of MH's mobility and the handoff blocking probability. Finally, in section 4, we study on the relationship between the reduction of buffer overhead and the blocking probability through simulations.

2 VARIOUS HANDOFF METHODS

Generally handoff methods presented in the literature can be divided into two categories : pont-to-point and point-to-multipoint. As in Figure 1, point-to-point handoff methods can be divided into three categories: path extension, forward-and-reroute and anchor-based handoff methods. (1) In case of path extension method (label 1) (Keeton,1993), when a MH moves into a new cell, it informs the new BS of its old BS information and its connection information. The new BS sends a path extension request message to the old BS. The old BS extends the MH's connection to the new BS. After connection extension, the old BS resumes sending data to the MH through the newly extended path. The problems of this method are as follows: the path length becomes very long and path cycle may happen. Therefore a path optimization algorithm and special cycle elimination algorithm are required. (2) In the forward-and-reroute method (label 2) (Keeton,1993), after a MH greets to the new BS, the new BS sends a forward-data-request message to the old BS. The old BS setups a forwarding path to the new BS and forwards data to the MH. And then the new BS searches for a branch point which is closest to the new BS and reroutes the connection to the branch point. But they are time consuming jobs. (3) The anchor-based handoff (label 3) (Akyildiz,1996), is similar to the forward-and-reroute method with difference that the forward point is fixed at the initial BS chosen at the call setup. It has also the problem that the path length between the initial BS and the new BS may be long.

Multicast-based handoff methods can be subdivided into two categories according to the method of constructing a multicast distribution tree. (1) Static: a multicast tree is created at call setup. The multicast tree does not change for the call lifetime (Acampora,1994). (2) Dynamic: a multicast tree changes dynamically during the call lifetime according to the MH's mobility (Ghai,1994). In multicast-based handoffs, it is before handoff that a multicast tree is composed of neighboring basestations and data is multicasted to all member BSs. When the MH enters a new cell, it greets the new BS and receives data from the new BS immediately. And the new BS informs the RS of the MH's location change. Therefore there is no service break. But they cost lots of buffer space at each member BS.

In summary, point-to-point handoff methods are not fit for seamless connection service in picocellular networks because of their inherent characteristics of data forwarding and rerouting overhead. Multicast-based handoff methods are appropriate for the seamless service because there is no handoff delay. But they have a main drawback of large buffer overhead because the preallocated buffers in member BSs excepts for the current BS are not used for the dwell time of the MH in the current cell. So we propose a multicat-based handoff method which can minimize buffer overhead.

3 THE PROPOSED SGMH HANDOFF METHOD

In this section, we will explain about the proposed SGMH handoff method in details : a picocellular network architecture, the concept of SGMH, and the selection of multicast subgroups.

3.1 A picocellular network architecture

We assume that a picocellular network, as shown in Figure 3, is composed of three layers: mobile host layer at the bottom, basestation layer in the middle and region server layer at the top. As shown in Figure 3, MH reports the received signal strengths to RS periodically. BS plays a role of mobile network access point and manages radio channels and buffer space for seamless connection service. RS covers one service area and performs connection-related functions: selection of multicast subgroups, setup multicast tree and data multicasting and so on. For seamless connection service, each layer must maintain connection-related information. We will not describe about the information here.

Compared with conventional cellular networks, picocellular network has following features: small cell size, high rate of frequency reuse, small-sized handset with low power consumption, high throughput, frequent handoffs and so on. Particularly frequent handoff is a main drawback. For example, if a mobile user moves with the average speed of 2 meters per second and picocell's diameter is 10 meters, handoff happens every 5 seconds. The mobile user experiences frequent service breaks. Therefore, for seamless connection service, we need a new handoff method fit for picocellular networks.

3.2 Concept

Here we will explain about the concept of the proposed SGMH method. The smaller cell size causes frequent handoffs in a picocellular network (Nanda,1993). The processing delay of frequent handoffs is an obstacle to provide mobile user with seamless service. But if the mobile system performs necessary jobs before the handoff, such as connectin pre-setup to the candidate BSs and transfer of data to the BSs, it can support mobile user with the seamless service.

In SGMH, RS predicts the MH's mobility and chooses the multicast group of the MH based on the velocity of the MH and the distances from the MH's current position to candidate basestations and sends data to member BSs before handoff. When the MH enters a new member BS, the MH can receive data from the new BS immediately. Therefore, SGMH method can eliminate handoff delays : the time to forward the transit data to the MH, the time to reroute the connection to the new BS and the time to resume sending

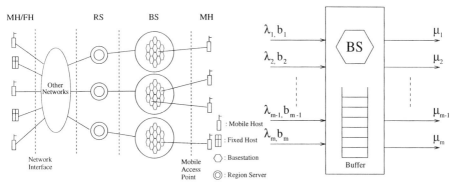

Figure 3 An architeture of picocellular network

Figure 4 Traffic model of a BS

data. In addition to the elimination of the handoff delay, becaue each member BS stores differnt amount of data according to its distance from the MH, SGMH can minimize the total buffer overhead and lowers the blocking probability of handoff calls and newly arriving calls due to the shortage of buffer space.

3.3 Selection of multicast subgroups

In order to estimate the location and the speed of a mobile user in a cellular system, it is reasonable to use the information about field strength and distance from the BS, both of which are supplied by the system. Various methods to obtain this information are broken down into three basic categories: radiolocation, dead-reckoning, and proximity methods (Kennenmann,1995). In order to estimate the position of the MH, they in general use trilateration and require installation of additional devices in the MH and the BS.

Here we do not consider the location methods, but we use the porposed methods to select the multicast subgroups of the MH. As shown in Figure 3, RS serves one service area and maintains the tables on the information about the cell layout and the signal propagation model. RS chooses the subgroups of the MH based on those informations. The procedure is as following. The MH informs the RS of the signal strengths received from neighboring BSs periodically. By using this information, RS selects the candidate BSs which the MH may move into in the near future. RS divides the BSs into several subgroups according to the expected handoff times. Because the earlier selection influences on the network performance, the selection time, if possible, is delayed to the latest time that the delay does not effect on the seamless connection service.

Let us introduce several variables. T_{sg} denotes the time taken to select the candidate basestations and divide them into several subgroups. T_{join} represents the sum of the time taken to setup a multicast distribution tree and the longest time among the joining times of member BSs: i.e., $T_{join} = Max(T_{join}(i)$ for each member BS). Then the total time taken to setup the subgroups of the MH is equal to $T_{gs} = T_{sg} + T_{join}$. Therefore, the group selection can start at the time which is calculated by subtracting the group selectin time T_{gs} from the expected handoff time T_{ho} : i.e., $T_{ho} - T_{gs}$. In case that there is not enough time to setup the subgroups, SGMH can not provide MHs with seamless service and thus the conventional handoff method is used.

3.4 Computation of buffer overhead

We define several parameters for calculating buffer overhead of SGMH.

- G : the number of member BSs in a multicast group
- m : the number of subgroups classified by their distances from the MH's current position
- SG_i : the number of BSs of the ith subgroup
- V : the estimated speed of the MH
- D_i : the distance from member BS$_i$ to the MH
- T_d : the real dwell time of the MH in the current cell
- T_{p_i} : the expected arrival time of the MH at BS$_i$
- N : the number of packets arrived at current cell during T_d
- K_i : the number of packets arrived at a BS of the ith subgroup between T_{p_i} and T_d

The expected arrival time of the MH at a member BS$_i$, T_{p_i}, is calculated by dividing the distance D_i by the speed V: i.e., $T_{p_i} = D_i/V$.

The buffer overhead exists only if $T_{p_i} < T_d$, because for seamless connection service other BSs except for the current BS must store packets arrived at between the expected handoff time T_{p_i} and the real sojourn time T_d. In case of the wrong expectation $T_{p_i} > T_d$, there is no buffer overhead. Thus, the buffer overhead of SGMH is equal to $B = \sum_{i=1}^{m} K_i \times SG_i$.

Let us take an example that the multicast group of the MH is composed of G={a, b1, b2, b3, c1, c2, c3, d2}.

(1) In case of the simple multicast-based handoff in (Keeton,1993), the packets to the MH are multicasted to all member BSs and are stored in buffer space for the near future usage. Thus, the total buffer overhead B is equal to the product of the multicast group size and the number of packets transmitted for the real dwell time : i.e., $B = N \times (|G| - 1) = 7 \times N$.

(2) In case of the method proposed in (Ghai,1994), the buffer overhead exists only if the expected handoff time is before the real handoff time : i.e., $T_p < T_d$. All member BSs except for the current BS must store packets transmitted for the time interval between the expected time T_p and the real handoff time T_d. The buffer overhead is $B = (|G|-1) \times K = 7 \times K$.

(3) In SGMH, the multicast group is divided into several subgroups according to the distances of BSs from the current MH's location. If the multicast group G is divided into G = { { a }, { b1, b2, b3 }, {c1, c2, c3 }, {d2} }, in case of $T_{p_i} < T_d$, the total buffer overhead is equal to $B = 3 \times K_2 + 3 \times K_3 + 1 \times K_4$ where $K_1 > K_2 > K_3$.

Therefore SGMH method has the lowest buffer overhead among three multicast-based handoff methods.

3.5 Analytic model

We consider a picocellular network as shown in Figure 3 with a suffciently large number of cells with the same geometric shape. Each cell is surrounded by the same number of neighboring cells. Each BS can belong to many subgroups and for the seamless connection service, the BS must reserve buffer space per each subgroup request. Therefore, the BS can be modeled as shown in Figure 4.

It is assumed that the total buffer capacity available to a BS is denoted by C and the buffer requests are divided into m classes and each class requires different amount of buffer units. And the buffer request arrivals follow a Poisson process and the buffer holding times are distributed exponentially.

We define several parameters for each class $i = 1, 2, 3, ..., m$ as followings:

- λ_i : the request arrival rate
- $1/\mu_i$: the mean buffer holding time
- b_i : the number of required buffer units
- P_{B_i} : blocking probability
- $P_{B_i,max}$: maximum permissible blocking probability
- $n = (n_1, n_2, n_3, n_4, ..., n_m)$: the current state of a BS denoted by the number of requests of each class in progress

The number of busy buffer units is given by $n \times b = \sum_{i=1}^{m} n_i \times b_i$. The problem is to provide each class with blocking probabilities less than or equal to maximum permissible blocking probabilities.

Now let us compute the blocking probability of a handoff request. The blocking probability is dependent on the strategy allocating the buffer space. The allocation policies can be divided into three categories: *complete sharing*, *fixed allocation* and *hybrid allocation* (see (Tekinay,1993) for more details).

Here we consider only the blocking probabilities under the complete sharing policy because the key issue of this paper is not the buffer allocation methods but the evaluation of the proposed SGMH method.

From the viewpoint of each BS, the traffic of buffer requests can be divided into m classes. The BS, therefore, is considered as a heterogeneous system and can be modeled as an m-dimensional Markov Chain. The transition probabilities along the ith axis are dependent on the arrival rate and the service rate of requests of the ith subgroup: i.e., $P_{n,n+1} = \lambda_i \cdot \Delta t$ and $P_{n,n-1} = \mu_i \cdot \Delta t$. The blocking probability of the ith subgroup is the sum of the probabilities of the blocking states of the ith subgroup. A blocking state of the ith subgroup is the state from which a transition to a higher state along the ith axis is not allowed. For example, if the state is $n = (n_1, n_2, n_3, ..., n_i, ..., n_m)$ and the next higher state $n = (n_1, n_2, n_3, ..., n_i + 1, ..., n_m)$ is not allowed, the current state is a blocking state.

The traffic of the ith subgroup in a BS can be modeled as $M/M/k/C$ Erlang's loss system. The probability of having k calls of the ith subgroup is equal to Equation(1).

$$P_k = \frac{(\lambda_i/\mu_i)^k}{k!} \times P_0 \tag{1}$$

And the probability of the current state of the BS, $n = (n_1, n_2, n_3, ..., n_i, ..., n_m)$, is obtained as Equation(2).

$$P(n) = P(0) \times \prod_{i=1}^{m} \frac{(\lambda_i/\mu_i)^{n_i}}{n_i!} \tag{2}$$

Therefore, the probability of a request of the ith subgroup becomes to Equation(3) where B_i is the set of blocking states for ith subgroup.

$$P_{B_i} = \sum_{n \in B_i} P(n) \tag{3}$$

4 SIMULATION RESULT

Here we will investigate how much the reduction of buffer overhead of SGMH method has an effect on lowering the blocking probability of the handoff calls and the newly arriving calls through simulations.

Simulation runs in two steps. Firstly, we search for the total buffer capacity C which is sufficient to satisfy for the given maximum permissible blocking probabilities of each subgroup traffic requests. And then for the selected buffer capacity, we measure the blocking probability with the variation of traffic intensity.

It is assumed that the number of subgroups is two. The simulation parameters in Table 1 are followings: C denotes the total buffer capacity of each BS, B means the requested buffer units of each subgroup handoff, $1/\lambda$ means the mean arrival time and $1/\mu$ means the mean buffer holding time respectively. And the arrival times of handoff calls and the buffer holding times are distribued exponetially. The values of parameters in Table 1 are not real values but relative values.

The first two simulation sets describe the cases that the movement of MHs is affected by the surroundings such as city streets and buildings. They are different by the expectation accuracy of the handoff time; the first one is more accurate than the second one. Figure 5 shows that the increase of buffer capacity of the BS lowers the blocking probabilities of each subgroup handoff calls and the blocking probability of the second subgroup calls is lower than that of the first subgroup calls. Figure 5 shows that the total buffer capacity with the maximum permissible blocking probability of less than 5 % is about 20 buffer units.

Figure 7 and Figure 8 show the blocking probability with the variation of the traffic intensity of each subgroup calls. The blocking probability of the first set is lower than the second set because it requires less buffer units than the second set. The simulation results show that the blocking probability of SGMH method is much lower than that of the simple multicast-based handoff method.

And the last two simulation sets have different subgroup sizes, that is, they describe the environment that the mobility of the MH is not affected by the surroundings. In these cases, the arrival rate of the first subgroup calls is higher than that of the second subgrouop calls. As in Figure 6, we choose the total buffer capacity of 30 buffer units. Figure 9 and Figure 10 show the blocking probability of the first subgroup calls and the second subgroup calls for the simulation set 3 and set 4 repectively. They shows that SGMH is better than simple multicast-based handoff in the blocking probability. And the simulation set 3 with the high expectation accuracy also has better performance than the simulation set 4.

In summary, the proposed SGMH method reduces the total buffer overhead and owing to the buffer reduction the blocking probability of handoff calls also decreases.

5 CONCLUSION

In this paper we proposed a multicast-based handoff method, so-called SGMH, which minimizes the service disruption delay owing to the frequent handoffs in the picocellular networks with the cell size of the order of ten meters or more. Because SGMH method takes consideration on the MH's spending time before the handoff and the distance from the MH's current location to a candidate handoff BS, it eliminates the handoff delay and reduces the buffer overhead significantly. And SGMH method also has the effect on dropping the handoff blocking probability. We evaluated the performance of SGMH method through the analysis and the simulation in terms of blocking probability and compared with the multicast-based handoff method. The simulation result shows that SGMH method is superior to the multicast-based handoff method. In the future, we will analyze the proposed SGMH method in various performance measures and design the protocol of SGMH method.

Table 1 Simulation parameters

Set	C	Multicast Method						SGMH Method					
		Subgroup 1			Subgroup 2			Subgroup 1			Subgroup 2		
		B	$1/\lambda$	$1/\mu$	B	$1/\lambda$	$1/\mu$	B	$1/\lambda$	$1/\mu$	B	$1/\lambda$	$1/\mu$
1	20	3	2	5	3	2	5	3	2	5	1	2	5
2	20	3	2	5	3	2	5	3	2	5	2	2	5
3	30	3	1	5	3	2	5	3	1	5	1	2	5
4	30	3	1	5	3	2	5	3	1	5	2	2	5

C: Total buffer capacity of BS
B: Requested buffer capacity

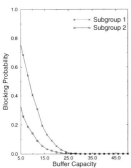

Figure 5 Set 1: Blk Prob vs Buffer Capacity

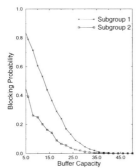

Figure 6 Set 3: Blk Prob vs Buffer Capacity

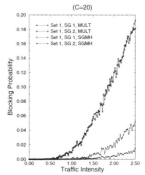

Figure 7 Set 1: Blk Prob vs Traffic Intensity

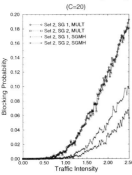

Figure 8 Set 2: Blk Prob vs Traffic Intensity

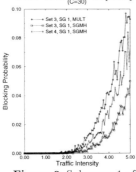

Figure 9 Subgroup 1 of Set 3,4 : Blk Prob vs Traffic Intensity

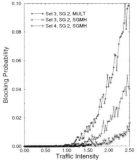

Figure 10 Subgroup 2 of Set 3,4 : Blk Prob vs Traffic Intensity

REFERENCES

Acampora, A.S. and Naghshineh, M. (1994) An Architecture and Methodology for Mobile-Executed Handoff in Cellular ATM Networks. *Journal on Selected Areas in Communications*, 12(8),1365–1375, Oct.

Akyildiz, I.F., Ho, J.S.M. and Ulema, M. (1996) Performance Analysis of the Anchor Radio System Handover Method for Personal Access Communications System. In *Proceedings of IEEE INFOCOM '96*, pages 1397–1404.

Ghai, R. and Singh, S. (1994) An Architecture and Communication Protocol for Picocellular Networks. *IEEE Personal Communications*, pages 36–46, Third Quater.

Chih-Lin I, Greenstein, L.J. and Gitlin, R.D. (1993) A Microcell/Macrocell Cellular Architecture for Low- and High-Mobility Wireless Users. *Journal on Selected Areas in Communications*, 11(6):885–891, Aug.

Keeton, K., Mah, B.A., Sehan, S., Karz, R.H. and Ferrari, D. (1993) Providing Connection-oriented Network Services to Mobile Hosts. In *Proceedings of the USENIX Symposium*, Aug.

Kennenmann, O. (1995) Locating Mobiles in Non-Flowing Traffic. In *Proceedings of PIMRC'95*, pages 274–278.

Nanda, S. (1993) Teletraffic models for urban and suburban microcells: Cell sizes and handoff rates. *IEEE Transaction on Vehicular Technology*, Nov.

Tekinay, S., Jabbari, B. and Kakaes, A. (1993) Modeling of Cellular Communication Networks with Hetrogeneous Traffic Sources. In *Proceedings of ICUPC'93*, pages 249–253.

Yeung, K.L. and Nanda, S. (1995) Optimal Mobile-Determined Micro-Macro Cell Selection. In *Proceedings of PIMRC'95*, pages 294–299.

Yu, O.T.W. and Leung, V.C.M. (1995) B-ISDN Architectures and Protocols to Support Wireless Personal Communications Internetworking. In *Proceedings of PIMRC'95*, pages 768–772.

Eunyong Ha received the B.S. and M.S. degrees in computer engineering from Seoul National University, Seoul, Korea, in 1986 and 1988 respectively. He is a Ph.D. student of Department of Computer Engineering at Seoul National University. His current research interests include wireless/mobile communication networks and multimedia systems.

Yanghee Choi received B.S. in electronics engineering from Seoul National University, M.S. in electrical engineering from KAIST , and Doctor of Engineering in Computer Science from ENST, Paris , in 1975, 1977 and 1984 respectively. He is an associate professor at the Department of Computer Engineering at Seoul National University. He has been with ETRI during 1977-1991. He is also associate director of the University Computing Center. He is associate editor for the Journal of Korea Information Science Society, and chairman of the Special Interest Group on Information Networking. His research interest lies in the field of multimedia systems and high-speed networking.

Chongsang Kim received the Ph.D degree in Electronic Engineering from Seoul National University. He is a professor at the Department of Computer Engineering at Seoul National University, where he has been since 1977. Dr. Kim was the president of the Korea Information Science Society from 1986 to 1988. He was also the chairman of the IEEE Korea Council during 1995. and director of the Research Institute of Advanced Computer Technology from 1993 to 1994. His interests are computer architecture and computer networks.

16
Impact of mobility in mobile communication systems

M. Zonoozi and P. Dassanayake
Department of Electrical & Electronic Engineering,
Victoria University of Technology,
PO Box 14428 MCMC, Melbourne, VIC 8001, AUSTRALIA.
Email: mahmood@cabsav.vut.edu.au
Fax: 613 6884908 - Tel: 613 6884767

Abstract
A new method is developed for systematic tracking of the random movement of a mobile station in a cellular environment. It incorporates mobility parameters under most generalized conditions, so that the model could be tailored to be applicable in most cellular environments. This mobility model is used to characterise the cell residence time of both new and handover calls occurring in a cellular mobile communication system. It is shown that the cell residence time can be described by the generalized gamma distribution.

Keywords
Handover, mobility, cell residence time

1 INTRODUCTION

In a cellular mobile communication network, depending on whether a call is originated in a cell or handed over from a neighbouring cell, two different cell residence times can be specified and they are the new call cell residence time and the handover call cell residence time, respectively. *New call cell residence time* is defined as the length of time a mobile terminal resides in the cell where the call was originated before crossing the cell boundary. Similarly, the *handover call cell residence time* is defined as the time spent by a mobile in a given cell to which the call was handed over from a neighbouring cell before crossing to another cell, (Figure 1). New call cell residence time, T_n, and the handover call cell residence time, T_h, are two random variables whose distributions have to be found. A literature survey shows that a relatively few in-depth papers have been published on this subject. Moreover, most of these are restricted to simple mobility cases, and often based on assumptions made

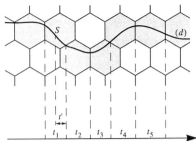

Figure 1. Cell residence times for a mobile travelling on a path of (d), S is the start time of a call. $T_h \in \{t_1, t_2, t_3, t_4, t_5\}$ $T_n \in \{t'\}$

without justification.

Hong and Rappaport [1] have obtained the probability density function (pdf) of a simplified case of mobility where there is no change in speed or direction of the mobile. Further, the initial speed of the mobile was assumed to follow a uniform distribution. Del Re, Fantacci and Giambene [2] have assumed that mobiles, before crossing a cell travel a distance uniformly distributed between 0 and $2R$, where R is the hexagonal cell side. They also assume a constant speed with uniform distribution and conclude that the pdf of cell residence time is different to that shown in [1]. Inoue, Morikawa and Mizumachi [3] have applied the procedure of [1] for a case of non-uniform speed distribution. However they end up with a set of unsolved integral equations. Yeung and Nanda [4], Xie and Kuek [5], Xie and Goodman [6] have shown that contrary to the assumption made in [1], the speed and direction distributions of the in-cell mobiles are different from those of the cell-crossing mobiles. A more precise distribution for the speed and direction can be obtained using their *Biased Sampling* formula.

While Sanchez Vargas [7], and Lue [8] have assumed cell residence time to be uniformly distributed over the call duration, Nanda [9], Lin, Mohan and Noerpel [10] have taken a general distribution for the cell residence time. For the sake of simplicity, in the absence of any proved probability distribution, many authors dealing with the mobility problem have assumed cell residence time to be an exponentially distributed random variable either explicitly or implicitly [11]-[17].

2 MOBILITY MODELLING OF THE SIMPLIFIED CASE

For a simplified case of mobility where there is no change in speed and direction of the mobile, cell residence time distribution can be obtained analytically. In this simplified case, the direction of the mobile at the starting point, α_0, is taken to be uniformly distributed in the range $(0, 2\pi)$, and it is assumed to remain constant along its path. Moreover, the initial speed of the mobile is also taken to be a uniformly distributed random variable in the range $(0, V_m)$, and it is assumed to remain constant along the mobile path. Users are assumed to be independent and uniformly distributed over the entire region. Initial location of a mobile is represented by its distance ρ_0 and direction θ_0 from the base station which is located at the centre of the cell. Let $f_{T_n}(t)$ denote the pdf of the new call cell residence time. Then, according to [1],

$$f_{T_n}(t) = \begin{cases} \dfrac{8R}{3\pi V_m t^2}\left\{1-\left[1-\left(\dfrac{V_m t}{2R}\right)^2\right]^{3/2}\right\} & 0 \leq t \leq \dfrac{2R}{V_m} \\ \dfrac{8R}{3\pi V_m t^2} & t \geq \dfrac{2R}{V_m} \end{cases} \qquad (1)$$

A handover call starts from the boundary of a cell with mobile having a direction α_0 uniformly distributed over $(-\pi/2, \pi/2)$. The pdf of the handover call cell residence time, $f_{T_h}(t)$ can be calculated in a similar manner to (1),

$$f_{T_h}(t) = \begin{cases} \dfrac{4R}{\pi V_m t^2}\left\{1 - \left[1 - \left(\dfrac{V_m t}{2R}\right)^2\right]^{1/2}\right\} & 0 \leq t \leq \dfrac{2R}{V_m} \\ \dfrac{4R}{\pi V_m t^2} & t \geq \dfrac{2R}{V_m} \end{cases} \qquad (2)$$

In [5]-[6], it is shown that the speed and direction distributions of the in-cell terminals are different from those of the cell-boundary crossing terminals. Let $f_{V_0}(v_0)$ denote the pdf of the speeds of all terminals and $f^*_{V_0}(v_0)$ denote the pdf of the speeds of cell-boundary crossing terminals. Based on the Biased Sampling [18], the speed pdf of the cell-boundary crossing terminals can be derived [5] as,

$$f^*_{V_0}(v_0) = \dfrac{v_0 f_{V_0}(v_0)}{E[V_0]} = \begin{cases} \dfrac{v_0}{V_m E[V_0]} & 0 \leq v_0 \leq V_m \\ 0 & \text{otherwise} \end{cases} \qquad (3)$$

Similarly, let $f(\alpha_0)$ be the pdf of the directions of all terminals, which has uniform density in the range $(0, 2\pi)$. Based on the Biased Sampling, the pdf of the directions of the cell-boundary crossing terminals, $f^*(\alpha_0)$, can be obtained as,

$$f^*(\alpha_0) = \begin{cases} \dfrac{1}{2}\cos(\alpha_0) & -\dfrac{\pi}{2} \leq \alpha_0 \leq \dfrac{\pi}{2} \\ 0 & \text{otherwise} \end{cases} \qquad (4)$$

Equation (4) shows that the pdf of the direction of the cell-crossing terminals is not uniform, but has a direction bias towards the normal. Considering (3) and (4), the relations for $f_{T_n}(t)$ and $f_{T_h}(t)$ could be modified accordingly.

3 MOBILITY MODELLING FOR GENERAL CASE

Eqs. (1)-(2) represent new and handover call cell residence time distributions for the simplified case of mobility where there is no change in speed or direction and no biasing in speed or direction of boundary crossing mobiles. In the general case, the mobility modelling should include changes in direction and speed of the mobile. Moreover, it is unrealistic to assume that the speed is uniformly distributed and remains constant. Extension of the analysis of the simplified case to cover the general case is virtually impossible, and simulation appears to be the only way out. The simulation model is aimed at obtaining statistical estimates of the mobile cell boundary crossings in a cellular environment where the mobile is allowed to move freely with randomly varying velocities and directions within realistic bounds. A uniform distribution is assumed for spatial location of the users. This assumption is valid, since throughout a cellular network, the relative orientation of streets and cells varies somewhat randomly, giving on the average a nearly uniform distribution of possible directions [6]. Since the destination point of the mobiles can be any place in the coverage area, mobiles are allowed to move away from the starting point in any direction with equal probability. Therefore, a uniform distribution in the range is suitable for the initial mobile direction. Depending on the structure of the cellular mobile coverage area, a mobile may move towards the destination point via

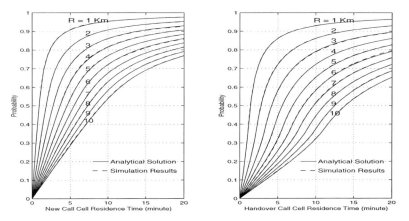

Figure 2. Comparison between simulation and analytic solution for the special case

different paths. However, in any case, the mobile direction is biased towards destination to prevent it from circling around. In the simulation model, the initial mobile direction is taken to be uniformly distributed in the cell area, and the directions at successive steps are allowed to change within a set bound referred herein as *drift*.

The probability distribution of the variation of the mobile direction along its path is taken to be uniform and the variation is taken to be in the range $(-\alpha, +\alpha)$ degrees with respect to the current direction. The value of α is chosen, depending on the street structure of the cell area, to be a low value for the cells with more straight streets and a high value for the cells with less straight streets. The effect of α on the probability of boundary crossing can be verified by comparing different values of α with respect to a reference. The initial speed of a mobile unit, at the instance the call is initiated, is taken as a random variable with truncated Gaussian pdf, $f_{V_0}(v_0)$, having a mean and standard deviation of μ_v and σ_v, respectively. The choice of such a distribution seems reasonable, since the more extreme the speed value, the less likelihood of its occurrence. Also, it is unlikely that the speed exceeds a certain maximum value.

At any time instant, the mobile speed is a random variable correlated with the previous speed, v_p. The current speed, v_c, of each mobile is taken to be a uniformly distributed random variable in the range ±10% of the previous speed. Any increase in the speed above 100 Km/h is not allowed, and the minimum speed is taken to be $0 Km/h$.

In order to check the validity of the proposed simulation model a test run is made for the simplified case described in Section 2, with the same assumptions held. The probability distribution function of the cell residence time is calculated through (1)-(2) and compared with the results obtained by the simulation. As shown in Figure 2 the simulation results are in good agreement with the analytical results.

4 CELL RESIDENCE TIME DISTRIBUTION

What is of important here is not the actual mobile trajectories, but the distribution of the users' cell residence time. With this in mind, we wish to test the hypothesis that the new call and handover call residence time data follows a particular probability distribution. Following [19], we proceed with the generalized gamma distribution which provides a series of pdf of the form

$$f_T(t; a,b,c) = \frac{c}{b^{ac}\Gamma(a)} t^{ac-1} e^{-\left(\frac{t}{b}\right)^c} \qquad t, a, b, c > 0 \qquad (5)$$

where $\Gamma(a)$ is the gamma function, defined as $\Gamma(a) = \int_0^\infty x^{a-1} e^{-x} dx$, for any real and positive number a. The parameters a, b, c can be classified on the basis of their physical or geometric interpretation, as one of the three types, namely location, scale and shape. The evaluation of the agreement between the distributions obtained by simulation and the best fitted generalized gamma distribution is done by using the Kolmogorov-Smirnov goodness-of-fit test. Given the generalized gamma distribution as the hypothesized distribution, the values of the parameters a, b, c are found such that the maximum deviation to be a minimum. The maximum deviation shows the biggest divergence between the observed and the hypothesized distributions. The results show that the values of a and c are constant and are independent of cell size, while b varies with the cell size according to

$$b \approx \begin{cases} 1.84R & \text{new call} \\ 1.22R & \text{handover call} \end{cases} \qquad (6)$$

5 MEAN CELL RESIDENCE TIME

The mean cell residence time for the new and the handover calls can be found by

$$E[T_n] = \int_0^\infty t \cdot f_{T_n}(t) dt \qquad (7)$$

$$E[T_h] = \int_0^\infty t \cdot f_{T_h}(t) dt \qquad (8)$$

Yeung and Nanda [4], have shown that for an arbitrary speed pdf and zero drift the mean cell residence time can be obtained through the following equations

$$E[T_n] = \frac{8R \; E[1/V]}{3\pi} \qquad (9)$$

$$E[T_h] = \frac{\pi R}{2} \frac{1}{E[V]} \qquad (10)$$

where R is the cell radius and V is the speed of the mobile in the cell. A comparison of the results obtained from (9)-(10) with (7)-(8) assuming generalized gamma pdf for $f_{T_n}(t)$ and $f_{T_h}(t)$ shows that the maximum difference (error) is less than 0.0007% in the case of new calls and 0.0021% in the case of handover calls.

6 EFFECT OF CHANGE IN DIRECTION AND SPEED

Depending on the street structure, a mobile can move in different paths and may possess different speeds. The extent of mobile change in direction (drift) and change in speed are the two parameters that govern its mobility. The effect of change in direction or speed of mobiles can be considered as equivalent to a change in an average distance travelled or time spent in the cell, before moving out. Any increase in mobile's drift can be treated as contributing to an effective increase in the cell radius. Similarly, any increase in speed of the mobile can be treated as contributing to a decrease in the cell residence time which can be interpreted as an effective decrease in the cell size. Therefore, cells with a broad variety of mobility parameters can be replaced by an equivalent reference cell with an effective radius. A reference cell is defined as a cell with the following mobility parameters, (a) mobile moves in a straight path, i.e. $\alpha = 0°$, (b) initial speed of a mobile follows a truncated Gaussian pdf with an average of $\mu_v = 50\ [Km/h]$ and standard deviation of $\sigma_v = 15[Km/h]$.

Our aim is to relate cells with given mobility parameters (i.e. drift α and average speed v) to the reference cell. We consider two different cases

case i.) cells having mobility parameters defined by a uniformly distributed drift pdf in the range $(-k°, k°)$ and speed pdf similar to the reference cell.

case ii.) cells having mobility parameters defined by zero drift (similar to the reference cell) and a truncated Gaussian speed pdf with an average value of $v \neq \mu_v$ and standard deviation of $\sigma_v = (v-5)/3[Km/h]$.

Consider a cell with the radius of R_α where its mobility parameters are according to case i. The radius of the equivalent cell (which has the same residence time) with mobility parameters of the reference cell, $\Re_\alpha$, is given by $\Re_\alpha = R_\alpha + \Delta R_\alpha$, where ΔR_α is the excess cell radius required to replace R_α (i.e. radius of a cell where its mobiles can move around with a uniformly distributed drift pdf in the range $(-k°, k°)$) with $\Re_\alpha$ (i.e. radius of an equivalent cell where its mobiles can move on a straight line). The data obtained by simulation satisfy the empirical equation of (11) in a least mean square sense.

$$\Delta R_\alpha = 0.0038 k R_\alpha \tag{11}$$

Therefore the equivalent cell radius will be

$$\Re_\alpha = K_\alpha R_\alpha \tag{12}$$

where K_α is the proportionality factor and is equal to $(0.0038k + 1)$. In the same manner, consider a cell with the radius of R_v where its mobility parameters are according to case ii. The radius of an equivalent reference cell, $\Re_v$, which has the same cell residence time is given by $\Re_v = R_v + \Delta R_v$, where ΔR_v is the excess cell radius required to replace R_v with $\Re_v$. The data obtained by simulation satisfy the empirical equation of (13) in a least mean square sense.

$$\Delta R_v = \left(\frac{\mu_v}{v} - 1\right) R_v \tag{13}$$

Therefore the equivalent cell radius will be

$$\Re_v = K_v R_v \tag{14}$$

where K_v is the proportionality factor and equals to (μ_v/v). In a case where both drift and speed are different from those of the reference cell, the equivalent value of the cell radius, $\Re_{\alpha v}$, for a cell of radius, $R_{\alpha v}$, can be calculated by the following relation, ($R_{\alpha v}$ is the cell radius for a cell which supports mobility parameters of α and v)

$$\Re_{\alpha v} = K_v K_\alpha R_{\alpha v} \tag{15}$$

Therefore, in a cell of radius $R_{\alpha v}$, the gamma distribution parameter b for a mobile with an average speed v and a drift α in the range $(-k° < \alpha < k°)$ can be described as

$$b \approx \begin{cases} 1.84\mathfrak{R}_{av} & new\ call \\ 1.22\mathfrak{R}_{av} & handover\ call \end{cases} \tag{16}$$

7 CONCLUSIONS

This paper presented a methodology appropriate for mobility modelling of users in a cellular mobile communication system. The proposed model traces mobiles systematically in a cellular environment where they are allowed to move in a quasi-random fashion with assigned degrees of freedom. This model enables the development of a computer simulation algorithm that provides statistical estimates of the cell boundary crossing features and hence the characterisation of cell residence times. Results show that the generalized gamma distribution is a good approximation for the cell residence time distribution of both new and handover calls. It is also shown and that the negative exponential distribution is a good approximation for the channel holding time distribution in cellular mobile systems.

It was also shown that an increase in mobile drift can be treated as contributing to an effective increase in the cell radius. Similarly, it was shown that an increase in the speed of a mobile can be treated as contributing to a decrease in the cell size, and vice versa. Taking this excess cell radius into account for different values of drift and speed, a broad variety of cell areas with different street orientations and traffic flows can be handled by this mobility model.

8 ACKNOWLEDGEMENTS

The authors gratefully acknowledge the support provided by the Australian Telecommunications and Electronics Research Board (ATERB) for this project. The first author also gratefully appreciates the support provided by the Telecommunication Company of Iran.

9 REFERENCES

[1] D. Hong, S. S. Rappaport, 'Traffic model and performance analysis for cellular mobile radio telephone systems with prioritized and nonprioritized handoff procedures', *IEEE Transactions on Vehicular Technology*, vol. 35, no. 3, pp. 77-92, 1986.

[2] E. Del Re, R. Fantacci, G. Giambene, 'Handover and dynamic channel allocation techniques in mobile cellular networks', *IEEE Transactions on Vehicular Technology*, vol. 44, no. 2, pp. 229-237, May 1995.

[3] M. Inoue, H. Morikawa and M. Mizumachi, "Performance analysis of microcellular mobile communication systems," *44th. IEEE Vehicular Technology Conference*, Stockholm, pp. 135-139, Jun. 1994.

[4] K. L. Yeung and S. Nanda, "Optimal mobile-determined micro-macro cell selection," *6th. IEEE International Symposium on Personal, Indoor and Mobile Radio Communications (PIMRC'95)*, Toronto, pp. 294-299, Sep. 1995.

[5] H. Xie, S. Kuek, "Priority handoff analysis," *43rd. IEEE Vehicular Technology Conference*, New Jersey, pp. 855-858, May 1993.

[6] H. Xie, D. J. Goodman, "Mobility models and biased sampling problem," *2nd. IEEE International Conference on Universal Personal Communications*, pp 804-807, 1993.

[7] J. H. Sanchez Vargas, "Teletraffic performance of cellular mobile radio systems," Ph.D dissertation, University of Essex, England, 1988.

[8] X. Lue, 'Investigation of traffic performance in mobile cellular communication systems' Master thesis, University of Melbourne, Australia, 1991.

[9] S. Nanda, "Teletraffic models for urban and suburban microcells: cell sizes and handoff rates," *IEEE Transactions on Vehicular Technology*, vol. 42, no. 4, pp. 673-682, Nov. 1993.

[10] Y. B. Lin, S. Mohan, A. Noerpel, "Queuing Priority channel assignment strategies for PCS handoff and initial access," *IEEE Transactions on Vehicular Technology*, vol. 43, no. 3, pp. 704-712, Aug. 1994.

[11] C. Purzynski and S. S. Rappaport, "Multiple call hand-off problem with queued hand-offs and mixed platform types," *IEE Proceedings on Communications*, vol. 142 no 1, pp. 31-39, Feb. 1995.

[12] C. Purzynski and S. S. Rappaport, "Traffic performance analysis for cellular communication systems with mixed platform types and queued hand-offs," *43rd. IEEE Vehicular Technology Conference*, New Jersey, pp. 172-175, May 1993.

[13] M. Naghshineh and A. S. Acampora, "Design and control of micro-cellular networks with QOS provisioning for real-time," *3rd. IEEE International Conference on Universal Personal Communications*, San Diego, pp 376-381, Sep. 1994.

[14] S. S. Rappaport, "The multiple-call handoff problem in high capacity cellular communications systems," *IEEE Transactions on Vehicular Technology*, vol. 40, no. 3, pp. 546-557, Aug. 1993.

[15] W. M. Jolley, R. E. Warfield, "Modelling and analysis of layered cellular mobile networks," *13th International Teletraffic Congress*, pp. 161-166, 1991.

[16] H. Jiang and S. Rappaport, "Handoff analysis for CBWL schemes in cellular communications," *3rd. IEEE International Conference on Universal Personal Communications*, San Diego, pp 496-500, Sep. 1994.

[17] S. Rappaport, L. R. Hu, "Microcellular communication systems with hierarchical macrocell overlays: traffic performance models and analysis," *Proceedings of IEEE*, vol. 82, no. 9, pp. 1383-1397, Sep. 1994.

[18] D. R. Cox and P. A. W. Lewis, *The statistical analysis of series of events*. London: Chapman and Hall, 1978.

[19] A. M. Law, W. D. Kelton, *Simulation modelling and analysis*. New York: McGraw-Hill, 1991.

BIOGRAPHIES

Mahmood Zonoozi received the B.Sc. and M.Sc. degrees in electrical engineering from K. N. Toosi University, Iran. He has worked in satellite and microwave communication departments of Iran Telecom for several years. He is currently working towards the PhD degree at the Victoria University of Technology, Melbourne, Australia. He is attached to the Mobile Communications and Signal Proccesing Research Group of the university, and his project deals with the issues involved in handover of cellular mobile communication systems.

Prem Dassanayake received his B.Sc. Eng degree from University of Sri Lanka, M.Sc. and PhD degrees from University of Wales, Cardiff, U.K. At present he is attached to the Department of Electronics and Electrical Engineering of the Victoria University of Technology, Footscray, Australia. Prior to joining Victoria University Dr. Dassanayake has worked at the University of Moratuwa, Sri Lanka and the University of Bahrain, Bahrain. He has also been a visiting researcher at the Bureau of Medical Devices, Health and welfare, Ottawa, Canada and at Telstra Research Laboratories, Clayton, Australia.

PART SEVEN

Mobile Communication Architectures

17
Mobile computing based on GSM: The Mowgli approach

Timo Alanko, Markku Kojo, Heimo Laamanen, Kimmo Raatikainen, and Martti Tienari
University of Helsinki
Department of Computer Science, P.O. Box 26 (Teollisuuskatu 23),
FIN-00014 University of Helsinki, Finland.
Telephone: +358-0-70851. Fax: +358-0-70844441.
email: {timo.alanko,markku.kojo,heimo.laamanen,
kimmo.raatikainen,martti.tienari}@cs.Helsinki.FI

Abstract
Modern cellular telephone systems extend the usability of portable personal computers enormously. A nomadic user can be given ubiquitous access to remote information stores and computing services. However, the behavior of wireless links creates severe inconveniences within the traditional data communication paradigm. In this paper we give an overview of the problems related to wireless mobility. We also present a new software architecture for mastering them and discuss a new paradigm for designing mobile distributed applications. The key idea in the architecture is to place a mediator, a distributed intelligent agent, between the mobile node and the wireline network. A prototype implementation of the architecture exists in an environment consisting of Linux and Windows platforms and the GSM cellular telephone network.

Keywords
Mobile clients, intelligent agents, nomadic users, personal communications, GSM Data Service.

1 INTRODUCTION

Developments in mobile communication and personal computer technology have created a new interesting platform for information processing. A modern portable computer gives remarkable processing power, always at hand for a nomadic user. A mobile telephone system gives ubiquitous access to remote information stores and computing services.

One of the most widely available cellular telephone systems is the Global System for Mobile Communications (GSM); see Rahnema (1993). It covers vast areas in Europe, and it is rapidly expanding in Asia and in Australia. It is also becoming available in the US. In addition, other cellular telephone systems already exist, and still more are emerging. This development implies a tremendous increase in the accessibility of information services. In principle, a mobile computer can always be connected to its home network through a cellular telephone system. Thus, the users will have the impression of working with their "normal desktop computers" wherever they happen to be. In practice, however, several reasons still prevent this scenario from working to the full satisfaction of the user. The problems are due to the different natures of the wireline and wireless worlds: the former is efficient and reliable, the latter is slow and vulnerable. Control methods tuned for one are not necessarily suitable for the other.

In Kojo et al. (1994) we introduced the Mowgli* approach to alleviate these problems. The key idea is to separate the behaviorally different wireline and wireless worlds. On the border of these two worlds we add a new component, a mediator. It contains intelligence, and it is able to cope with both worlds. In essence, this is a shift of paradigm: the traditional "client-server" paradigm is replaced with a new "client-mediator-server" paradigm.

At the same time similar principles were independently introduced for wireless LANs; see Yavatkar and Bhagawat (1994), Bakre and Badrinath (1995), Balakrishnan et al. (1995). However, in the wireless LAN the problem is treated at the transport layer. In Mowgli it is seen as a problem concerning all layers up to the application layer and the user interface.

In Section 2 we address the fundamental problems related to mobility and wirelessness and in Section 3 we outline our mediator-based Mowgli solution. The benefits of Mowgli are discussed in Section 4. In Section 5 we summarize the essentials of the Mowgli approach.

2 PROBLEMS RELATED TO MOBILITY AND WIRELESSNESS

Under favorable radio conditions a wireless telephone link behaves almost like a traditional PSTN link. Therefore, it can be expected that existing applications still work when the conditions are good. However, there are differences between wireline and wireless links that must be taken into account:

- Wireless telephone links have low throughput, high latency, and a long connection establishment time.
- Wireless links are expensive. Today, the GSM offers only connection oriented services. Thus, customary habits of working with the link open over the whole session are not economical.
- Transmission delays may be highly variable. When error rates on radio links are high, the strong error correction mechanisms of the GSM data service increase significantly the delay variation. In unfavorable conditions the delays may extend to tens of seconds.
- The wireless link is vulnerable. The mobile workstation may move through an uncovered area, or the radio conditions may temporarily deteriorate. In both cases the link becomes non-accessible for some period of time.

Current communication architectures, such as the TCP/IP protocol suite, usually offer improper performance over networks with wireless links. In addition, neither they nor the

---

*Mowgli is the acronym of the project name **Mobile Office Workstations using GSM Links**.

Mobile computing based on GSM: the Mowgli approach 153

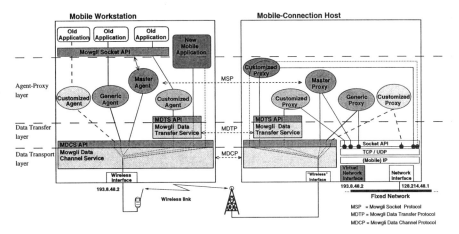

Figure 1 The Mowgli communication architecture for connecting mobile nodes to fixed networks

existing applications are designed to react in an appropriate way to the behavior characteristics of wireless WAN environments. Hence, even minor disturbances in connections usually result in failures at the application level as explained in Kojo et al. (1994) and in Alanko et al. (1994).

The user confronted with these circumstances would benefit from having new functionalities provided by the infrastructure: information about expected performance, direct control over the transfer operations, possibility to specify a condition-based control policy, capability to work in disconnected mode, advanced error recovery methods.

3 A MEDIATOR-BASED APPROACH: THE MOWGLI COMMUNICATION ARCHITECTURE

The key idea in the Mowgli architecture is to separate the behaviorally different wireline and wireless worlds. The essential component for doing this is the mediator, a node located in the fixed network.

A mobile node is connected to the fixed network through a wireless telephone link. The mediator is the node which provides the mobile node with a connection point to the wireline Internet. In Mowgli the mediator node is called the mobile-connection host (MCH).

The Mowgli system offers to the applications on the mobile node a rich set of communication and control services. Figure 1 depicts the conceptual organization of the service architecture. Services are provided in three different layers: in the agent-proxy layer, in the data transfer layer, and in the data transport layer.

The basic communication services are available through an API called the Mowgli socket interface. This interface includes the socket operations of the traditional BSD socket interface. Hence, existing applications that use TCP or UDP sockets can be executed on the mobile node without modifications. In addition to this, the Mowgli sockets provide new features not available in conventional TCP/IP sockets. These features include assignment of priorities to the data delivered through the socket, establishing the wireless link, and ordering an automatic recovery from unexpected link-level disconnections.

The Mowgli socket interface binds the existing applications to the services available in the agent-proxy layer. These services are implemented as agent-proxy teams. A proxy on the MCH and an agent on the mobile node cooperate to act as the mediator for all the data delivered between an application on the mobile node and its peer in the Internet. The proxy has a special mission: it plays the role of the mobile node while communicating on its behalf with the peer.

In the agent-proxy layer we have three different kinds of teams:

1. **Master agent - Master proxy.** Master agent and proxy provide the semantics of the traditional socket operations. In order to accomplish this the Mowgli socket layer informs the master agent whenever an application on the mobile node invokes a socket operation. The master agent then sends a message to the master proxy which performs the corresponding operation on behalf of the application. The masters are able to create other agents and proxies. They also take care of delivering UDP datagrams.
2. **Generic agent - Generic proxy.** Once a TCP connection with a remote peer has been established on a TCP socket the masters create a generic agent-proxy team, which will take care of delivering the data for that socket.
3. **Customized agent - Customized proxy.** The generic agent and proxy can be replaced with a customized agent and proxy which are tailored for a specific application protocol, for example for the HTTP protocol used in the World-Wide Web. The Mowgli WWW software, see Liljeberg et al. (1995,1996), is a good example of an agent-proxy team taking advantage of its knowledge about application semantics.

The functionality of a customized agent can be integrated into the client software when new applications are implemented for mobile users. Such a mobile client can cooperate directly with the customized proxy on the MCH. This approach amounts to splitting the traditional client program into two parts — one part on each side of the wireless link. The communication services available in the data transfer layer and in the data transport layer are specifically designed to support application-specific agent-proxy teams to be used in communication over the wireless (telephone) link.

In the data transfer layer, the Mowgli Data Transfer Service (MDTS) is able to take over the responsibility of transferring structured user data over the wireless link. The basic element of information for transfer operations is *Information eXchange Unit* (IXU). In general, an IXU is something that the user considers as an independent unit of information, the transfer of which he or she may want to control. Examples of IXUs include mail messages, files, print jobs, and inline images of WWW pages. Each IXU can have a set of attributes, which are used in controlling the transfer of the IXU. The MDTS provides an API for invoking the transfer of IXUs, for managing transfer queues, and for changing attributes of specified IXUs. Due to the attribute system, the MDTS is able to operate independently, without a direct user control. According to user specifications, expressed by attributes of IXUs, the MDTS can make its own decisions about invoking operations when conditions are favorable, about postponing transfers when conditions deteriorate, about trying to recover from failures, and about cancelling operations. In certain circumstances, the MDTS can even be allowed to make telephone calls.

In the data transport layer, the standard TCP/IP protocol stack is replaced with the Mowgli Data Channel Service (MDCS), which takes care of transmitting data over the wireless link between the mobile node and the MCH. All communication above the data transport layer uses the transport services offered by the MDCS. The MDCS is designed to cope with the special characteristics of the cellular telephone links. The communication is based on data channels with a priority-based scheduling system. Furthermore, each data channel has a set of additional attributes for controlling the behavior of the channel in case of exceptional events. The programming interface to the MDCS is very similar to the standard BSD socket interface. A detailed description of the MDCS can be found in Kiiskinen et al. (1996).

4 BENEFITS OF THE MOWGLI APPROACH

The mediator-based architecture of Mowgli affects the behavior of applications in several ways. Below we briefly summarize various benefits observed in experiments with Mowgli prototypes.

4.1 Working in normal conditions

The performance of distributed applications is essentially improved. The major benefits can be attributed to the agent-proxy team which makes it possible to divide the end-to-end data communication path into two autonomous parts. The agent-proxy team can also exploit the application semantics leading, for example, to type-specific data compression and to application-specific error recovery. We have also implemented several technical improvements including reduction of round trips, avoidance of unnecessary data transmissions, error recovery from short-term disconnections, and compression of headers. The increase in performance is remarkable as reported in Liljeberg et al. (1995,1996) and in Kiiskinen et al. (1996).

In a wireless WAN environment the user's possibility to control the communication is important. We allow the user to specify the order of data transmissions, to query the expected transmission times, and to decide if the data should be sent immediately. We have also allowed the user to delegate (with given advice) the decision making to the agent-proxy team.

4.2 Working in disconnected mode

There are three primary reasons why the user may want to work without a connection. Firstly, the wireless link is usually quite expensive. Secondly, the link quality may temporarily be below a required Quality-of-Service level. Thirdly, the user may be in an area not covered by the cellular telephone system.

In the disconnected mode a Mowgli agent in the mobile node can take the role of the remote component in the application. The basic principle is to transform the user data into Information eXchange Units. When the connection is later established, the Mowgli Data Transfer Service takes care of transferring the IXUs without any explicit user intervention.

The agent-proxy approach gives rise to functionality that may, in the future, turn out to be the most important one: The proxy can take the role of the mobile component and proceed autonomously. For example, the agent delegates time-consuming operations to the proxy to be executed asynchronously. The independently working proxy can control remote operations and receive incoming messages. If needed, it can reestablish the connection with the mobile node.

4.3 Working in almost disconnected mode

Traditionally the nodes in distributed systems work either in a connected mode or in a disconnected mode. The mobile telephone systems give another possibility: Components of the application will only connect to each other at the moment when cooperation is really needed. Typical examples include cache validation and notification of incoming messages.

An alternative usage for proxies: An application–specific proxy may act as an intelligent agent for the mobile node. It works independently, in a disconnected mode, to fulfill a task specified by the mobile application. For example, it may filter incoming data, and when something of interest arrives, it sends a notification to the mobile node.

The implementation of notifications can be based on the Short Message Service of GSM, which is a low–bandwidth but light–weight form of communication over the wireless link. An important implication is that the connection setup can be implemented in an almost transparent manner: The agent and the proxy communicate according to predefined rules.

4.4 Working in changing conditions

The disconnected operations can be exploited further. The user may be willing to specify minimum conditions under which the data transfer should take place. For example, large files are transferred only during cheap night time or during periods of high link quality, but urgent messages are always sent as soon as possible. The MDTS invokes the operations when the specified conditions are satisfied. In other words, the operations of the MDTS can be controlled like guarded commands.

5 ESSENTIALS OF MOWGLI

The original challenge in the Mowgli project was to combine two data communication systems of very different characteristics under a single control. This was accomplished through a mediator, consisting of an agent-proxy team and the associated infrastructure.

The role of the agent-proxy team is pronounced: Together the agent and proxy form a model for a distributed intelligent agent. This agent is "Janus-faced" in two ways. Horizontally it presents the client-interface on the fixed-net side and the server-interface on the mobile node, or vice versa. Vertically, it knows both the application semantics and the problems in wireless communication. This general approach gives a rich choice of opportunities for the application developer:

- hide or show,
- make the control "goal-based" or "explicit",
- solve the conflicts between the user's wishes and skills — or let the user do it.

An intelligent agent can take several different roles. The possibilities include the roles of an advisor, a filter, a booster, and a representative.

Today the research in mobility is primarily concentrated on the problems of physical mobility: The systems perceive how the nodes move. In the Mowgli environment the cellular telephone system hides certain issues in the terminal mobility like paging, hand-overs, and location updates. Therefore, we have been able to focus on the next set of problems: when to connect, how to control disconnected agents, how to handle the multiple-copy-update problem when the peers are only probably accessible, and — in general — how to improve the dependability of applications comprising wireless data communications.

Traditionally middleware solutions have been regarded as a promising approach in implementing distributed applications on heterogeneous platforms. According to our experience we would like to extend the concept: when implementing distributed applications in heterogeneous data communication environments the middleware should be replaced by a functionally more powerful "mediatorware".

ACKNOWLEDGEMENTS

This work was carried out as a part of the Mowgli research project funded by Digital Equipment Corporation, Nokia Mobile Phones, Nokia Telecommunications, and Telecom Finland. The authors are thankful to the rest of the Mowgli team, particularly Heikki Helin, Petteri Kaskenpalo, Jani Kiiskinen and Mika Liljeberg, for the fruitful comments and discussions during the Mowgli project as well as during the preparation of this paper. Last but not least the authors want to express their gratitude to the software engineering student groups who have participated in the implementation of the system.

REFERENCES

Alanko, T., Kojo, M., Laamanen, H., Liljeberg, M., Moilanen, M. and Raatikainen, K. (1994) Measured Performance of Data Transmission over Cellular Telephone Networks. *Computer Communications Review*, **24**, 5, 24–44.

Bakre, A. and Badrinath, B.R. (1995) I-TCP: Indirect TCP for Mobile Hosts, in *Proceedings of the IEEE 15th International Conference on Distributed Computer Systems*, IEEE Computer Society, Los Alamitos.

Balakrishnan, H., Seshan, S., Amir, E. and Katz, R. (1995) Improving TCP/IP Performance over Wireless Networks, in *Proceedings of the First ACM International Conference on Mobile Computing and Networking (Mobicom '95)*, ACM, New York.

Kiiskinen, J., Kojo, M., Liljeberg, M. and Raatikainen, K. (1996) Data Channel Service for Wireless Telephone Links, in *Proceedings of the Second International Mobile Computing Conference*, National Chiao Tung University, Hsinchu.

Kojo, M., Raatikainen, K. and Alanko, T. (1994) Connecting Mobile Workstations to the Internet over a Digital Cellular Telephone Network. Report C-1994-39, University of Helsinki, Department of Computer Science. Revised version in *Mobile Computing* (eds. Imieliński, T. and Korth, H.F. 1996), Kluwer, Boston.

Liljeberg, M., Alanko, T., Kojo, M., Laamanen, H. and Raatikainen, K. (1995) Optimizing World-Wide Web for Weakly-Connected Mobile Workstations: An Indirect Approach, in *Proceedings of the 2nd International Workshop on Services in Distributed and Networked Environments (SDNE)*, IEEE Computer Society, Los Alamitos.

Liljeberg, M., Helin, H., Kojo, M. and Raatikainen, K. (1996) Enhanced Services for World-Wide Web in Mobile WAN Environment, to appear in *Proceedings of the 3rd International Conference on Communicating by Image and Multimedia (Image'Com 96)*, ADERA, Bordeaux.

Rahnema, M. (1993) Overview of the GSM System and Protocol Architecture. *IEEE Communication Magazine*, **31**, 4, 92–100.

Yavatkar, R. and Bhagawat, N. (1994) Improving End-to-End Performance of TCP over Mobile Internetworks, in *Proceedings of the IEEE Workshop on Mobile Computing Systems and Applications*, IEEE Computer Society, Los Alamitos.

BIOGRAPHY

Timo Alanko received in 1983 his Ph.D. (Computer Science) from the University of Helsinki. He is an Assistant Professor in Computer Science at the University of Helsinki; currently the project leader of Mowgli. His research interests include distributed operating systems, mobile computing, and performance analysis. **Markku Kojo** received in 1995 his M.Sc. (Computer Science) from the University of Helsinki. Currently he is Ph.D. student in Computer Science at the University of Helsinki. His research interests include mobile computing, data communications, and distributed systems. **Heimo Laamanen** received in 1982 his M.Sc. (Computer Science) from the University of Helsinki. Currently he is researcher in Digital Equipment Corporation and Ph.D. student in Computer Science at the University of Helsinki. His research interests include nomadic computing, wireless data communications, and distributed systems. **Kimmo Raatikainen** received in 1990 his Ph.D. (Computer Science) from the University of Helsinki. Since 1990 he has been Assistant Professor in Computer Science at the University of Helsinki. He is a member of IFIP TC6 Special Interest Group of Intelligent Networks. His research interests include nomadic computing, telecommunications software architectures, and real-time databases. **Matti Tienari** received in 1962 his Ph.D. (Mathematics) from the University of Helsinki. Since 1969 he has been Full Professor in Computer Science at the University of Helsinki; currently Chairman of the Department. He has been the representative of Finland at IFIP General Assembly 1987–96, IFIP Trustee 1989–95 and member of IFIP WG 6.1 (Architecture and Protocols for Computer Networks). His research interests include modeling of concurrency, computer networks, and distributed systems.

18
An adaptive data distribution system for mobile environments

Sascha Kümmel, Alexander Schill, Karsten Schumann, Thomas Ziegert
Dresden University of Technology, Department of Computer Science,
Institute for Operating Systems, Databases, and Computer Networks
D-01062 Dresden, Germany, Tel.: +49 351 4575 457, Fax.: +49 351 4575 251, { kuemmel, ziegert }@ibdr.inf.tu-dresden.de

Abstract
Common transport systems lack the appropriate mechanisms to deal with the problems of mobile computing systems (e.g. temporary inaccessibility, transient network addresses of mobile hosts, along with varying quality of service parameters of physical network connections). Therefore, a need for new adaptive data distribution mechanisms.
This paper discusses specific features of GISMO's (**G**eneric **I**nfrastructure **S**upport for **M**obile **O**bjects) infrastructure; in particular, a mobile queuing service - which distributes data in an adaptive manner. In the remainder of this paper we introduce and motivate the need for adaptive data distribution mechanisms, describe our concept, as well as some implementation details and an overview of our first experiences with a prototype implementation.

Keywords
adaptation, data distribution, disconnected operation, mobile computing, mobile transport system, queuing systems

1 INTRODUCTION

During the last few years the fields of mobile computing and mobile communications has make considerable progress which has lead to a strong trend towards integrating mobile hosts within existing data networks. Mobile hosts appear at different locations at different times. During moves and connections over wireless links we have a situation where frequent disconnections occur. Furthermore, the bandwidth of common wireless communication infrastructures is limited (Mello, 93), (Davies, 94). Other problems are: resource heterogeneity, security and the management of location dependent data.

One way to overcome these problems is to implement new applications from scratch. Due to the similarity of a significant number of applications concerning the above mentioned issues, we outline the importance of a support platform for mobile applications giving generic assistance in a reusable manner. Considering this, we are currently developing and implementing a **G**eneric **I**nfrastructure **S**upport for **M**obile **O**bjects (the GISMO-Project). The basic design of our support platform is presented by (Schill, 95). This paper discusses specific features of this infrastructure, in particular a mobile queuing service.

2 MOTIVATION

Common transport systems lack the appropriate mechanisms to deal with temporary inaccessibility, transient network addresses of mobile hosts, along with varying quality of service parameters of physical network connections.

There are some considerable efforts to solve the problem of disconnected operations and temporary inaccessibility ((Huston, 93), (Huston, 95), (Kistler, 92), (Satyanarayanan, 93) and (Satyanarayanan, 94)). These solutions mainly focus on file systems, based on caching, operation logging and reintegration. A recoverable queuing service for distributed transaction processing as a solution for reliable operation handling in case of inaccessibility is described in (Dietzen, 92) and (Transarc, 94). But they did not deal with service mobility and QoS adaptation. (Balakrishnan, 95) shows a way for the improvement of TCP/IP performance over wireless networks and (Bakre, 95) enhances the RPC-mechanism with mobility awareness.

While working on GISMO, the need of a new adaptive data distribution mechanism arose. This mechanism has to cope with postponed data transfer by intermediate persistent storing, „mobile addressing" supported by special locating mechanisms, detection of QoS parameters and appropriate adaptation during the transfer.

QoS adaptation can be achieved by varying the data packet size depending on the average error rate and by data conversion or compression. The parallel transfer of data packets by multiple threads in case of long delay connections obviously decreases the transmission time, see (Kümmel, 95).

3 THE CONCEPT

In this paper we focus on the data transport layer and the disconnected operation handling within the GISMO-architecture (for a detailed description see (Schill, 95)). For better understanding we use an e-mail system as an application scenario below (details can be found in (Schill, 96)).

We present a transport system according to the queuing principle, the queuing service (QS). The QS is designed to support transfer of data units in any size, so called databodies. It is possible to describe dependencies between databodies. This allows the transfer of complex data structures with explicit access to each element by the QS. The explicit access enables data format detection and conversion in advance (before the transmission). A suitable example therefore is a multimedia e-mail, composed of text documents, audio and/or video and other attachments. By building an e-mail using single databodies for each type of information, the user is able to select only the databodies he really wants in case of a low-bandwidth connection.

It is possible to assign the QS to reduce the size of a bodypart in advance. This allows the reduction of costs and time during a transmission over a low bandwidth link,. The reduction may cause a loss of quality, so we believe that user interaction during the selection of the rather subjective compression parameters is very desireable. For an automatic adaptation 'on the fly' we suggest another solution. The data source should deliver further information about the maximum possible compression, or on the other hand which parts of the information are vital to be transmitted in any case. The QS data structures enable the transfer of additional data, so this kind of operaion is supported.

For the evaluation of QoS parameters, the assessment and conversion respectively compression of different media types, external services within the GISMO System will be deployed. The QS is only aware of the fact, that the need for data conversion and QoS-detection exists.

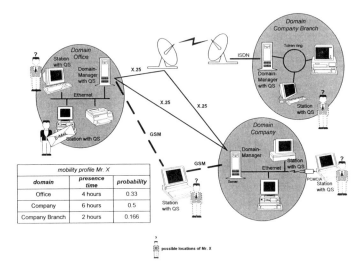

Figure 1 a sample environment

Figure 1 shows a sample environment. Queuing systems run on every station within the distributed environment. There is no central queuing server, due to the fact that redundancy and optimum routing are supported. Target addressing is not only limited to network addresses, it is also possible to use a list of target descriptions in order to meet the mobile aspects. In the example e-mail application a description is derived from a user identifier and an application identifier (we use in our prototype universal unique identifiers). Target locating is achieved by an external service - the Application Data Mobilizer and Manager (ADMM; for further explanation see (Schill, 95)).

The following short description of one distribution cycle is based on figure 2. The source application enqueues the databody together with a destination description (1). The ADMM locates the QS which is currently assigned to the target application and the respective user (for mechanisms for distributed location of mobile objects see (Dasgupta, 94)) and assembles a routing path (2). In case a user is not connected and hence it is impossible to locate him, the ADMM supports *mobility profiles* which include possible locations of the users. Users usually relocate to a lazy changing set of locations, see also figure 1. Therefore the QS supports multicast data transfer to all probable destinations. The distribution is controlled by the routing path obtained from the ADMM and data is sent hop by hop. There can be several multicast distribution hops in the path for optimum exploitation of available bandwidth. Upon reception of a databody at the QS it is possible to contact the ADMM (3). So adaptation to changing network topology and very high target mobility are supported. We call it „adaptive source routing".

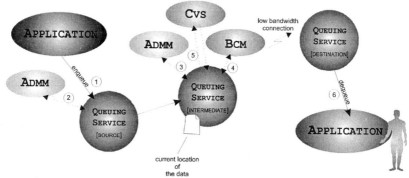

Figure 2 the queuing service

On every intermediate QS the physical connection to the next QS is evaluated by the bandwidth and cost manager (BCM) (4). A conversion service (CVS) converts the databody into another format with a significant reduction of its size in case of a low bandwidth connection or high costs and if the databody is indicated as convertible (5). The modified databody is dequeued by the destination application (6).
The transfer between two QS is covered by a transaction-like protocol and optimized accordingly to the currently available QoS parameters. All received data is stored persistent to ensure reliability and recovery in case of a system crash or network breakage. To support multicast distribution and to protect the system from data overflow we implemented lifetime control and garbage collection mechanisms. Each databody has a maximum lifetime. After expiration, the databody and all its copies will be removed from the whole distributed system. If a databody is dequeued by an application, a garbage collection process is started which will remove all copies within the system and inform the source application the successful operation has completed.
In order to consider security requirements, security tags (to support various security techniques) will also be transferred with every databody. All distribution steps could be verified based on these tags using an external security component.

4 THE IMPLEMENTATION

The basic architecture of the queuing service is shown in figure 3. It is made of five core components and multiple in/out data queues to support priority queuing (we distinguish between hierarchy and priority rather than (Athan, 93)). The QS acts as a single process with multiple threads. It uses three external services to obtain information about addressing, current QoS parameters and data conversion in advance of transmission.

The distribution control component monitors the data distribution within the mobile environment. All data transfer actions will be logged. Therefore, the component is informed by the four other components in case of data enqueuing or dequeuing and data transfer to or from a remote QS. The distribution control component itself communicates via RPC with other distribution control components running on other QS elsewhere. The logged transfer actions allow tracing the data distribution within the system. This accomplishes garbage collection, lifetime control and back reporting of distribution results.

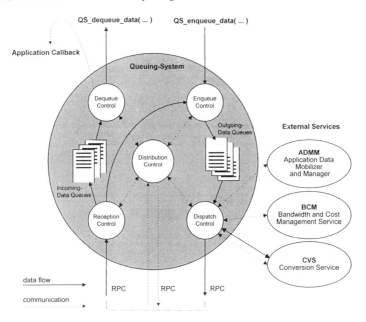

Figure 3 The basic architecture of the Queuing Service

Applications, which want to use the QS have to implement a callback procedure, so the QS can inform them about data arrival or the completion of a dequeuing operation at a remote site. Assuming a long time between the enqueue and dequeue operations there is a high likelihood that the sending application has already gone. So what will happen with the back reported message about data arrival? Applications can specify a file, so the system can log any reports about successful arrivals or failures during delivery (timeouts).

5 EXPERIENCES AND FUTURE WORK

We have implemented a first prototype queuing system based on the Windows NT/95 operating systems and Microsoft RPC. The sources will be ported to UNIX platforms on top of OSF/DCE soon. So the system can run on mobile clients based on Windows95 while the servers are running on more powerful Unix workstations. We successfully built a mobile aware multimedia X.400 e-mail system on top of our prototype (see (Schill, 96)).

Currently we are able to present the first results of performance measurements with the system prototype. Figure 4 shows the average transfer times between two QS running on a PC Pentium 100 MHz, 32 MB memory each, connected by a HP 100 VG AnyLAN. The transfer time is measured between the moment an application has completed it's enqueuing operation on one QS until an other application completes its dequeuing operation on the other. The application will be informed by a callback function called by the QS if addressed data is available.

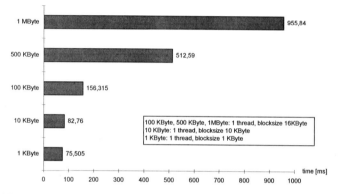

Figure 4 throughput measurement

We currently working on an enhanced transfer protocol. The protocol performs optimized asynchronous multithreaded packet transport and continuous recovery abilities for aborted transfers in case of network disconnection, disruption or station switch off. There are some further enhancements to do. Considering the worst case of a low bandwidth and cost intensive connection to a remote site and assuming a few messages in the queue waiting for delivery, we must decide, which message will be delivered first. Of course the message headers own a higher priority than other databodies within a container, but sending all headers first involves a combination of multiple operations, if an error occurs lots of things have to set up again, so it may be better to send some containers in one trial.

Future work will focus on packet transport mechanisms without using RPC respective TCP/IP to maximize throughput for low-bandwidth connections found in wireless networks, because of the known TCP problems with large and also very short delay times due to the window mechanism (for details see (Balakrishnan, 95), (Thekkath, 93) and (Kay, 93)).

We also plan to implement mobile database access and mobile-RPC (Winkler, 95) facilities on top of our prototype. An overall transaction-management and enhanced security facilities within the system are also under development.

Acknowledgments

We would like to thank all involved colleagues and students for their significant efforts in building the implementation of the described prototype. Moreover, we would like to thank Digital Equipment GmbH (EARC Karlsruhe for sponsoring and for supplying the X.400-Infrastructure).

6 REFERENCES

Athan, A. and Duchamp, D. (1993) Agent-Mediated Message Passing for Constrained Environments, in *Proceedings of the USENIX Mobile and Location-Independent Computing Workshop*, Cambridge MA, 103-7

Bakre, A. and Badrinath, B.R. (1995) M-RPC: A Remote Procedure Call Service for Mobile Clients, in *Proceedings of the 1st ACM Mobicom Conference*, 2-11

Balakrishnan, H., Amir, E. and Katz, R.H. (1995) Improving TCP/IP Performance over Wireless Networks, in *Proceedings of the 1st ACM Mobicom Conference*, 124-31

Dasgupta, P. (1994) Resource Location in Very Large Networks, in *Proceedings of the IEEE Computer Society First International Workshop on Services in Distributed and Networked Environments (SDNE'94)*, 156-63

Davies, N., Pink, S. and Blair, G. S. (1994) Services to Support Distributed Applications in a Mobile Environment, in *Proceedings of the IEEE Computer Society First International Workshop on Services in Distributed and Networked Environments (SDNE'94)*, 84-9

Dietzen S. (1992) *Distributed Transaction Processing with Encina and the OSF DCE*, Transarc Corporation

Huston L.B. and Honeyman P. (1993) Disconnected Operation for AFS, in Technical Report No. *CITI 93-3*, University of Michigan

Huston L.B. and Honeyman P. (1995) Partially Connected Operation, in Technical Report No. *CITI 95-5*, University of Michigan

Kay, J. and Pasquale, J. (1993) The Importance of Non-Data Touching Processing Over-heads in TCP/IP, in Computer Communications Review

Kümmel, S. and Schill, A. (1995) Leistungsanalyse und Vergleich von RPC-Systemen für heterogene Workstation-Netze, in *PIK - Praxis der Informationsverarbeitung und Kommunikation*, Issue No. 3, 148-53

Mello, J. and Wayner, P. (1993) Wireless Mobile Telecommunications, *Byte*, Februar 1993, 147-54

Satyanarayanan M. et. Al. (1993) Experience with Disconnected Operation in a Mobile Computing Environment, in Technical Report No. *CMU-CS-93-168*, Carnegie Mellon University

Satyanarayanan M. and Noble D.B. (1994) Coda, An Empirical Study of a Highly Available File System, in Technical Report No. *CMU-CS-94-120*, Carnegie Mellon University

Kistler, J. J. and Satyanarayanan M. (1992) Disconnected Operation in the Coda File System, in *ACM Transactions on Computer Systems*, No. 10

Schill, A. and Kümmel, S. (1995) Design and Implementation of a Support Platform for Distributed Mobile Computing in *Mobile Computing Special Issue of Distributed Systems Engineering,* 128-41

Schill, A., Kümmel, S. and Ziegert, T. (1996) Mobility aware Multimedia X.400 email: A Sample Application Based on a Support Platform for Distributed Mobile Computing, in *Proceedings of the IMC '96 Workshop for Information Visualization & Mobile Computing*

Transarc Corporation (1994) *Encina RQS Programmer's Guide and Reference*

Thekkath, C.A. (1993) Limits to Low-Latency Communication on High Speed Networks in *ACM Transactions on Computer Systems*; Vol.11, No.2

Winkler, M. and Kümmel, S. (1995) Mobile RPC - Eine Erweiterung des DCE Remote Procedure Call, in *Proceedings of the GUUG-Jahrestagung 1995*, 116-22

7 BIOGRAPHY

Prof. Dr. Alexander Schill is chair of the institute of Operating Systems, Data Bases and Computer Networks at Dresden University of Technology. He received his Ph.D. in Computer Science in 1989 at Karlsruhe University. In 1990/91 he worked at IBM T.J. Watsom Research Center, Yorktown Heights. His main research areas are distributed systems, high performance networking and mobile computing.

Sascha Kümmel received his diploma in computer science in 1994 from the Dresden University of Technology. He's currently working as research assistant at the institute of Computer Networks at Dresden University of Technology. His main research areas are mobile computing and high performance networking. His work is supported by Digital Equipment Corporation.

Karsten Schumann is a student of Computer Science at Dresden University of Technology since 1991. Karsten is student member of the GISMO Group. He is interested in Mobile Computing and Distributed Systems.

Thomas Ziegert received his diploma in computer science in 1995 from the Dresden University of Technology. He is currently working towards his Ph.D. at the Dresden University of Technology with a special interest in efficient location mechanisms in mobile environments. His work is supported by the German Science Foundation (DFG). Thomas is a member of the GISMO Group.

19
Object oriented system architecture and strategies for the exchange of structured multimedia data with mobile hosts

J. Bönigk, U. von Lukas
Computer Graphics Center Rostock (ZGDV)
Joachim-Jungius-Str. 9, D-18059 Rostock, Germany
Phone.:+49 381 4024 150   Fax:+49 381 446088
E-mail: {joerge,uvl}@rostock.zgdv.de
URL: http://www.rostock.zgdv.de

Abstract
In this paper an architecture is presented which enables applications on mobile computers to transparently exchange multimedia objects with applications on stationary servers via the *Object Bus*. The components of our architecture face the common problems of mobile computing like limited bandwidth, end systems with limited resources and frequent disconnections. Therefore they are designed to efficiently use the available resources and to minimize these problems. This is done, e.g. by using object specific methods of data reduction and level-of-detail, by adaptation of transfer to given Quality-of-Service parameters, and by data compression. All these methods are influenced by contexts like local resources, available communication channels, and user preferences. Finally the abstract architecture is mapped to OMG CORBA. The architecture and a selection of the exchange strategies are used to build a prototype of a mobile information system.

Keywords
mobile computing, mobile communication, detail-on-demand, data reduction, mobile system architecture, distributed systems, object orientation, CORBA

1 INTRODUCTION

The main task of the underlying system architecture and communication model for the aim of mobile visualization within the MoVi project[*] (Heuer et al., 1995) is to enable the use of complex visualization applications and large sets of multimedia data in a mobile environment.

---

[*] The work of the project „Mobile Visualization" (MoVi) is supported by the German Research Association (DFG) under contract Schu 887/3-1.

One of our typical scenarios is the access to public or private data stored at several stationary data servers (SDS) for visualization on a mobile end system (MES) using GSM[1].

However, there are the well-known problems of mobile computing (Forman and Zahorjan, 1994) (Imielinski and Badrinath, 1994) (Weiser 1993) which should be considered when designing an architecture: limited bandwidth, resource poor mobile terminals, variations in quality of transfer and possible disconnections. Our architecture addresses the first two problems by reducing the amount of data to be transferred by compression and by filtering of relevant data according to the resources of the mobile system. The transfer of data is adapted to varying transfer characteristics and our communication model is designed to cope with frequent disconnections in all stages of the transfer.

The remainder of this paper is structured as follows: In the next section the components of our architecture will be introduced as well as the exchange strategies they make use of. In section 3 a mapping of the architecture and the communication model to OMG CORBA is presented. A first prototype of a mobile information system, which is based on the architecture and uses a selection of the exchange strategies is explained in section 4. After an overview of related work in section 5 a conclusion of our work and an outlook on future plans regarding the architecture and the prototype are given in section 6.

2 SYSTEM ARCHITECTURE

The system architecture has to serve as a flexible platform for the efficient exchange of user data and applications between stationary data servers and mobile end systems. Our main concept is an object oriented approach with the *Object Bus* (OBus) as one central feature. This Object Bus serves as a transparent layer for mobile communication and is responsible for the delivery of messages. The second main feature is the introduction of *Message Handlers* (MH) that act in place of the communicating processes when exchanging structured objects. They notify each other about transfer procedures and transfer the objects (see Figure 1).

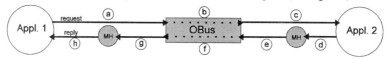

Figure 1: Exchange of messages via the Object Bus.

That means, that the whole communication between two processes A and B (a) is divided into the following main steps (see Figure 1):
(1) Applications generate requests that are handed over to the local Object Bus for transportation (a).
(2) The following transfer inside the Object Bus is carried out by several components (Network Scheduler, Transfer Manager) which reside between application and network layer on every computer (b).
(3) The receiver gets the message (c) and generates reply objects.
(4) The reply objects are handed over to appropriate Message Handler processes (d). The Message Handlers use knowledge about the context to enrich the reply with additional information, that is necessary for an effective scheduling. Additionally they perform

---
[1] Global System for Mobile Communication

operations on the reply objects like data reduction that are influenced by the resources of the mobile host. These operations are necessary in order to save the limited bandwidth of wireless communication channels.
(5) The Message Handlers on both sides transfer the objects with suitable methods via the Object Bus (e/f/g).
(6) The application receives the reply objects (h).

The third main component is a *Context Manager* (see Figure 2), which covers all aspects connected with contexts; e.g., context sensitive trading, context based modification of requests, and the management of contexts. It intervenes in the communication flow (a),(c),(d), and (h) between the components in Figure 1. In addition, it provides Message Handlers and the Object Bus with context information that are necessary for their work.

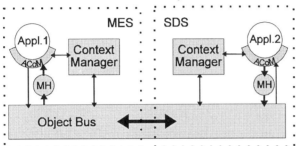

Figure 2: Overview: Components of the System Architecture on MES and SDS.

The communication between applications and Object Bus and Message handlers respectively is done via the Application Communication Manager (ACoM). The named components will be explained in the next sections in more detail.

2.1 Object Bus

In the context of mobile computing with slow, unreliable, and possible expensive wireless links, asynchronous calls (messages) are preferred to synchronous calls. The aim of the *Object Bus* (OBus) is to manage and efficiently transfer these messages on all kinds of networks. Components of the Object Bus are a *Network Scheduler*, a *Request/Reply Cache*, a *Transfer Manager* as well as a *user interface* on mobile hosts. The OBus is linked with the local Context Manager where it can access information that is necessary for its work, e.g. the current context information or the address of a requested service.

The *Network Scheduler* (NS) is the main component of the Object Bus. It is responsible for the efficient transfer of messages. To achieve this aim it manages a list of all requests that it has received and of all replies that have to be sent to other NSs together with their context information. Based on this list, the NS determines which request or reply should be send next in an existing connection with the best use of the channel. This message is then handed over to the Transfer Manager for transportation. The decision is not only based on information describing the request, like size, type, priority and requested Quality of Service (QoS) parameters, but also on information from the Transfer Manager about the state of connections and general context information obtained from the Context Manager; e.g., information about available channels and their parameters. It depends on the intelligence of the Network Scheduler to what extend all this information is used to make the decision.

For incoming new messages the NS decides whether to start an appropriate MH that has to handle the message or to deliver the message directly to the receiver's ACoM. The latter case is true for messages that do not need a special processing; e.g., error messages, acknowledgements, or messages where their content is negligibly small or does not support special techniques (see section 2.3). Typical exchange strategies that are performed by this component are scheduling with priorities and with the influence of Quality-of-service parameters as well as data compression initiated by the NS.

The *Request/Reply Cache (Bus cache)* stores requests and replies at the local host to facilitate the immediate delivery of replies to the application when receiving an identical request later on. Since local applications also communicate via the Object Bus its cache can also be employed for acceleration of the local communication. Due to changing resources it is necessary to store also additional information about the current contexts. As a base for minimizing inconsistencies and for decisions which cached items have to be deleted serves a combination of several methods (Cate, 1992)(Chankhunthod et al., 1995). In our approach the time since the last modification as well as information about least recently accessed cached items and information about the type of the items are used for cache management functions in case that the cached items do not have timestamps that explicitly define their validation time.

On mobile end systems the Object Bus includes an *user interface* as a separate application independent component. In addition to functionality that is offered by applications and their respective ACoMs the user interface of the Object Bus provides the user with information about the current state of all active requests and pending replies and allows the interactive control of exchange parameters. The user has for example the possibility to modify request priorities, to cancel transfers, and to configure the handling of special message types.

The *Transfer Manager* carries out the lower level network functions like opening and closing of connections as well as the actual send and receive functions. In order to have the possibility to evaluate QoS parameters this component monitors the traffic of current connections. These traffic data are handed over to the Context Manager. So the other components like the Network Scheduler can make use of information about the last connections to a special host as a base for planning their work.

2.2 Context Manager

Another basic component of our abstract architecture is the *Context Manager*. The *context* is our main concept for describing all dynamic and static characteristics of the mobile and stationary entities (user, resources, information, location and time). Since the contexts represent the parameters of the mobile environment, they have an effect on all architectural components. This includes, how to locate, to access, to transfer, and to present information. Main tasks of the Context Manager include:

Context sensitive trading:

The Context Manager serves to deliver the addresses of desired information and services. This process is influenced by contexts. If no address information is locally available by the use of this technique, a request will be sent to the next Context Manager on a remote site that can in turn involve further Context Managers to obtain the required information.

Modification of requests and replies according to context criteria:
The Context Manager is in addition responsible for the adaptation of requests to the mobile environment by evaluation of contexts in order to save the mobile resources. That includes substitution of environment variables like location and time with their current values and the modification of requests, for example including information about resources (e.g. free disk capacity, display properties) and user preferences (e.g. maximum costs for a transfer, priorities for certain data types).

Management of all relevant context information:
This point includes the management of all current context parameters including statistics as well as context information from some other frequently contacted SDS and MES. Context information on the one hand will be inquired not only by the ACoM but also by all other components of the architecture. On the other hand, contexts will be inserted from all components of the architecture. Therefore, the Context Manager allows all components of the architecture to access the contexts in a specified manner. The Context Manager stores the contexts in its context database or it uses built-in functionality to inquire them directly on demand. For example, resource information with a high dynamic is always requested directly, since their storage in a database would consume too many resources for its continuos updating.

2.3 Message Handler

Message Handlers are built as processes for the optimized exchange of structured information. They can be launched by the Application Communication Manager of an application that has generated a reply as well as by the Network Scheduler that receives a suited reply. Due to their *object specific* design and their knowledge about type and structure of the data to be transferred they are able to use type and structure dependent methods for minimizing network traffic and response times. One typical group of methods are level-of-detail concepts; e.g., *Successive Refinement* and *Detail-on-demand* (see Figure 3).

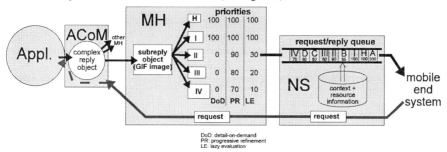

Figure 3: Level-of-detail methods: cooperation of Message Handler and Network Scheduler on a SDS.

Another method is *Data Reduction* which can be achieved by the modification of requests (see Heuer and Lubinski, 1996) and by reducing reply objects according to available resources, for example reduction of color images to the color depth of the mobile display, no transfer of objects where there are no viewers on the MES configured for, no transfer to a MES without sufficient resources, and adaptation of data to user defined constraints regarding the network connection.

2.4 Application Communication Manager

The *Application Communication Manager* is an application specific interface to the components of our architecture. It can be realized as a built-in interface in especially designed mobile applications or as a separate component that uses a defined interface of an existing application (see Figure 2). Tasks of the ACoM include the management of a list of all requests and related replies of an application together with applied contexts and the decomposition/composition of complex compound replies. The sending ACoM divides compound replies into several subreplies of certain defined types and hands them over to MHs for type specific exchange. These MHs have to be started by the ACoM. They transfer the subreplies to their counterpart at the receiving site (started by its NS) and hand them over to the receiving ACoM that will reassemble the reply object.

2.5 Message Exchange

The protocol that we use for the communication over the Object Bus is a combination of an efficient protocol for coding header information and the possibility to express the contents of messages in an application dependent format, like SQL, HTTP, MIME, or KQML. Since the content of a message should not be evaluated by the Network Scheduler all data that are necessary for planning the transfer, e.g. priority, size, and QoS demands, have to be part of the header. Our protocol allows splitting of larger messages into smaller pieces and sending them separately. The headers of these pieces are provided with offset values so that a transfer can be continued at this offset without a full retransmission after a disconnection has occurred.

3 USE OF CORBA

The acceptance of a new approach is often measured by its integration in the "real world". Due to this, the use of widely known and accepted standards and the integration of legacy systems into the new environment is crucial. We propose the Common Object Request Broker Architecture (CORBA) of the Object Management Group (OMG, 1991) as a platform for the design of object oriented, distributed systems – even in the context of mobile visualization.

We show, how this standard can be used to support resource poor mobile hosts and minimize the amount of data by optimization of the data flow. To do this we map the abstract architecture of the OBus to CORBA and point out which work is necessary to gain interoperability between standard CORBA and the modified broker architecture implementing the OBus.

3.1 CORBA Motivation

The design of software systems in general and of software for special circumstances like mobile environments in particular is a non trivial and expensive challenge. The problems of today's software engineering can only be faced by the use of modern paradigms and powerful tools that support the user by hiding great parts of the complexity one has to cover. The role of CORBA in this context is a platform for integration and distribution: objects can be used in a comfortable way independent of their location in the network, their implementation language and platform. The inter process communication and the marshalling of parameters are done by this middleware (the CORBA implementation).

All the functionality which should be available to local or remote clients resides at objects in the server. Servers are described by the signature of their interface. This interface is specified in the Interface Definition Language (IDL). An IDL compiler maps the description to a specific implementation language and generates marshalling code for client (stub/proxy) and server (Basic Object Adapter, BOA) side.

In a mobile environment, we have to deal with resource poor mobile terminals that are not able to host applications with high memory or processing demands. These parts should reside at a stationary server and be coupled with the mobile application. The location of external services and the communication among these components is managed by CORBA. The configuration, which services to provide local or remote is specified at runtime. Further more, CORBA accelerates the development of applications by using existing OMG defined services like persistence, transaction, or security.

But CORBA in its actual state is not prepared to solve the special problems in a mobile environment. It is not only the lack of special purpose CORBA implementations but also some fundamental insufficiencies of the architecture itself: CORBA hides a lot of details and prevents the application to deal with a variety of parameters. Buffer sizes, coding of parameters, location of objects, choice of protocols, marshalling etc. are not under control of the developer. This total information hiding is inadequate for mobile computing. Also, there is no way to deal with disconnections but to completely restart the transfer. Finally, CORBA does not have a stream concept at the moment and CORBA does not support any form of QoS.

3.2 Implementation of the OBus as a modified ORB

The concept of the Object Bus can be implemented on top of a request broker core. In this section we will map the abstract OBus based architecture presented above, to the more concrete architecture proposed by the OMG.

The "normal" way to invoke CORBA servers is by a synchronous call of a method of a remote object. This call can have in and out parameters and also an optional return value. This call blocks the client until the remote function returns. In our scenario where the transfer and the generation of the result may be very slow, this synchronous mode is not acceptable. We only support oneway calls that cannot have out parameters or return values. Synchronous calls must be transformed into oneway calls by the ACoM. This transformation is necessary to facilitate the use of all the existing and standardized services with their asynchronous interface.

First we have to decide, at which level mobility support should be inserted. The application level could be used for most of the concepts but is not very fast and would not be transparent for the users. The design for a mobile application should not differ significantly from a stationary solution.

Stubs and skeletons are generated by the compiler. This code is linked with also implementation specific client or server libraries for inter process communication. Existing CORBA implementations cannot deal with priorities and disconnections. They only have the functionality of a "dumb Transfer Manager". For optimal support we have to insert the functionality of the Network Scheduler at this level: It holds a list with submitted calls, extracts the context information and manages the choice of physical network. The priority driven transfer is done by calling functions of the chosen Transfer Manager. Figure 4 gives a sketch of the architecture with the modified/additional components. As we can see there, the broker also must be adapted to make best use of the introduced features. The broker now includes the functionality of the Context Manager and uses dynamic as well as static context information to bind to an appropriate server. It also supports the negotiation of QoS parameters for stream connections between client and server.

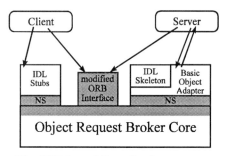

Figure 4: Modifications of the ORB for mobile applications

One concept that is not available in CORBA today is the concept of streams. They are not only useful for multimedia applications with realtime demands but also to implement successive refinement concepts. With streams we can think of an image transmission as a stream from the source (e.g. data base at SDS) to the sink (application at mobile terminal). This stream can be interrupted if the necessary level of detail is reached. But this is only possible when streams are introduced beside attributes and operations, as a new feature in CORBA. A stream is characterized by a frame type, a direction, some QoS characteristics, source and sink callbacks.

The modified communication bus has an additional interface. This is used for query and manipulation functionality at request level. The application or a special control application for the OBus can use it to facilitate a higher level of controllability for the user.

3.3 Access to standard CORBA services

Interoperability of the modified broker and standard CORBA is reached by two half bridges which translate their format to the standardized UNO protocol defined by CORBA 2.0 (OMG, 1995).

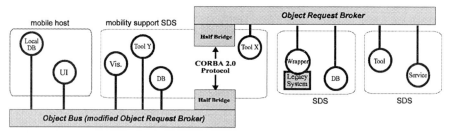

Figure 5: Bridging between modified and standard ORB

Using this approach we lose performance, but win flexibility and lots of existing and future services using CORBA. The optimized communication can only be gained for objects on the mobile host, contacting services at the mobility support SDS. Remote calls over the bridge are limited to standard features. The resulting structure with the mobile host and several stationary servers connected, is presented in Figure 5.

4 MOBILE INFORMATION SYSTEM

The presented Object Bus architecture and a selection of the exchange strategies are used to build a first prototype of a mobile information system (MIS). This MIS is intended to serve as integrating framework to provide a user with necessary information while he/she is mobile. First data types that are supported by Message Handlers include textual information and GIF images. They are used when a mobile user makes a query to a remote database to get the brand-new weather information regarding his/her current location. That means that database queries are modified to include the current location as an additional selection criterion. Information about the resources of the mobile end system and user preferences are used to achieve a further reduction of data; e.g., the user can specify that images should match the display capabilities of the mobile system.

The application consists of several modules: the user interface, a huge database holding actual weather data and visualization tools that produce still images. All these modules are implemented as CORBA clients and servers that communicate via the OBus. The complete database is stored at the stationary mobility support server at the office or can be accessed via this server. Due to the poor resources of the mobile terminal, we split the application: Only the user interface and the main control reside on the terminal. The computing intensive visualization and the data are located at the workstation. A typical request would be generated by the user, who is asking for a forecast for the actual location six hours later. The ACoM of the database decomposes the compound reply into several subreplies of individual types; e.g., textual descriptions and satellite images. The subreplies are handed over to Message Handlers. There the image and the text objects are treated separately for optimization issues; for example, the image is reduced in resolution and color depth adequate for the display capabilities of the notebook. At the client side, the objects are decompressed if necessary and are handed over to the UI to present it to the user. In case of an interlaced GIF image, streams can be used to implement a successive refinement that can be interrupted if the level of detail is acceptable.

5 RELATED WORK

There are some other projects that focus on some problems that arise, when mobile end systems access globally distributed information on stationary servers. Mobisaic (Voelker and Bershad, 1994) is an information system for a mobile wireless computing environment, which provides users with documents that are dependent on dynamic environment variables such as location. Mowser (Joshi et al., 1995), a smart Web browsing application, performs transactions based on the user's available resources like display resolution and sound capabilities. Other projects try to find solutions for context-aware applications (Adams et al., 1994) (Satyanarayanan et al., 1994) and for the message exchange with mobile hosts; e.g., the agent-mediated message passing (Athan and Duchamp, 1993).

Two groups also use standardized distribution platforms in their projetcs. The MOST project (Friday et al., 1996) relies on ANSAware. A group at the Dresden University of Technology use OSF DCE as the platform for distribution (Schill et al., 1995). A research group at BBN deals with Quality of Service for object communication in a more general manner (Zinky et al., 1995).

6 SUMMARY

We proposed an architecture, where applications transparently exchange objects via the Object Bus. It enables the development of mobility aware applications with state of the art support for disconnections and bandwidth minimization. Message Handler processes and a Network Scheduler, the main components of this architecture, were described as well as the methods that they perform to make the best use of slow and unreliable communication channels when acting in mobile environments. These methods include level-of-detail techniques, data reduction, caching, data compression and scheduling with priorities and influence of Quality-of-Service parameters. The given abstract architecture was finally mapped to OMG CORBA. The architecture and a selection of the exchange strategies were used to build a first prototype of a mobile information system.

The following table sums up the presented components of the Object Bus, the objects/data types treated by them with typical operations and gives the equivalent component of the CORBA mapping.

Table 1: Logical components of the architecture and their CORBA equivalents

component	operating on	typical operations	CORBA equivalent
ACoM	request, reply (complex objects)	decomposition, reassembly	application level
Context Manager	request, context data	address resolving, enrich request	broker, trader
NS	request/ reply messages	compression, fragmentation, QoS-handling	core
Message Handler	homog. data (simple objects)	disconnection-handling, reduce color, drop frames	BOA + stub (proxy)
TM	package	send, multiplex	BOA + stub (proxy)

7 ACKNOWLEDGEMENT

We would like to thank Thomas Kirste and Astrid Lubinski for fruitful and valuable discussions regarding the presented object bus architecture and exchange strategies. Additionally, we would like to thank Steffen Nowacki for cooperation regarding the integration with CORBA.

The described work is part of the workpackages "Exchange strategies" and "Distributed functionality" of the MoVi project funded by the German Research Association (DFG).

8 REFERENCES

Adams, N.; Gold, R.; Schilit,B.; Tso,M.; Want, R.(1994) An Infrared Network for Mobile Computers. *Proc. of the 1st Usenix Symposium on Mobile & Location Independent Computing*, August 1994.

Athan,A.; Duchamp,D.(1993) Agent-Mediated Message Passing for Constraint Environments *Proc. Mobile and location-Independent Computing, USENIX,* Cambridge, August 1993.

Cate, V.: Alex - a Global Filesystem. *Proceedings of the USENIX File Systems Workshop, pp. 1-11, May 1992.* ftp://alex.sp.cs.cmu.edu/doc/intro.ps

Chankhunthod,A., Danzig,P.B., Neerdaels,C., Schwartz,M.F., and Worrell,K.F.(1995) A Hierarchical Internet Object Cache. *University of Southern California, University of Colorado,* Boulder, 1995.

Forman, G.H. and Zahorjan, J.(1994) The Challenges of Mobile Computing. *IEEE Computer,* 27(4), April 1994.

Friday, A.J., Blair, G.S., Cheverst, K.W.J., Davies,N. (1996) Extensions to ANSAware for advanced mobile applications. *Proceedings of the IFIP/IEEE International Conference on Distributed Platforms,* Dresden, 1996.

Heuer, A. and Lubinski, A. (1996) Mobile Information Access - Challenges and possible Solutions. *Workshop "Information Visualization and Mobile Compuing" (IMC '96),* Rostock, February 1996.

Heuer, A., Kehrer, B., Kirste, T., Schumann, H., and Urban, B. (1995) Concepts for Mobile Information Visualization - The MoVi-Project. *Proc. Eurographics Workshop on Scientific Visualization 1995,* May 1995.

Imielinski, T. and Badrinath, B.R (1994) Mobile wireless computing. *Communication of the ACM,* Oct. 1994, Vol.37, No.10.

Joshi,A., Weerasinghe,R., McDermott,S.P., Tan, B.K., Bernhardt,G., and Weernawarna,S. (1995) Mowser: Mobile Platforms and Web Browsers. *Purdue University, 1995.*

Object Management Group (1995) CORBA 2.0/Interoperability - Universal Networked Objects. *OMG TC Document 95.3,* March 20, 1995.

Object Management Group (1991) The Common Object Request Broker: Architecture and Specification. *OMG TC Document 91.12.1 Revision 1.1,* 1991.

Satyanarayanan, M., Noble, B., Kumar, P., Price, M. (1995) Application-aware adaption for mobile computing. *Operating Systems Review,* 29(1) January 1995.

Schill,A., Bellmann,B., Böhmak,W., Kümmel,S. (1995) System Support for Mobile Distributed Applications. *IEEE Workshop on Services in Distributed and Networked Environments (SDNE),* Vancouver, June 1995.

Voelker, G.M., Bershad, B.N. (1994) Mobisaic: An Information System for a Mobile Wireless Computing Environment. *University of Washington,* Sept. 1994.

Weiser,M. (1993) Some computer science issues in ubiquitous computing. *Communications of the ACM,* 36(7), July 1993.

Zinky,J, Bakken, D.E., Schantz, R. (1995) Overview of Quality of Service for Objects. *Proceedings of the Fifth IEEE Dual Use Conference,* May 1995.

9 BIOGRAPHY

Jörg Bönigk studied computer science at University of Rostock. 1993 he received his diploma and then became a member of the scientific staff of the Computer Graphics Center Rostock. His research topics are scientific visualization and mobile computing. He is currently involved in the project MoVi (Mobile Visualization).

Uwe von Lukas graduated in computer science from Darmstadt University of. He became a research assistant at Fraunhofer Institute for Computer Graphics in 1994 and is now a member of the scientific staff of the Computer Graphics Center Rostock. His research interests are in the area of distributed platforms, CSCW and the application fields of CAD.

PART EIGHT

QoS-Management and Resource Discovery for Mobile Communication

20

Agents skills and their roles in mobile computing and personal communications

M. Mendes[*], *W. Loyolla*[*], *T. Magedanz*[+], *F. M. Assis Silva*[+,1], *S. Krause*[+]
[*]*Pontifícia Universidade Católica de Campinas (PUCCAMP)*
Campinas, S.P. Brazil, Email: mendes@dca.fee.unicamp.br
[+]*Research Institute for Open Communication Systems (GMD FOKUS)*
Berlin, Germany, Email: [magedanz\krause\flavio]@fokus.gmd.de

Abstract
In the context of the emergence of a global service market enabling the ubiquitous access to information and communication services, the new paradigm of mobile intelligent agents is attracting strong interest throughout the research community, as well as the information and telecommunication industries. After giving a brief account of the development of the agent paradigm and the basic terminology, this paper analyzes different application fields of this highly promising technology. Among the reviewed application areas are electronic markets, virtual enterprises, flexible information systems, and the domain of mobile and personal communications. The analysis leads to the identification of a number of roles agents may take on. The capabilities that an agent exhibits in the course of performing its role can be expressed as levels of agent intelligence, mobility, communicability, autonomy and cooperation ability, i.e. the agents skills.

Keywords
Agents, agent roles, agent skills, electronic markets, flexible information systems, mobile computing, personal communications, virtual enterprises

1 INTRODUCTION

The vision for future telecommunications and information networking environments is often described as "information at any time, at any place, in any form". The envisioned ubiquitous information access will result in an open "electronic" service market, where an unlimited spectrum of communication and information services will be offered to the customers, ranging from simple communication services to complex distributed applications. This vision is based on both, the society's increasing demand for "universal connectivity" and on the technological progress in the emerging areas of mobile computing and mobile/personal telecommunications. The ultimate target is to support customer-controlled wired and wireless global access to information and telecommunication services, regardless of the customer's current

---

[1] The work of this author is partially supported by Conselho Nacional de Desenvolvimento Científico e Tecnológico (CNPq), Brazil

location, or terminal and network capabilities. The realization of this target depends on the provision of an intelligent platform/environment, which is capable of adapting available information and telecommunication services, and the underlying systems and network resources according to the customers demands.

In this context the notion of Agents has evolved over the last years, as a new problem-solving paradigm, suited to a wide range of problem domains. The number of products, projects, and new applications using this paradigm is constantly growing, calling attention of the research community to study it and to propose solutions for several of the already detected problems.

The term *agent* is original of the Artificial Intelligence (AI) community, and its notion is central to the definition of AI itself. A common definition for AI is '*the sub-field of computer science that aims the construction of agents that exhibit aspects of intelligent behavior*' (Wooldridge, 1994a).

Since the early 1970s, the AI planning community has been closely concerned with the design of artificial agents. Nevertheless, until the middle of the 1980s, researchers from mainstream AI had given relatively little consideration to the issues surrounding agent synthesis. However, from this time there has been an intense growing of interest in the subject {Wooldridge, 1994a}.

Nowadays several research projects and developing efforts concerning agent technology are on-going and cover a multitude of other areas, other than AI, but there is an agreement in almost all of them about the nature of the environment where agents exist, and this comprehends the massively distributed system.

Generally, an agent can be any entity (physical or logical) that is responsible for the execution of a set of tasks, delegated to it, directly or by means of other entities, by a user human-being. The agent represents the user while performing its cores, acting on the user's behalf.

Analyzing the R&D projects, most of them developed by the AI community, it is possible to perceive that there are currently at least five major research sub-fields:

- *Physical agents* (robots), employing AI techniques to conceive them with human behavior. Most of the surveyed projects lie in this context. The emphasis in this sub-field is in the intelligent agent architecture, so these projects are closer related to IA. Although the architectures proposed are to robots, it seems that they can be modified to be employed in software agents, as was done in the Sumpy project {Franklin, 1996}

- *Multi-agent systems,* made up of multiple heterogeneous intelligent agents, where competition or cooperation is possible between them. There is no common application domain or global goal to be achieved. Projects in this second group cover the development of agents written by different people, in different languages, at different times for different purposes {Dalmonte, 1995}. This sub-field is directly concerned with the area of DAI (Distributed Artificial Intelligence);

- *Task-specific agents*, with the use of AI techniques to create software that performs information filtering and other autonomous tasks for users. "Intelligent" agents of this sort have been developed at Xerox PARC, at MIT's Media Lab, and by researchers at others companies and universities. These agents may or may not display any explicit anthropomorphic features;

- *Interface Agents*, as an interface metaphor that aids the user. These agents may or may not incorporate new AI techniques; their essential function is to act as effective bridges between a person's goals and expectations and the computer's capabilities. The agent metaphor is used to make the interface more intuitive and to encourage types of interactions that might be difficult to evoke with a GUI. Agents of this sort need not to be explicitly anthro-

pomorphic, although this is the arena where the expressive qualities of 'character' are being explored;
- *Mobile agents*: It seems that these projects do not intent to use high levels of intelligence skill. (Tele)Communication industries are the main interested in this research field. In this context agents will be used for customization of existing services and configuration of new services {Magedanz, 1996c}.

In accordance to the environment where the agent inhabits, there can be physical agents, such as robots, or software agents, such as programs running on a computer. The scope of the present survey and resulting analysis is to use the term agent technology concerning to *software agents*. In Section 2 of the paper some agent applications are reviewed in order to better characterize the numerous roles that agents may play. Issues related to multi-agent systems, intelligent agents, interface-agents, and mobility should not be treated as separate aspects, but yet as complementary aspects that, if combined, can result in very powerful software agents for the support of an unlimited set of tasks. Thus those issues are conceptualized and worked on in Section 3, in order to allow a broader and more generic definition of what an agent is. In Section 4 some conclusions are drawn.

2 AGENT APPLICATIONS

In the following analysis applications are briefly described in order to better understand the extent of agents potential use. To this end we outline some agent applications in Virtual Environments, with emphasis in two high level application areas, and three supporting areas. The two high level applications areas are Electronic Commerce, and Virtual Enterprises, and the supporting areas Flexible Information Systems, Mobile Computing and Mobile Communication.

Electronic Market was chosen since it is an area of current explosive growing interest, not just to the commercial community but also to the academic community. Virtual Enterprise is a current and important research and standardization area which offers multiple opportunities to the use of agents.

Flexible Information Systems, which are the basis of all information manipulation, and Mobile Computing and Communication, by their very nature, are basic support facilities not just to the analyzed high level application areas but also to almost anyone collaborative application in the near future.

Mobile Communication and Mobile Computing are considered, since these are the key aspects of an emerging ubiquitous computing environment, where universal access to information, communication and services is provided.

2.1 Electronic Market

The term "electronic market" has been used in various ways, frequently referring just to few parts of commercial transactions. Electronic markets may be considered as hybrid systems, in which people and computers are working together in order to reach certain commercial results. In order to represent a real market in an electronic form, the "electronic market system" has to offer support to all possible transactions in a marketing process.

A generic market transaction is {Klein, 1994} "a finite sum of interaction processes between members of different roles". The participants try in a certain period of time to settle their interests and expectations amongst themselves. The aim of a market transaction is a trading agreement between the actors, supplier and customer, and the following settlement of goods and/or services. In some markets, their may exist intermediaries like dealers, brokers or market makers who perform some supporting functions in order to improve the efficiency of the transaction process.

Interaction processes in a market transaction can be combined in three phases: *Information* phase, *Trading/matching* phase and the *Settlement* phase. Supporting these phases requires an Electronic Market Information System who may be defined {Klein, 1994} as "an information system, embedded in the context of an electronic market, in order to automate or to support electronically the interaction processes between the actors in a market".

The use of agents in the realm of electronic commerce may have a explosive growth in the next few years. Currently, some emblematic experiences have been already done by offering the work of some agents in the WWW. These experiences cover aspects as finding people who have their own tastes (Agents Incorporation's Firefly) {Agent Incorporation, 1995}; filtering news information to receive a personalized newspaper (IBM's Infosage) {IBM, 1995}, or even searching for the cheaper site where to buy an CD.

Agents may be used in all phases of a market transaction as:

- *Information* - the main purpose is to search for, find, filter and present information required both by potential customers and potential suppliers. These tasks may be performed with different aspects as:
  - directly searching in the net for potential suppliers on behalf of a buyer, accordingly to some personal rules or conditions. Agents could visit firm's virtual presence sites, searching for offerings or for desired goods and services;
  - searching potential buyers (marketing research) in the net by placing agents in the sites where advertisements are presented or by asking people (or another agents) about their preferences. In this way, agents could acquire not just the buyers' preferences profile from keeping track their advertisement site navigation but also their addresses.
  - indirectly searching for potential suppliers or buyers by contacting information traders or by purchasing information from special commercial sites who maintain commercial information.
- *Trading/matching* - the main purpose is to support the tasks involving negotiations. Agents, on behalf of the buyer, could negotiate with suppliers the terms and conditions of delivery and payment, warranties, and the execution or offering of additional services. In this case, agents could act at least in the following ways:
  - traveling through the selected suppliers sites and asking them for a set of marketing conditions which could be sent to the buyer. In this case, the agent could:
    - just pick the required information up and send it to the buyer;
    - pick the required information up and provide some kind of information filtering, accordingly to some pre-defined rules established by the agent's owner. After that, the prepared information could be sent to the buyer.
  - locally receiving invoices from selected suppliers, and, accordingly to pre-defined rules, to analyze them and to suggest a preferential supplier.
  - performing some part of the negotiation. The results of preliminary negotiation would be submitted to both supplier and buyer for posterior formal approval.
  - performing the whole negotiation. Both buyer and supplier are notified about final negotiation set.
- *Settlement* - agents could act both by:
  - providing electronic payment;

- following the occurrence of derived transaction processes related to the deal as packaging, storage, shipping, insurance, customs clearance, etc.

Despite not fulfilling the above electronic market system definition, the Internet and particularly the World Wide Web have been used as a commercial medium where firms are learning new ways of marketing to consumers in computer-mediated environments {Hauser, 1995}. In the next future it is expected that these experiences drive the evolution of complete Open Electronic Market Systems where almost all phases of market transactions can be electronically supported, probably running upon the Internet or new communication environments. Currently, many categories of commercial sites have been already defined and implemented in the Web {Hoffman, 1996}.

2.2 Virtual Enterprises

Virtual Enterprises correspond to a temporary alliance of companies formed to share costs and skills in order to address fast-changing commercial/industrial opportunities by providing competitive products, services and solutions cost-effectively and in a timely manner, regardless of size, organization, geographic or technical boundaries. Its members carry out their tasks as if they all belong to the same organization, under one roof, using a very powerful computing system to access and manage all information needed to support the product or service life-cycle. Yet, in reality they are employees of different companies, located abroad in the world, using a variety of computing systems and applications {NIIIP, 1996b}. The powerful computing system which offers support for this cooperation is called a Virtual Computing Environment.

Agents may be useful in all existence phases of a Virtual Enterprises as:
- *Formation of a Virtual Enterprise* - an agent owned by a virtual enterprise initiator could
  - search for potential virtual enterprise partners by traversing the net and asking (directly or consulting some kind of yellow pages) for companies who may fulfill some desired production requirements or perform some required services;
  - negotiate partners admittance after analyzing technical or commercial responses to participation requests accordingly to initiator limits and rules;
  - negotiate workflow templates to be executed during the virtual enterprise existence,
    - search and install appropriate software supporting the needed protocols to facilitate the operation of the virtual enterprise (workflow execution);
    - scheduling and completing projects - agents could be used to monitor remote status of resources and services execution in order to reschedule local or global workflows;
  - sharing information - remote placed agents could receive requests for product or service information and perform
    - security verification for accessing information, so saving the transmission of unauthorized requests,
    - data request format conversion in order to translate data requests to an understandable format for data repositories.
    - data format translation in order to provide data understandable to requester.
- *Collaborating on solving problems* - agents could
  - offer support on enabling groupwork such as controlling the use of groupware tools as whiteboards, shared sessions, and meeting schedulers,

- offer support on controlling particular responsibilities due to virtual enterprise participant in a specific workflow,
- offer support on controlling the enactment of particular workflows;
- *Communicating with organizations outside the VE* - agents could be useful in:
  - filtering incoming communications and routing them to the right partners' sites,
  - offering information translation (mediation) between different application's information format;
- *Negotiating, mediating and resolving disputes* - agents could perform activities as
  - searching information or services related to a specific matter of discussion,
  - offering the owner's position related to a specific matter of discussion,
  - summarizing voting related to a specific matter of discussion;
- *Dissolving a VE* - agents owned by each virtual enterprise participant could act:
  - searching for unfinished tasks related to the participation in a virtual enterprise,
  - negotiating with other agents the responsibilities due by the participant after the dissolution of the virtual enterprise.

2.3 Flexible Information systems

The demand for tools to manage the vast amount of information has come with the explosive growth of network information systems, such as the one that can be found in the World Wide Web (WWW). The provision of information is a key prerequisite for a successful distributed information system. There is the need to concentrate on what can be done to supply the user with the required information, so that communications problems, location of the data, the amount of data to be manipulated, among other issues do not interfere with an adequate provision. Agent technology could play a major role in this context allowing the construction of Flexible Information Systems that can operate autonomously or in cooperation with other systems in order to satisfy the specific goals of users and applications supporting distributed information services over an heterogeneous distributed processing environment.

Information agents could act as intelligent and active front-ends, in the sense of being based on user preferences or patterns and on environmental conditions. Information agents could be created and launched by the user's computational environment in order to travel through the information network and bring information, or they could be created by an information server and launched to the user computational environment in order to bring information, after a process of offer and negotiation.

An information agent is an agent that, by itself or asking for cooperation, has access through network services to at least one, and potentially many information sources, and is able to collate and manipulate information obtained from these sources in order to answer queries posed by users and other agents. It can be added to this definition the ability of an information agent to sell and deliver information that was not initially of interest to the user.

Basically an information agent could be seen as an information trader, forager, cataloger, presenter, or combination of them with some degree of intelligence and mobility, and that is part of a cooperative information system.

Agent skills can produce very flexible cooperative information systems allowing that, for example:

- *agents with knowledge* - Foraging is one of the central issues in agent programming {Harrison, 1994} and the ideal system posses knowledge specific to the domain in which it operates and specific to the user's interests, as well as the ability to filter data based on this knowledge. This knowledge is also obtained through the exposure to a lot of current data related in the proper area of expertise. The difference between real semantic foraging and simple keyword searches is the amount of information that can be passed through the system to allow to it become a real expert in its field and to have high quality, current information at its disposal.

- *mobile agents* - To the extend that mobility allows the agents to get closer to the actual data source, mobility becomes a real advantage. It is far more efficient for a program extracting knowledge to go to the source of the data instead of sending the data to the user, especially since the program can primarily filter and summarize the data.
- *dependency* - Distributed intelligent agents, which are resident at the same local as the data sources and persist there of their own accord have a real advantage in this respect over centralized, more static systems. Thus an agent can do automatic indexing of documents, which will include identifying a small number of interesting documents from among a large number of uninteresting ones. It might also identify documents potentially interesting to other agents and inform them of this fact. Distributed indexes could be built up hierarchically, geographically, and by subject.

2.4 Mobile/Personal Telecommunications

In the light of an increasing number of wired and wireless communication services available for accessing and exchanging information, the vision for future communications, i.e. *"information anytime, any place in any form"*, is becoming reality in the near future. However, besides the traditional aspect of information transport, the fundamental issue for new telecommunication services is information processing, i.e. the filtering, transformation and presentation of multimedia information in accord to customer needs, available communication networks and end systems currently used by the customer.

In addition to user and terminal mobility support, it will be the aspects of service customization and service interworking that are of fundamental importance for future telecommunications, enabling customers to define when, where, for whom and in what form they will be reachable or not. Consequently, adequate means are required to support end users in configuring their communications environment in accord to individual preferences.

Looking at today's mobile/personal communications systems, one will identify the Intelligent Network (IN) and the Telecommunications Management Network (TMN) standards as the basis for controlling and managing the user/terminal locations, user preferences and network resources. Current IN and TMN architectures are based on highly centralized approaches for the location of service control and network (and service) management "intelligence", where the related protocols rely on the traditional client/server paradigm. This means that centralized nodes, known as IN Service Control Points (SCPs) and TMN Operations Systems (OSs) are used for hosting service and management programs, respectively.

However, agent technology is considered as an emerging technology in this domain. The driving force for this view is the assumption that agents will provide better support for distribution of control and management in telecommunication systems in general {Magedanz, 1995a, Magedanz 1996b, Magedanz, 1996c}. This is due to the fact that the agent concept can be seen as an evolution step beyond the object-oriented paradigm, since agents enhance objects by distributing control and goals to them. This means that agents, i.e. their identified skills, such as intelligence, cooperation, asynchronous operation and agent mobility, allow to perform control and management tasks at the most appropriate locations within the telecommunications environment in contrast to existing architectures. This may have significant impacts on the architecture and the related (signaling) protocols of existing (mobile) telecommunication systems.

In principle, the following chances of emerging agent technology can be identified in the context of mobile communications {Magedanz, 1996a}:

- Customization and configuration of services: In the light of an electronic market place, agent technology allows to instantly provide new services either by customization or (re)configuration of existing services. In this case agents act as "service adaptors" and could be easily installed.
- Instant service usage and active trading: Mobile agents realizing service clients travel to potential customers providing spontaneous access to new services. This feature enabling easy distribution of service clients can be exploited to perform active trading.
- Decentralization of management: Mobile agents allow to decrease pressure on centralized network management systems and network bandwidth by delegating specific management tasks from central operations system to dispersed management agents. Mobile agents representing management scripts enable both temporal distribution (i.e. distribution over time) and spatial distribution (i.e. distribution over different network nodes) of management activities {Magedanz, 1996a}.
- Intelligent communications: Agents provide the basis for advanced communications. They support the configuration of a user's communications environment, where they perform control of incoming and outgoing communications on behalf of the end user. This includes communications screening, intelligent adaptation of services (i.e. conversion of information formats) to network access arrangements and end user devices, as well as advanced service interworking and integration {Magedanz, 1996b}.
- Information retrieval and support of dynamic information types: Mobile agents provide an effective means for retrieving information and services within a distributed environment and support for dynamic information types within electronic mail and advanced networked information systems.

In accord to the above given list two basic approaches of agent-based mobile/personal telecommunications can be identified:

- "Smart network" approach:

    Agents are stationary entities in the network, providing the necessary intelligence, and are able to perform specific predefined tasks autonomously (on behalf of a user or an application). The basic attributes of this type of agent are their ability to act asynchronously, to communicate, to cooperate with other agents, and to be dynamically configurable. In order to enable this dynamic reconfiguration of the agent, the dynamic downloading and/or exchange of corresponding control scripts represents a fundamental issue. This means that this kind of static agent, such as user agents or management agents, could be considered more likely as a specialized agent execution environment, which executes scripts (i.e. remote execution type agents). Examples for such fixed agents are given in {Eckardt, 1996}, {Ida, 1995}, {Mohan, 1995}, {Lauer, 1995}, and {Rizzo, 1995}.

- "Smart message" approach:

    Agents are mobile entities, which travel between different computers/systems and perform specific tasks at remote locations. In respect to the task to be performed either remote execution agents or migration agents may be deployed. Consequently, a mobile agent contains all necessary control information (i.e. the service logic of a particular telecommunications or management service), instead of the corresponding end systems and /or nodes within the network. This means that corresponding agent execution environments have to be provided by the potential end user systems and within the network, in order to perform the execution of agents and thus the realization of the intended services. Architectures promoting mobile agent-based services are AT&T´s PersonaLink and IBM´s Intelligent Communications {Reinhardt, 1994}.

2.5 Mobile computing

The evolution of computing and telecommunications are bringing new ways of their use. The main shift is in the direction of nomadic computing and communications. The "nomadic" property refers to the system support needed to provide services to the nomad as he/she moves from place to place. This new paradigm is already manifesting itself as users travel to many different locations with laptops, PDAs , cellular telephones, and so on.

Mobile Computing Systems are a combination of the following three elements {Mazer, 1995}:

- Stationary computer and network devices, which we will call the "infrastructure";
- Mobile computers, of varying physical properties (such weight, size, and power) and computational capabilities, which will likely be "resource-poor" relative to the infrastructure;
- Wireline and wireless communications channels of varying bandwidth, cost, reliability, and service properties.

There are many motivations for using mobile agents in this mobile environment {Chess, 1995}

- Support to lightweight devices;
- necessity of asynchronous methods of searching for information or transaction services;
- the reduction of overall communication traffic over the diverse communication channels that a mobile device may encounter;
- the ability of the agent to engage in a efficient communication with remote servers;
- the ability of the simple mobile computer to interact with complex applications, without necessarily knowing the remote server capabilities;
- the ability to create "personalized services" for the user, by tailoring the agents that move to the server and respond to the user's requests.

Two examples are now considered, one of an intelligent nomadic Information Systems {Bgrodia, 1995}, and the other from ubiquitous computing {Want, 1995}.

The vast number of clients and servers interconnecting in a nomadic information system cause new problems in respect to the diversity of data formats, data consistency requirements, data schema and bandwidth limitations. Data servers need to have improved resource discovery facilities with a set of intelligent agents to provide searching and adaptive matching of resources. Based on context, user profile and current operating environment the CoBase Project {Chu, 1996} introduces *adaptive agents*. Agents in this complex situation must present several abilities:

- Different agents interact to achieve a common task and may use the output of other agents as input;
- they communicate with each other through some Agent Communication;

- new concepts and algorithms need to be developed for performing negotiations, recovery, commitments;
- agents may be replicated and reused, thus needing a scaleable infrastructure.

The philosophy of Ubiquitous Computing aims to enrich processing by embedding into the environment many computing devices of different types, both personal and anonymous, fixed and mobile, large and small, and by using the virtual and physical context to provide more flexible personal access to information.ParcTab {Want, 1995} describes a representative project under development at the CSL from XEROX PARC. TAB is shorthand for "Small Tablet Computer" . A multilayer system architecture integrates the ParcTab devices into the PARC Office Network so that network applications can easily control and respond to the mobile devices. The TABs act primarily as input/output devices that rely on workstation-based applications for most computation. For each ParcTab there is exactly one agent which acts like a switchboard to connect applications. The agent performs four functions:

- it receives requests from applications to deliver packets to the mobile ParcTab that it serves;
- in the reverse direction, it forwards messages from its tab to the applications;
- it provides the source of Tab location information;
- It manages application communication channels.

2.6 An Agent Role Classification

Software agents may be further subdivided in the following classes:
- *Group enabling agents*: these agents support short-lived or long-lived collaborations. Several examples may be considered as:
  - meeting schedulers;
  - conferencing tools administrators;
  - notificators;
  - task activators;
  - mediators;
  - negotiators;
  - facilitators: enable the joining of agents in cooperating groups of agents and provide the sharing of operations among them;
- *User Agents:* they represent and act on behalf of the users and operators (individual or groups of users). User agents will assist the user by representing the characteristics and the preferences of the user, and are further characterized by the roles which the user plays. Examples of these agents are:
  - Personal Assistants: assists the user in his tasks and responsibilities, working in cooperation with him
  - Administrative assistant: relieves the user from the burden of routine and predictable tasks, such as filtering and conveying information to the user (e-mail, mobile computer, faxes, voice-mail,...) and scheduling meetings
  - Role-specific agents: they provide high-level assistance to the user, specialized according to the role of the user (e.g. designers, engineers, supervisors,...).
- *Task-oriented Agents:* designed to the execution of specific tasks, not necessarily on behalf of system resources, as for example,

- *Organizational agents*: normally other agents report to the organizational agent, which itself may represent a department, project, division of the enterprise or the enterprise itself. They may have resources such as money, own a "timeline schedule", approve work and make settlements
- *Service Agents*: may represent high-level services, generally in client/server architectures, located at the server as well as at client side As it will be discussed later on this work they may be great importance for legacy systems, and reduced in the implementation to so called proxy objects. Examples of those agents may be:
  - *client agents*: at client or server side, representing the interests of the client
  - *server agents*: at the server or client side, representing the interests of the server
  - *license server;*
- *Resource agents:* these agents represent exhaustible assets (i.e. resources) of system or enterprise. One specific commodity among every agent managing a resource is that it has to be exclusive when it comes to access the variable representing the quantity of the resource. Otherwise some resource agents may represent the interests of multiple resources. Examples:
  - *Equipment agent*: represents a specific physical resource (a "life-entity" in the real world). Such resources are usually characterized by having sensors and actuators and may be associated with humans operators. These agents can be so smart that they assume an "anthropomorphic" nature in respect to the physical resource they supervise
  - *Computational agent:* representing computing resources
  - *Personal resource agents*: because people are often treated as resources
- Information *agents:* information agents are primarily concerned with the managing, gathering and interpreting of information. Several kinds may be considered :
  - *Trading Agents* are responsible for the negotiation of an information service because of the interest demonstrated by a user or because of the selling intention of a server; they receive a request from an application by some type of information service or they are asked to offer some specific information service to a community of users.
  - *Foraging Agents* can be:
    – *Brokering Agents* are responsible for managing system information objects on behalf of the applications that share them; they receive a request from an application, search the network object repositories for the information agent or server, and deliver a reference to it; they act as "Yellow Page" servers.
    – *Localization Agents* are responsible by the exploration of the network in order to localize documents according to a user criteria. This user criteria could be simple as a pattern or very complex as a set of preferences constrained by established rules and dynamic conditions.
    – *Retrieval Agents* are responsible for the retrieval of document content, along with information about its characteristics (which could be both spread through the network), at adequate data rates, in order to allow proper network transportation and presentation to the user. They are responsible by QoS (quality of service) negotiation and communication binding, recovery, and commitments of data flows.
- *Cataloging Agents* react to data events or other type of events in order to create, update, or destroy information in network data banks.
- *Presentation Agents* are responsible for the proper exhibition of information that arrives at the network node where the application has asked; proper exhibition means a combination

of decompression, filtering, format conversion, language translation, synchronization, and user perception.
- *Focusing Agents* are briefed with a search profile, but the agent is looking for specific data items, like when a value threshold is reached; they are like database triggers, without the necessity to reside and depend on a database system. Once triggered the agent can deliver the information to the application.

3 AGENT SKILLS

3.1 Agent Skills

One way to achieve a general characterization for agents is through the use of the *skill* concept, borrowed from Minsky {Minsky, 1994}. An agent skill is a capacity that an agent presents (in some degree) that sometimes allows or facilitates the agent to play its role(s) and accomplish its goals.

We propose five major skills:
- *intelligence* or the degree of reasoning and learning ability of an agent;
- *mobility* or the degree of migration power of an agent;
- *communicability* or the degree of interaction ability of an agent;
- *autonomy* or the degree of auto-control ability of an agent and
- *cooperation* or the degree of an agent ability to collaborate and to do some joint work with other entities.

Each agent acting on some specific role has different graduations of each of these skills. Each of the skills is also related to a subset of the properties commonly cited in definitions of agents of various authors. Table 1 lists a set of such properties.

Table 1 Agent characterization properties

Property	Other Name	Meaning
activeness	self-start	ability to **decide** when to act {Hayes-Roth, 1995, Russel, 1995, Franklin, 1996}
perceptiveness	sensing	sensibility to (ability to **receive information** about) changes in its environment {Maes, 1995, Hayes-Roth, 1995, Russel, 1995, Franklin, 1996}
reactivity		ability to **respond** to changes in its environment at just the right time in order to be successful {Etzioni, 1994, Wooldridge, 1995}
reflection		**behavior** in a quickly asynchronous stimulus-response fashion {Goodwin, 1993}
autonomy		**exclusive control** over its own actions {Maes, 1995, Smith, 1994, Brustoloni, 1991, Gilbert, 1995, Franklin, 1995, Wooldridge, 1995, Franklin, 1996}

Agent skills and their roles in mobile computing 193

Property	Other Name	Meaning
pro-activity	goal-directed purpose	do not simply act in response to the environment, ability to *take the initiative* to satisfy its own goals {Maes, 1995, Smith, 1994, Brustoloni, 1991, Franklin, 1995, Wooldridge, 1995, Goodwin, 1993}
consideration		do not blindly *obey commands*, but has the ability to modify requests, *ask* clarification questions, or even refuse to satisfy certain requests {Genesereth, 1994, Wooldridge, 1995}
independence	asynchronous operation	may execute its tasks totally *decoupled* from its user or other agents {Gilbert, 1995}
communication	socially able interaction	*communicates* with applications, facilities, services, other agents and perhaps people, in order to obtain information or enlist their help to accomplishing its goals {Genesereth, 1994, Wooldridge, 1995, Coen, 1995, Lingnau, 1995}
rule-based		possession of statements (rules) with preferences and ability to *choose what actions* to take based on these rules {Gilbert, 1995}
ability to plan	reasoning about actions	maintenance of some internal model of the world and *ability to reason* about the results of the actions taken. Tentative choose of actions that more likely lead it to accomplish the task {Hayes-Roth, 1995, Gilbert, 1995}
ability to learn	adaptability	collection of information about the environment and its *use as knowledge* in order to perform its task {Gilbert, 1995}
ability to predict	rationality	the model of the world, if sufficiently accurate, allows the correct *prediction* of what actions more likely lead to accomplish the task successful {Gilbert, 1995}
interpretation		*interpretation* of sensor readings {Maes, 1995, Wooldridge, 1995}
soundness		if it is predictive, interpretive and rational {Wooldridge, 1995}
mobility		ability to transport itself or be transported from one machine to another, independently of the system architecture and platform {Heilmann, 1995}

Intelligence

Intelligence is the degree of reasoning and learning behavior that an agent contains. An agent employs some knowledge or representation of the user's goals and desires, in order to operate on behalf of the user.

An agent can present one of the degrees of intelligence listed below. The degrees are presented in order of increased complexity. Each degree includes its predecessors:

- *preference* - the agent acts according to some statements of the user's preferences. The user's preferences can be represented, for example, directly in the instructions (code) of the

agent or by a set of rules. The agent can have an internal inference mechanism to process the rules or this mechanism can be provided by the platform;
- *planning* - the agent maintains some internal model of the world and of what the user wants to be done, and is able to choose what actions are more likely to lead it to accomplish its goals;
- *reasoning* - the agent reasons about the results of its actions and about the world, in order to choose the more appropriate actions to complete successfully its tasks. The agent presents some capacity for knowledge inference, data gathering, and extrapolation;
- *learning* - the agent learns and becomes able to adapt to a dynamic environment, both in terms of the user's objectives as in terms of the resources available to the agent. Such an agent may, for example, discover new relationships, connections, or concepts independently from the human user. Exploiting the acquired knowledge, the agent can make decisions, anticipating and satisfying user needs.

The intelligence of agents is associated with properties such as: *rule-based, ability to predict, ability to plan, ability to learn, interpretation* and *soundness* (see Table 1). An agent exhibits such properties depending on the level of intelligence it has or on the way the agent perform its tasks.

Mobility

Mobility is the degree of migratory power an agent exhibits. Agents can be dispatched from a source computer and roam over a set of networked servers until they accomplish their tasks. The list of migratory ability levels below is presented in order of increased complexity:
- *static* - the agent does not migrate. Although executing only in the platform of its user, such an agent may access not only local resources but also remote resources, for example via some kind of FTP command;
- *remote execution (mobile without state)* - remote execution of agents is realized through a remote procedure call, not necessarily synchronous, that asks for the execution of a program code existent in a remote node, or through a program, that may be composed on and dispatched from a client computer and whipped and transported to a remote server computer for execution. It can be seen as the distribution of a static (not initialized) agent to be executed at remote nodes;
- *migration (mobile with state)* - the agent migrates with information about its execution state. It can move from machine to machine in the middle of execution, and carries the accumulated information with it, in order to complete its tasks. When such an agent arrives at a remote node, it can present its credentials and so obtains access to services and data managed by the local software platform. Such a migration capability enables an agent to progressively accomplish its task while moving from place to place.

Very important to note is that a mobile agent decides to move by itself. That makes agent mobility different from the concept of migrating processes in some batch systems or distributed operating systems. In the latter case, processes are moved by some system coordination component, which tries to achieve, for example, load balancing.

Communicability

The communication skill represents the complexity degree of interactions an agent can perform with an application, a service, a facility, an user, or another agent. Communicability can be measured qualitatively by the nature of such interactions. The levels of communicability an agent can exhibit are listed below. The levels are presented in order of increased complexity:

- *isolated* - the agent exhibits no capacity for interacting with any other entity;
- *simple data exchange* - the agent can interact through data exchange with other entities, with minor semantics in this interaction. Service calls with minor control information and some input data are characteristic of data exchange. Service responses are also communicated through data exchange. An agent can also monitor and respond to changes in the environment. Most of these changes are notified by asynchronous event driven communications, that is basically a data exchange communication;
- *semantic data passing* - the agent exchanges data with other entities but needs to interpret the semantics of the data in order to accomplish its tasks. An application domain vocabulary is needed to allow message passing. Semantic data can be exchanged between agents and other entities through message passing { Pitt, 1995} or the blackboard {Magedanz, 1995a}. Messages exchanged between agents can represent, for example, an *assertion* (when it states a fact), a *directive* (if it includes a command, a request or a suggestion), or a *declaration* (when an agent sends an information on his internal capacities);
- *conversation* - the highest level of interaction between agents requires more than a conventional communication scheme in order to address the following issues: heterogeneity between agents, knowledge exchange, control localization and organizational structure {Wong, 1993}. Agents talk with their peers by exchanging messages in an expressive communication language with agent-independent semantics (it can not vary from one agent to another). The communication language allows for the exchange of data and logical information, individual commands and scripts. The conversation may involve several rounds of data exchange.

Depending on the level of its communication ability agents can present properties such as *reactivity, perceptiveness* and *reflection* (see Table 1). These properties result from the level of communication ability of the agent or the way this ability is used to perform the tasks.

Autonomy

Autonomy has been commonly used in the context of agents to refer to the principle that agents operate with or without the intervention of humans or others {Wooldridge, 1994, Nwana, 1996, Castelfranchi, 1994}.

The autonomy of an agent reflects in the behavior of the agent in relation to external requests. If an agent acts only following orders, it does not exhibit much autonomy. If the agent has "its own goals" and can, therefore, accept or refuse external requests or do something that was not requested, then the agent exhibits more autonomy. This notion of autonomy is called *Execution Autonomy* by Castelfranchi {Castelfranchi, 1994}. Two major levels of autonomy can be observed:

- *(restricted by) delegation:* the agent executes tasks in behalf of the user, but the tasks it executes are constrained (or defined) by a delegation by the user (other agent or others);
- *(based on) negotiation:* when the agent has the ability to act on requests (accepting, refusing, negotiating) taking into consideration its own goals and capacities.

Autonomy is related to properties such as *activeness, pro-activity, independence* and *consideration* (see Table 1).

Cooperation ability

The cooperation ability of an agent relates to the degree with which the agent can establish and maintain a relationship to a set of other entities in order to perform a joint work. We identify the following degrees of cooperation ability. The list is presented in order of increased complexity:

- *isolation* - the agent executes its task by itself, it does not have ability to cooperate with other entities;
- *static cooperation* - the agent can cooperate with other entities, but in a static environment. The term *static* in this context means that the environment does not modify frequently. The rules and entities with which the agent can cooperate are relative stable and the agent has previously knowledge of them;
- *dynamic cooperation* - the agent is inserted in a dynamic environment, where components and rules constantly change. So agents are required to evaluate and compromise based on a range of preferences. Typically this type of cooperation involves a negotiation phase before the agent begins cooperating.

Figure1. Levels (degrees) of agents skills exhibited by different types of agents

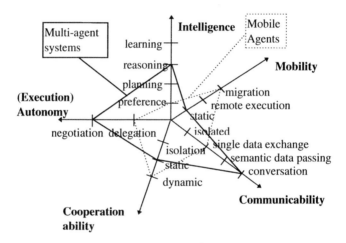

Figure 1 presents a graphical representation of the agent skills and their levels (degrees). The figure exemplifies also the degree of each skill that agents in different contexts (mobile agents and multi-agent systems) exhibit.

4 CONCLUSIONS

The future massive distributed systems bring together a set of new challenges for software engineering which cannot be suitably fulfilled by conventional distributed systems concepts. The support for mobile users and devices; the vast geographical extension; the huge dynamism allowing new information sources and servers to be added and removed without formal control and registration; the increasingly easier access to enormous amount of information; and high competition are only some of the characteristics of forthcoming systems.

The technology of agents converges developments in different software development areas and is emerging as the technology to fulfill those advanced requirements. Associated with notions of intelligence, mobility, autonomy, cooperation and communication (identified in the text as major *agents skills*), agents represent the building blocks of a new problem-solving paradigm which makes possible the development of very flexible, dynamic and adaptable systems. The exploration of the capacities of this emerging technology conducts us towards the prospect of how the future distributed systems will be.

This text aims at providing a compilation of the current research on the subject, trying to make clear what an agent is and what we can expect from them. Much effort has been expended in the development of theories and concepts to build the basis of the agents paradigm. An explosion of different types of contexts where agents are used is currently taking place. Efforts, however, must still be concentrated on the development of the technological support to make agent-based systems reality. That includes a more extensive investigation of the impact of agent technology in all the aspects of software engineering, from design up to implementation.

5 ACKNOWLEDGEMENTS

The authors are very grateful to Paulo Sérgio da Silva, Carlos Miguel Tobar Toledo, Ivo Fernandes, Carlos Raul Arias Méndez and Orandi Mina Falsarella from the UNICAMP and PUCCAMP Universities in Brazil. Much of this text resulted from their dedicated work.

6 REFERENCES AND BIBLIOGRAPHY

Note: Cited references are clearly marked by an asterisk after the author's name(s), e.g. *name**. Unmarked references are included for bibliographical purposes.

Agent Concepts and Theory

___FAQ (1996) URL http://www.ee.mcgill.ca:80/~belmarc/agent_faq.html

___OMG (1995) OMG Common Facilities RFP3, OMG TC Document 95-9-40, October, 1995.

Belgrave, M.[*] (1995) The Unified Agent Architecture: A Platform for Intelligent Software Agents, Working Draft, revision 1.0, October. Available in http://www.ee.mcgill.ca/~belmarc/agent_root.html.

Brustoloni, J.C[*] (1991) Autonomous Agents: Characterization and Requirements, Carnegie Mellon Technical Report CMU-CS-91-204, Pittsburgh, Carnegie Mellon University.

Castelfranchi, C.[*] (1994) Guarantees for Autonomy in Cognitive Agent Architecture, Intelligent Agents, Lecture Notes in Artificial Intelligence, M.J.Wooldridge, N.R.Jennings (Eds.), Springer-Verlag, pp.56-70

Chess, D.M. et. al.[*] (1995) Itinerant Agents for Mobile Computing, IEEE Personal Comm. Magazine, Vol.2 ,No. 5.

Coen, M.[*] (1995) The Sodabot Slide Show, URL http://www.ai.mit.edu/people/sodabot/slideshow/P001/html.

Dalmonte, A. and Gaspari, M.[*] (1995) Modeling Interaction in Agent Systems. http://www.cs.umbc.edu/kqml/papers/gaspari-ijcai95.ps.

Eckardt, T. and Magedanz, T.[*] (1996) From IN towards TINA-based Personal Communications Support, IEEE Int. Conference on Universal Personal Communications (ICUPC), Cambridge, Boston, September 29 - October 2, 1996.

Etzioni, O. and Weld, D. S. (1995b) Intelligent Agents on the Internet: Fact, Fiction and Forecast, *IEEE Expert*, 4:44-49, August.

Etzioni, O. and Weld, D. S.* (1994) A Softbot-based Interface to the Internet, Communications of the ACM, 37(7):72:76.

Etzioni, O. et al. (1995a) The Softbot Approach to OS Interface, *IEEE Software*, pp.42-51, July.

Franklin, S and Graesser, A.* (1996) Is it an agent, or just a program?: A taxonomy for Autonomous Agents", http://www.msci.memphis.edu.

Franklin, S.* (1995) Artificial Minds, Cambridge, MA, MIT Press. Quoted in (Franklin, 1996).

Genesereth, M.R. and Ketchpel, S.P.* (1994) Software Agents, *Communications of the ACM*, 37(7):48-53.

Gilbert, D.* (1995), IBM Intelligent Agent Strategy White Paper, IBM Corporation, Research Triangle Park, NC, URL http://activist.gpl.ibm.com:81/WhitePaper/pct2.html.

Goodwin, R.* (1993) Formalizing Properties of Agents, Technical Report CMU-CS-93-159, School of Computer Science, Carnigie-Mellon University, Pittsburgh, PA.

Harrison, C.G. et al* (1994) Mobile Agents: Are they a good idea?, IBM Research Report, RC 19887, October 1994.

Hayes-Roth, B.* (1995) An Architecture for Adaptive Intelligent Systems, *Artificial Intelligence:* Special Issue on Agents and Interactivity, 72:329-365. Quoted in (Franklin, 1996).

Heilmann, K. et al (1995) Intelligent Agents: A Technology and Business Application Analysis, http://haas.berkeley.edu/~heilmann/agents/.

Kay, A. (1984) Computer Software, Scientific American", 251(3):191-207. Quoted in (Maes, 1994) and (Magedanz, 1995).

Lauer, G. et al.* (1995) Broadband Intelligent Network Architecture, Intelligent Network Workshop (IN'95), Ottawa, Canada.

Lingnau, A and Drobnik, O.* (1995) An Infrastructure for Mobile Agents: Requirements and Architecture, ftp://ftp.tm.informatik.uni-frankfurt.de/orlando.ps.

Maes, P. (1994) Agents that Reduce Work and Information Overload, Communications of the ACM, 37(7):30-40.

Maes, P.* (1995) Artificial Life Meets Entertainment: Life like Autonomous Agents, *Communications of the ACM*, 38(11), 108-114. Quoted in (Franklin, 1996).

Magedanz, T.* (1995a) Intelligent Agents: Concepts, Architectures and Applications, Part 1 Intelligent Agents- state-of-the-art Analysis: Part 2: Impacts of Intelligent Agents Concepts on the Telecommunications Environment, FOKUS Report, Berlin.

Magedanz, T.* (1995b) On the Impacts of Intelligent Agent Concepts on Future Telecommunication Environments", in: Lecture Notes on Computer Science 998 - Bringing Telecommunication Services to the People - IS&N'95, pp. 396 - 414, A. Clarke et al. (Eds.), ISBN: 3-540-60479-0, Springer Verlag, (Proceedings of the 3rd Int. Conference on Intelligence in Services and Networks (IS&N), Heraclion, Greece, October 16-20, 1995)

Magedanz, T. and Eckardt, T.* (1996a) Mobile Software Agents: A new Paradigm for Telecommunications Management, in: Proceedings of IEEE/IFIP Network Operations and Management Symposium (NOMS), Kyoto, Japan, April 15-19, 1996, pp. 360-369, IEEE Catalog No. 96CH35757, ISBN: 0-7803-2518-4, IEEE Press.

Magedanz, T. and Eckardt, T.* (1996b) Mobile Service Agents and their Impacts on IN-based Service Architectures, in: Proceedings of IEEE Intelligent Network Workshop, , Melbourne, Australia, April 21-24, 1996, IEEE Catalog No. 96TH8174, ISBN: 0-7803-3230-X, IEEE Press.

Magedanz, T. et al* (1996c) Intelligent Agents: An Emerging Technology for Next Generation Telecommunication?, Proceedings of IEEE INFOCOM, San Francisco, March 1996.

Minsky, M. and Riecken, D. * (1994) A Conversation with Marvin Minsky About Agents, Communications of the ACM, 37(7):22-29.

Mohan et.al. * (1995) A Personal Messenger Application Based on TINA-C, XV. Int. Switching Symposium, Berlin, Germany, April 1995.

Negroponte, N. (1970) The Architecture Machine; Towards a more Human Environment, MIT Press, Cambridge, Mass. Quoted in (Maes, 1994) and (Magedanz, 1995).

Nwana, H.S. (1996) Software Agents: An Overview, to appear in Knowledge Engineering Review.

Ousterout, J. (1995) Scripts and Agents: The New Software High Ground. Invited talk, USENIX Conference. Quoted in (Lingnau, 1995)

Pitt, J.* (1995) Normalized Interactions Between Autonomous Agents, http://medlar.doc.ic.ac.uk/

Reinhardt, A. * (1994) The Network with Smarts, *BYTE Magazine*, pp. 51-64, October.

Rizzo, M and Utting, I.A. * (1995), An Agent-based Model for the Provision of Advanced Telecommunication Services, pp 205-218, Proceedings of the 5th Telecommunications Information Networking Architecture (TINA) Workshop, Melbourne, Australia, February 1995.

Russel, S.J. and Norving P. * (1995), Artificial Intelligence: A Modern Approach, Englewood Cliffs, NJ, Prentice Hall. Quoted in (Franklin, 1996).

Selker, T. (1994), Coach: A Teaching Agent that Learns, *Communications of the ACM*, 37(7):92-99.

Smith, D.C. et al* (1994), KIDSIM: Programming Agents Without a Programming Language, *Communications of the ACM*, 37(7):55-67.

Virdhagriswaran, S. (1995a), URL http://www.crystaliz.com/logicware/mubot.html.

Virdhagriswaran, S. (1995b), Response to Request for Information (RFI) on Common Facilities - Common Intelligent Agent Services Specification (CIASS), OMG.

Wong, S.T.C. * (1993), COSMO: A Communication Scheme for Cooperative Knowledge-based Systems", IEEE Transac. on Systems, Man, and Cybernetics, 23 (3):809-824.

Wooldridge, M. and Jennings, N.R. * (1994a), Agents Theories, Architectures, and Languages: A Survey in Intelligent Agents, Proceedings of. ECAI-94, Workshop on Agent Theories Architectures, and Languages, Lecture Notes in Artificial Intelligence, 890.

Wooldridge, M. and Jennings, N.R. * (1995), Intelligent Agents: Theory and Practice, Submitted to Knowledge Engineering Review, October 1994, Revision of January 1995.

Survey of R&D Projects

___ (1995) *A Survey of Cognitive and Agent Architectures*. In URL http://ai.eecs.umich.edu:80/cogarch0/

Coen, M. (1996). *SodaBot: A Software Agent Construction System*. Draft Version, URL http://www.ai.mit.edu/people/mhcoen/sodabot.ps

Survey of AI Theory

Agre, P. and Chapman, D. (1987) PENGI: An implementationof the theory of activity, In *Proceedings of the Sixth National Conference on Artificial Intelligence (AAAI-87)*, pp. 268-272, Seattle, WA. Quoted in {Wooldridge, 1995}.

Almasi, G. and Jannathan, V. (1996) *Integrating the WWW and CORBA based Environments*, Concurrent Engineering Research Center, West Virginia University, http://webstar.cerc.wvu. edu/www4.html

Barry Kitson, (1995) *CORBA and TINA: The Architectural Relationships*, TINA Symposium , Australia , May

Brooks, R. A. (1986) A robust layered control system for a mobile robot, *IEEE Journal of Robotics and Automation*, 2(1):14-23. Quoted in {Wooldridge 1995}.

Burmeister, B. and Sundermeyer, K. (1992) Cooperative problem solving guided by intentions and perception, In *Descentralized AI 3 - Proceedings of the Third European Workshop on Modelling Autonomous Agents and Mutli-Agents Worlds (MAAMAW-91)*, Werner, E. and Demazeau, Y., editors, pp. 77-92, Elsevier Science Publishers B.V., Amsterdan, The Netherlands. Quoted in {Wooldridge, 1995}.

Chapman, D. (1987) Planning for conjunctive goals, *Artificial Intelligence*, 32:333-378. Quoted in {Wooldridge, 1995}.

Cohen, P.R. and Levesque, H. J. (1990) Intention is choice with commitment, *Artificial Intelligence*, 42:213-261. Quoted in {Wooldridge, 1995}.

Doyle, J.; Shoham, Y.; and Wellman, M. P. (1991) A logic of relative desire, *in Methodologies for Intelligent Systems - Sixth International Symposium, ISMIS-91* (LNAI Volume 542). Ras, Z. W. and Zemankova, M., editors. Spring-Verlag: Heidelbeg, Germany. Quoted in {Wooldridge, 1995}.

Ferguson, I. A. (1992b) Towards an architecture for adaptive, rational, mobile agents. In *Descentralized AI 3 - Proceedings of the Thrid European Workshop on Modelling Autonomous Agents and Mutli-Agents Worlds (MAAMAW-91)*, Werner, E. aand Demazeau, Y., editors pp. 249-262, Elsevier Science Publishers B.V., Amsterdan, The Netherlands. Quoted in {Wooldridge, 1995}.

Ferguson, I.A. (1992a) *TouringMachines: An Architecture for Dynamic, Rational, Mobile Agents*, PhD thesis, Clare Hall, University of Cambridge, UK. Quoted in {Wooldridge, 1995}.

Franklin, S., Graesser, A. (1996*), Is it an agent, or just a program?: A taxonomy for Autonomous Agents*, http://www.msci.memphis.edu, March

Georgeff, M.P. and Lansky, A.L. (1987) Reactive reasoning and planning, *In Proceedings of the Sixth National Conference on Artificial Intelligence (AAAI-87)*, pp. 677-682, Seattle, WA. Quoted in {Wooldridge, 1995}.

Haddadi, A. (1994) A hybrid architecture for multi-agent systems, *In Proceedings of the 1993 Workshop on Cooperating Knowledge Based Systems (CKBS-93)*, Deen, S. M. editor, pp. 13-26, DAKE Centre, University of Keele, UK. Quoted in {Wooldridge, 1995}.

Hintikka, J. (1992) *Knowledge and Belief*, Cornell University Press: Ithaca, NY. Quoted in {Wooldridge, 1995}.

Kiss, G. and Reichgelt, H. (1992) Towards a semantics of desires, *In Descentralized AI 3 - Proceedings of the Thrid European Workshop on Modelling Autonomous Agents and Mutli-Agents Worlds (MAAMAW-91)*, Werner, E. and Demazeau, Y., editors, pp. 115-128, Elsevier Science Publishers B.V., Amsterdan, The Netherlands. Quoted in {Wooldridge, 1995}.

Maes, P. (1989) The dynamics of action selection, In *Proceedings of the Eleventh International Joint Conference on Artificial Intelligence (IJCAI-89)*, pp. 991-997, Detroit, MI. Quoted in {Wooldridge, 1995}.

Maes, P. (1990) Situated agents can have goals, In *Designing Autonomous Agents*, Maes, P. editor, pp. 49-70, The MIT Press, Cambridge, MA. Quoted in {Wooldridge, 1995}.

Maes, P. (1991) The agent network architecture (ANA). *SIGART Bulletin*, 2(4):115-120. Quoted in {Wooldridge, 1995}.

Moore, R.C. (1990) A formal theory of knowledge and action. Readings in Planning, In Allen, J.F., Hendler, J., and Tate, A., editors, pp. 480-519, Morgan Kaufmann Publishers, San Mateo, CA. Quoted in { Wooldridge, 1995}.

Müller, J.P., Pischel, M., and Thiel, M. (1995), Modelling reactive behaivour in vertically layred agent architecutures, In *Intelligent Agents: Theories, Architectures, and Languages*, Wooldridge, M. and Jennings, N.R., editors, Lecture Notes in Artificial Intelligence, 890:261-276, August.

OMG (1993) *The Common Object Request Broker: Architecture and Specification*, r.1.2, OMG Documentation.

Rao, A.S. and Georgeff, M.P. (1991a) Asymetry thesis and side-effect problems in linear time and branching time intention logics, In *Proceedings of the Twelfth International Joint Conference on Artificial Intelligence (IJCAI-91)*, pp.498-504, Sydney, Australia. Quoted in {Wooldridge, 1995}.

Rao, A.S. and Georgeff, M.P. (1991b) Modeling rational agents within a DBI-architecture, In Proceedings of Knowledge Representation and Reasoning (KR&R-91), Fikes, R. and Sandewall, E., editors, pp. 473-484, Morgan Kaufmann Publishers: San Mateo, CA. Quoted in {Wooldridge, 1995}.

Rao, A.S. and Georgeff, M.P. (1993) A model-theoretic approach to the verification of situated reasoning systems, In *Proceedings of the Thirteenth International Joint Conference on Artificial Intelligence (IJCAI-93)*, pp. 318-324, Chambéry, France. Quoted in {Wooldridge, 1995}.

Rosenschien, S. and Kaelbling, L.P. (1986) The synthesis of digital machines with provable epistemic properties. In *Proceedings of the 1986 Conference on Theoretical Aspects of Reasoning About Knowledge,* Halpern, J.Y., editor, pp. 83-98, Morgan Kaufmann Publishers, San Mateo, CA. Quoted in {Wooldridge, 1995}.

Sacerdoti, E. (1974) Planning in hierarchy of abstraction spaces, *Artificial Intelligence*, 5:115-135. Quoted in {Wooldridge, 1995}.

Sacerdoti, E. (1975) The non-linear nature of plans, In *Proceedings of the Fourth International Joint Conference on Artificial Intelligence (IJCAI-75)*, pp.206-214, Stanford, CA. Quoted in {Wooldridge, 1995}.

Singh, M.P. (1994) *Multiagents Systems: A Theoretical Framework for Intentions, Know How, and Communications*, (LNAI Volume 79), Springer-Verlag, Heidelberg, Germany. Quoted in {Wooldridge, 1995}.

Thomas, D. (1995) Ubiquitous Applications: Embedded Systems to Mainframe, *Communication od the ACM*, October.

TINA (1995a) *Distributed Processing Environment (DPE)*, version 1.0, August.

TINA (1995b) *Object Definition Language (ODL) Manual*, version 1.3 June.

Tschammer, V. and Mendes, M.J. (1995): *Architecture and Distributed Support Environment for Open Distributed Applications and Telecommunications Services*, GMD FOKUS Berlin.

Wainer, J. (1994) Yet another semantics of goals and goal priorities, *In Proceedings of the Eleventh European Conference on Artificial Intelligence (ECAI-94)*, pp.269-273, Amsterdam, The Netherlands. Quoted in {Wooldridge, 1995}.

Wayner, P. (1995), Free Agents, *Byte*, March

Wooldridge, M. (1994b) Coherent social action*, In Proceedings of the Eleventh European Conference on Artificial Intelligence (ECAI-94)*, pp.279-283, Amsterdam, The Netherlands. Quoted in {Wooldridge, 1995}.

Agents applications

Agent Incorporation[*] (1995) *Firefly*, reached at http://www.ffly.com/

Athan, A. and Duchamp, D. (1993) Agent-Mediated Message Passing for Conatrained Environments, *Proc. Usenix Symp. Mobile and Location-Independen Comp.*

Bgrodia, R. et al* (1995) Vision, Issues, and Architecture for Nomadic Computing, *IEEE Personal Communications,* December 1995

Borenstein, N., Freed, N. (1992) *MIME (Multipurpose Internet Mail Extensions Part One: Mechanisms for Specifying and describing the Format of Internet Messages Bodies",* RFC 1341, June

Borenstein, N.S. (1992) Computational Mail as Network Infrastructure for Computer Supported Cooperative Work, *CSCW92 Proceedings,* Toronto, November.

Borenstein, N.S. (1994) Email with a mind of its own: The SafeTcl Language for Enabled Mail, *ULPAA,* Barcelona.

Borenstein, N.S. and Rose, M.T. (1993) *MIME Extensions for Mail-Enabled Applications: application/SafeTcl and Multiparty/enabled Mai",* October.

Chess, D.M. et al* (1995), Itinerant Agents for Mobile Computing. *IEEE Personal Comm. Magazine,* Vol.2 ,No. 5, October.

Chess, D.M. et al. (1994), *Mobile Agents: Are they a good idea?,* IBM Research Report, RC 19887, October.

Chu, W.W.* (1996) CoBase: A Scaleable and Extensible Cooperative Information System, *J. Intelligent Information Systems*

Connolly, J.H. (1994b) Artificial Intelligence and CSCW in International contexts, in {Connoly 1994a}, pp 141-159

Connolly, J.H. and Edmonds, E.A. (editors) (1994a), *CSCW and Artificial Intelligence,* Springer Verlag.

Goyal, S.K. (1991) Knowledge technologies for evolving networks, *Integrated Network Management II,* Krishnan & Zimmer (Eds), Elsevier Publishers.

Guilfoyle, C., Warner, E. (1994) *Intelligent Agents: The New revolution in Software,* Technical report, OVUM Limited.

Hauser, R.* (1995) *NetWatch Top Ten - Intelligent Agents / Information Agents,* reached at http://www.pulver.com/netwatch/topten/tt9.html

Hoffman, D. L.; T. P. Novak; P. Chatterjfe, (1996) Commercial Scenarios for the Web: Opportunities and Challenges, to appear *in Journal of Computer-Mediated Communication", Special Issue on Electronic Commerce.*

IBM* (1995) *Infosage,* reached at http://www.infosage.ibm.com/, 1995.

Ida, I. et al* (1995) DUET: An agent based Personal Communications Network, *IEEE Communications Magazine,* Vol. 33, No. 11, November.

ITU-T (1991) *ITU-T Draft Recommendation F.851, "Universal Personal Telecommunications- Service Principles and Operational Provision",* November.

ITU-T (1992*) ITU-T Recommendations Q.12xx Series on Intelligent Networks,* Geneva, March

Klein, S.; Langenohl, T.* (1994) Electronic Markets: An Introduction, *In Proceedings of Information and Communications Technologies in Tourism,* Wien, Springer-Verlag.

Le, M.T. et al (1995), Infonet: The Network Infrastructure of Infopad, *Proc. CompCom, California,* March

Liu,G., Maguire, G. (1995) Efficient mobility management support for wireless data services, *Proc. 45th IEEE Vehicular Tech. Conf., Chicago, IL,* July.

Magedanz, T. (1995c), On the integration of IN and TMN- Modeling IN based Service control capabilities as Part of TMN-based Service Management, *ISINM 95,* Santa Barbara, USA, May 1-5

Mazer, M. et al* (1995) Issues in mobile Computing Systems. Guest Ediors Note in *IEEE Personal Communications,* Dezember.

NIIIP (1996a) *NIIIP Reference Architecture*, National Industrial Information Infrastructure Protocols Consortium

NIIIP (1996c) National Industrial Information Infrastructure Protocols Consortium, *Reference Architecture: Concepts and Guidelines*, Stamford, CT, USA.

NIIIP[*] (1996b) National Industrial Information Infrastructure Protocols Consortium, *Virtual Enterprise Computing: the Competitive Edge for the 21st Century*, Stamford, CT, USA.

Open Applications Group,Inc. (1996), *Open Application Group Integration Specification*, document N°1996.02.02, version 1.002.0.

Ramjee, R. et al (1995), The use of Networked-Bsased Migrating User Agents for Personal Communication Services, *IEEE Personal Communications*, December.

Richardson, T. et al. (1994), Teleporting in an X Window system environment, *IEEE Personal Comm.*, 3rd Quarter.

Schuster, H.; S. Jablonski; T. Kirsche; C. Bussler, (1994) A Client/Server Architecture for Distributed Workflow Management Systems", reached from World Wide Web as http://www6.informatik.uni-erlangen.de:1200/publ/sjkb94.ps.Z

TINA (1993) Telecommunication Information Networking Architecture -TINA Consortium, Dic No. TB_B.HT.004_1.o_93, *Description of Telecom Service Examples*, December.

Want, R. et al[*] (1995) An Overview of the ParcTab Ubiquitous Computing Experiment, *IEEE Personal Communications* , December.

Technical requirements for the development of agents

Colusa Software (1996), *Colusa Software White Paper: Omniware Technical Overview*, Colusa Software Inc.

7 BIOGRAPHIES

Prof. Dr. Manuel de Jesus Mendes received his Dipl.Ing. degree in Electrical Engineering in 1965 and a Dr.Ing. degree in 1968 at the Technical University of Berlin. He was a researcher from 1968-1969 at the Heinrich Hertz Institut in Berlin, and from 1969 to 1973 at the Max-von-Laue Paul Langevin Institut in Grenoble. Since 1973 he is full professor at the Electrical Engineering Faculty, Universidade Estadual de Campinas (UNICAMP) Brazil, where he retired in 1994. He became since then an invited professor at UNICAMP and is currently full professor at the Instituto de Informática, Pontifícia Universidade Católica de Campinas (PUCCAMP). His current research area is Open Distributed Platforms focused to telecommunications.

Dr. Waldomiro Loyolla received a B.S. degree in Physics in 1976 and in Electrical Engineering in 1980, a M.S. degree in Electrical Engineering in 1987, and a Ph.D. degree in Electrical Engineering from the Department of Computer Engineering and Industrial Automation at UNICAMP in 1992. He was a lecturer at the State University of São Paulo (UNESP) in Bauru since 1977, and was the Head of the Electrical Engineering Department from 1993 to 1995. Since 1996 he is a full Professor at the Informatics Institute at Pontifícia Universidade Católica de Campinas (PUCCAMP). His current research is directed to Workflow Management and Organizational Computing in Open Distributed Processing Environments.

Dr. Thomas Magedanz is assistent professor at the Department for Open Communication Systems of the Technical University Berlin with focus on distributed computing and telecommunications. In addition, he is head of the "Intelligent Communication Environment" research department of the GMD Research Center for Open Communication Systems (FOKUS) in Berlin. He received his master's and Ph. D. in computer science from the Technical University of Berlin, Germany in 1988 and 1993, respectively. Since 1989 he has been involved in several international research studies and projects related to IN, TMN, mobile computing and personal communications. In 1996 he was named project leader of the TINA-C Auxiliary Project "Personal Communication Support in TINA".

Flávio Morais de Assis Silva received a B.S. degree in Computer Sciences from the Universidade Federal de Minas Gerais (UFMG), Brazil, in 1989, and a M.S. degree in Computer Sciences from the Universidade Estadual de Campinas (UNICAMP), Brazil, in 1993. He is currently a Ph.D. candidate at the Technical University Berlin, Germany. During 1995 he has contributed to the RACE Prepare Project. His areas of interest are Advanced Transaction Processing, Mobile Computing and Open Distributed Processing Environments.

Sven Krause received a Dipl.-Inform. degree in Informatics from the Technical University of Berlin in 1992. From 1993 until 1995 he worked at GMD FOKUS, the German National Research Centre for Information Technology's Research Institute for Open Communication Systems. His work there focused on system and service management for ATM-based VPN networks. In 1995 he joined the Institute for Open Communication Systems of the Technical University of Berlin. His research area is the application of agent technology for the rapid and customized provision of information and telecommunication services in dynamic environments characterized by user and end system mobility.

21

A Global QoS Management for Wireless Network

My T. Le & Jan Rabaey
Department of Electrical Engineering & Computer Sciences
Cory Hall, University of California, Berkeley, CA 94720, USA
Phone: (510) 643-9380
Fax: (510) 642-2739
E-mail: mtl@eecs.berkeley.edu

Abstract

The Infopad Downlink Power Control System (IDPCS) is proposed as a scheme for the assignment of mobile units to cells and the management of power control for the downlink of Direct Sequence-Code Division Multiple Access (DS-CDMA) wireless networks. Its aim is to maximize the number of mobile units connected to each cell on the downlink subjecting to three requirements: maximizing the signal-to-noise-ratio that each mobile unit receives, meeting the maximum power constraint imposed on each base station, and meeting the limit on the number of mobile units that can be assigned to each base station.

Key words

Power control, wireless network, Direct Sequence-Code Division Multiple Access.

1 INTRODUCTION

From the phenomenal successes of the Internet and the cellular radio network, two key user demands can be discerned: the ability to gain access to a wide variety of information and the freedom of movement while being connected. Based on these requirements, a new type of communication system has emerged, the wireless multimedia system. The goal of the InfoPad project at the University of California is to design and build a prototype of such a system. It consists of a high speed wireless network, low-power, portable displaying devices, and supporting software for media access and data delivery (Sheng, 1992).

There exist many types of wireless networks, designed for a variety of applications. Our work is based on a particular type of network with the following characteristics. The wireless network is comprised of a number of cells with each cell controlled by a base station. Base stations send data to mobile units over a downlink channel - in our case using a Direct Sequence-Code Division

Multiple Access (DS-CDMA) scheme over a radio frequency (Viterbi, 1995). Mobile units send data to the base stations using an uplink, but we do not consider this in our work. The wireless network is assumed to support multiple types of data such as text, audio, video and provides data rates in the order of megabits per second to each mobile unit.

Different applications require different Bit-Error-Rates (BER). The BER is related to the ratio of the received signal to the interfering noise, the Signal-to-Noise Ratio (SNR), although this relationship is not a simple one (Proakis, 1989). In a system with only one mobile unit, the SNR is simply the ratio of the power levels of the signal and the noise experienced by the receiver. If the base station transmits more power, the power of the received signal increases and hence the SNR increases. The situation becomes more complicated in a system with more than one mobile units. The signal destined to one mobile unit is experienced as interfering noise by other mobile units. By increasing the power to increase the SNR at one mobile, the noise levels at other mobile units in the system are increased causing a decrease in their SNRs.

This mutual interference creates an interesting power allocation problem. Allocating power efficiently requires us to solve an optimization problem - providing the necessary power level for each mobile unit to achieve its required SNR.

The proposed solution addresses the problem of allocating power by providing the minimum transmitted powers to meet the SNR requirements of all mobile units in the system. It is comprised of three parts (Figure 1):

1. *Global Minimization* Algorithm: each mobile unit is assigned to a cell such that the transmitted power level to the mobile unit is minimized.

2. *Local Minimization* Algorithm: a feasible solution is determined such that each base station transmits the minimum power levels needed to provide its mobile units with the required SNR.

3. *SNR Renegotiation* Algorithm: when the Local Minimization Algorithm does not yield a feasible solution, the base station negotiates with the mobile units to reduce their SNR requirements such that the power limitation can be met.

The overall objective of our solution is to minimize the transmitted power to each mobile unit in the network so as to minimize the interference to other mobile units. The transmitted power to each mobile unit can be written in terms of: (1) the transmitted powers to other mobile units of the network, (2) the noise experienced by the mobile unit, and (3) the required SNR. These equations form the matrix that represents the problem of determining the downlink transmitted powers of the network. In addition, there are two boundary conditions: (1) an upper bound on the maximum power that each base station can transmit and (2) an upper limit on the number of mobile units that can be connected to a base station.

With the goal of providing an intuitive and comprehensive treatment of the power control problem, we will derive the solution based on a step-by-step approach:

Step 1. Derive the equation representing the transmitted power from a base station to a mobile unit.

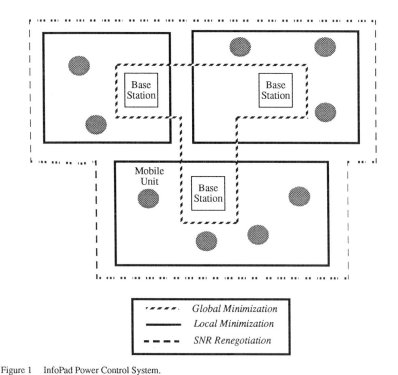

Figure 1   InfoPad Power Control System.

Step 2. Formulate power control as a linear programming problem based on the power matrix.

Step 3. Describe the algorithms that are used to solve the linear programming problem.

2   TRANSMITTED POWER EQUATION

Call the power level transmitted from base station i to mobile unit j, x_{ij}. The value of x_{ij} is determined by the power required at mobile unit j, x^r_{ij}, where the relationship between the two power levels is as follows. Signal arrives at the mobile unit with less power level than that was transmitted by the base station. The reduction in power level is due to the fact that the radio is a dispersive medium; transmitted signals experience reflection, refraction, scattering, etc., before reaching the receiver (Andersen, 1995). The relationship between the transmitted and received powers is well known:

$$\frac{x_{ij}^r}{x_{ij}} = \rho_{ij} = d^{-\lambda} 10^{-\varsigma/10} \qquad (1)$$

x_{ij}^r must be large enough to overcome the effects of noise and interference experienced by the mobile unit. In fact, the ratio of the receiver power over noise and interferences must be equal or greater than the required SNR:

$$\frac{x_{ij}^r}{(I_{ij}^{intra} + I_{ij}^{inter} + \sigma_j)} \geq SNR_j \qquad (2)$$

with I^{intra}, I^{inter}, and σ the intracell, intercell and background noise interference, respectively. Consider mobile unit j in cell i. In a CDMA scheme using PN codes for signal spreading, the *intracell interference* is comprised of two parts: the interference caused by the delayed paths of x_{ij}^r and the interference caused by other signals transmitted by base station i:

$$I_{ij}^{intra} = \left(\frac{1}{\kappa_j}\right) x_{ij} + \left(\frac{\rho_{ij}}{PG}\right) \sum_{l \neq j}^{N_i} \frac{\kappa_j + 1}{\kappa_j} x_{il} \qquad (3)$$

where N_i is the number of mobile units in cell i; κ is the ratio of the first path to the sum of all other paths (the Rician K ratio); PG is processing gain of the system; ρ_{ij} is the transmission coefficient from base station i to mobile unit j; and x_{ij} is the transmitted power from base station i to mobile unit j.

If the signal is modulated with a PN sequence followed by a Walsh code, the intracell interference is comprised of the delayed paths of all signal transmitted by base station i:

$$I_{ij}^{intra} = \rho_{ij} \left(\frac{1}{\kappa_j}\right)\left(\frac{1}{PG}\right) \sum_{l=1}^{N_i} x_{il} \qquad (4)$$

In the environment such as indoor where the first path is much larger than other paths, Walsh codes significantly reduce the intracell interference.

When multiple base stations transmit on the same channel, the mobile unit receives interfering signals from other base stations in addition to the signal transmitted from its connected base station. This type of interference is termed *intercell interference*. In a DS-CDMA system with frequency reuse of one, the same channel spectrum is used by every cell in the system. Thus, mobile units in each cell receive interference from all other base stations in the system.

Signal transmits from base station k and received by mobile unit j is subjected to the path transmission coefficient, ρ_{kj}. Furthermore, the intercell interference from base station k as experienced by mobile unit j (of cell i) is reduced by a factor of the processing gain due to the spreading effect of DS-CDMA.

$$I_{ij}^{\text{inter}} = \sum_{k \neq i}^{M} \frac{\rho_{kj}}{PG} \left(\sum_{l=1}^{N_k} x_{kl} \right) \quad (5)$$

where M is the number of base stations (cells) in system; N_k is the number of mobile units in cell k; PG is the processing gain; ρ_{kj} is the transmission coefficient from base station k to mobile unit j; and x_{kl} is the transmitted power from base station k to mobile unit l.

Two types of noises are considered in a wireless environment: thermal noise and external noise. Thermal noise is generated within the physical layer components and is quite small compared with other noises and interferences. Thermal noise is the baseline power level at a mobile unit because without any other types of noises or interferences, the mobile unit still needs a minimum power level to overcome the effect of this noise. All noises or interferences that are not classified as intercell interference, intracell interference, or thermal noise are considered external noise. The most common type of external noise is the power generated by equipment that operates in the same frequency bands, such as microwave ovens. Typically, all external noises are ignored.

Re-writing the equation of the transmitted power (2) using the results from (4) and (5) yields:

$$x_{ij} \geq \frac{SNR_j \left(\sigma_j + \rho_{ij} \left(\frac{1}{\kappa_j} \right) \left(\frac{1}{PG} \right) \sum_{l=1}^{N_i} x_{il} + \sum_{k \neq i}^{M} \frac{\rho_{kj}}{PG} \left(\sum_{l=1}^{N_k} x_{kl} \right) \right)}{\rho_{ij}} \quad (6)$$

3 BUILDING BLOCKS OF THE LINEAR PROGRAMMING PROBLEM

As we shall see in subsequent sections, IDPCS determines the transmitted power levels of the system by solving a set of linear programming problems (Luenberger, 1989). Before proceeding, we need to establish the blocks upon which the linear programming problems are built, the objective function and the constraints.

3.1 Objective Function

The overall objective of IDPCS is to minimize the transmitted power to each mobile unit in the network so as to minimize its interference to other mobile units:

$$\text{Minimize} \sum_{i=1}^{M} \sum_{j=1}^{N_i} x_{ij} \quad (7)$$

3.2 Constraints

The system has three overall constraints:

1. ***Power constraint***: A base station cannot transmit more than its maximum power level. An upper bound on the transmitted power guarantees that a base station does not cause excessive interference to other users in the same frequency band:

$$\sum_{j=1}^{N_i} x_{ij} \leq P_{max} \tag{8}$$

2. ***Single base station assignment constraint***: Each mobile unit is assigned to one base station:

$$\sum_{i=1}^{M} z_{ij} = 1, z_{ij} \in \{0, 1\} \tag{9}$$

3. ***Mobile unit constraint***: A base station cannot accept more than its maximum allowable mobile units. This upper bound is a system parameter that is set based on the capacity of the base station, such as its processing power or its number of DS-CDMA codes:

$$\sum_{j=1}^{N} z_{ij} \leq N_i^{max}, z_{ij} = 1 \text{ if } x_{ij} > 0, z_{ij} = 0 \text{ if } x_{ij} = 0 \tag{10}$$

4 PROPOSED SOLUTION

The goal of power control algorithms is to determine the power levels a base station needs to transmit so that all mobile units in its cell meet their required SNRs. This requirement is met by the proposed control system - the *InfoPad Downlink Power Control System (IDPCS)*. It is a global optimization system comprising of three algorithms: Global Minimization, Local Minimization, and SNR Renegotiation.

While IDPCS shares the same goal as other power control algorithms (Bock, 1964) (Foschini, 1995) (Gilhousen, 1991) (Hanly, 1995) (Yates, 1995) (Yun, 1995) (Zander, 1992), the proposed methods differ in a number of important ways:

1. All previously discussed algorithms determine whether the algorithm can converge and, if the algorithm converges, whether a solution exists and is optimal. The proposed system goes beyond this solution by considering additional algorithms required to change a non-convergent condition into a convergent one.

2. The proposed system represents a comprehensive way to implement power control instead of single algorithm that only deals with a particular aspect of power control.

3. All previously discussed algorithms operate distributively on a per-cell basis. This means that each base station determines the operating conditions that are optimized for its own cell. The proposed system describes an algorithm that is also distributed but has a global minimization goal. This means that each base station determines the operating conditions that optimized he entire system and not just its own cell.

4. All previously discussed algorithms assign mobile units to a base station according to a simple algorithm - the mobile unit is assigned to the base station where it experiences the least path loss. The proposed system employs a more sophisticated algorithm that guarantees the optimal assignment of each mobile unit.

4.1 Procedures

Consider a wireless system with M cells and N mobile units, $\sum_{i=1}^{M} N_i = N$. The implementation of IDPCS in this system can be divided into two parts: assignment of mobile units and allocation of power. First, the assignment of mobile units is solved using the Global Minimization algorithm. Once the mobile units have been assigned, the powers that base stations need to transmitted to their mobile units are determined by the Local Minimization and SNR Renegotiation algorithms.

1. **Assignment of Mobile Units**: Using the global minimization algorithm, the mobile units are assigned to the base stations where the total transmitted power of the system is minimum:

$$\min \left(\sum_{i=1}^{M} \sum_{j=1}^{N_i} x_{ij} \right) \qquad (11)$$

The solution of the algorithm can be represented as an assignment matrix where each column represents a mobile unit and each row represents a base station:

$$Z = \begin{bmatrix} z_{11} & z_{12} & \cdots & z_{1N} \\ z_{21} & z_{22} & \cdots & z_{2N} \\ \cdots & \cdots & \cdots & \cdots \\ z_{M1} & z_{M2} & \cdots & z_{MN} \end{bmatrix} \text{ where } \begin{matrix} z_{ij} = 1 & \text{if } x_{ij} > 0 \\ z_{ij} = 0 & \text{if } x_{ij} = 0 \end{matrix}.$$

2. **Allocation of Power:** The allocation of power can be accomplished in two steps: Using the local minimization algorithm, the minimum power level required by each base station is determined. If the total transmitted power by base station exceeds P_{max}, the base station negotiates with the mobile units to reduce the supported SNR level such that the power required can be reduced to meet the power limitation. There are two goals for the SNR Renegotiation algorithm: (a) meeting the power limitation at each base station and (b) providing an SNR that is closest to the desired SNR given constraint (a).

$$\min(desiredSNR_j - supportedSNR_j), \forall j \in i \qquad (12)$$

The solution of the algorithms can be represented as a power matrix where each column represents the power allocation of each base station:

$$X = \begin{bmatrix} x_{11} & x_{21} & \cdots & x_{M1} \\ x_{12} & x_{22} & \cdots & x_{M2} \\ \cdots & \cdots & \cdots & \cdots \\ x_{1N} & x_{2N} & \cdots & x_{MN} \end{bmatrix}$$

5 GLOBAL MINIMIZATION ALGORITHM

The Global Minimization algorithm determines the optimal assignment of a mobile unit to a base station. The optimal base station-mobile unit pair is defined as the one that requires the minimum transmitter power level by the base station. This is also called the assignment of a base station-mobile unit pair. In other power control algorithms, an optimal base station-mobile unit pair is defined as one where the transmission coefficient has the largest value among the transmission coefficients of the mobile unit: $\{z_{ij} = 0, z_{kj} = 1, \forall k \neq j : \rho_{ij} > \rho_{kj}, \forall k \neq j\}$. This policy does not always result in the minimum transmitted power and hence the Global Minimization algorithm is necessary to guarantee the objective of minimizing the total transmitted power levels of the system.

Consider two base station-mobile unit pairs (i, j) and (k, l). Let x_{ij} be the transmitted power level of each base station-mobile unit pair (i, j). This power level is experienced as an interference at mobile unit l. Base station k needs to provide mobile unit l with sufficient power to overcome this interference (Figure 2). If $k = i$, base station-mobile unit pair (k, l) experiences an intracell interference of the amount $\left(\frac{SNR_l}{PG}\right)\left(\frac{1}{\kappa_l}\right)x_{ij}$. If $k \neq i$, base station-mobile unit pair (i, j) generates an intercell interference of the amount $\left(\frac{SNR_l}{PG}\right)\left(\frac{\rho_{il}}{\rho_{kl}}\right)x_{ij}$ to (k, l).

Define y_{klij} as the portion of the transmitted power x_{kl} used to overcome the interference caused by x_{ij}. The relationship of y_{klij} and x_{ij} is an inequality

$$\frac{y_{klij}}{\left(\frac{SNR_l}{PG}\right)\left(\frac{1}{\kappa_l}\right)} \geq x_{ij} \text{ if } k = i, \quad \frac{y_{klij}}{\left(\frac{SNR_l}{PG}\right)\left(\frac{\rho_{il}}{\rho_{kl}}\right)} \geq x_{ij} \text{ if } k \neq i \qquad (13)$$

A global QoS management for wireless network

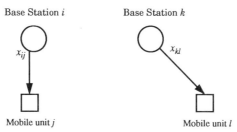

Figure 2 Base Station i outputs power x_{ij} to mobile unit j. Base Station i must provide power $y_{klij} = v_{kli} x_{ij}$ to overcome the interference from mobile unit l.

In the Global Minimization Algorithm, mobile unit l is (initially) assumed to be partially connected to all base stations. Its total transmitted power is $x_l = \sum_{k=1}^{M} x_{kl}$. Define y_{lij} as the portion of x_l used to overcome the interference caused by x_{ij}. y_{lij} can then be expressed as:

$$y_{lij} = \sum_{k=1}^{M} y_{klij} = y_{ilij} + \sum_{k \neq i}^{M} y_{klij} \qquad (14)$$

A *feasible activity* for $f = (i, j)$ is a vector of the form $[x_{ij}, y^f]$ where $y^f := (y_{klij})$ such that for each l

$$\frac{y_{ilij}}{\left(\frac{SNR_l}{PG}\right)\left(\frac{1}{\kappa_l}\right)} + \sum_{k \neq i}^{M} \frac{y_{klij}}{\left(\frac{SNR_l}{PG}\right)\left(\frac{\rho_{il}}{\rho_{kl}}\right)} \geq x_{ij} \qquad (15)$$

The power allocated to a mobile unit l must be large enough to overcome the interference caused by the base station-mobile unit pair (i, j):

$$-x_{ij} + \frac{y_{ilij}}{\left(\frac{SNR_l}{PG}\right)\left(\frac{1}{\kappa_l}\right)} + \sum_{k \neq i}^{M} \frac{y_{klij}}{\left(\frac{SNR_l}{PG}\right)\left(\frac{\rho_{il}}{\rho_{kl}}\right)} \geq 0 \qquad (16)$$

Furthermore, the net power of the base station-mobile unit pair (i, j) as received by the mobile unit j must be large enough to over come the noise experienced by the base station-mobile unit pair:

$$\rho_{ij}(x_{ij}) - \rho_{ij}\left(\sum_k \sum_l y_{ijkl}\right) - n_{ij} \geq 0 \tag{17}$$

where the total noise experienced by a mobile unit equals the sum of the noises of its base station-mobile unit pairs:

$$\sum_{i=1}^{M} n_{ij} = n_j \tag{18}$$

We can formulate the Global Minimization algorithm as a linear programming problem of the form:

$$\text{Minimize} \quad \sum_{i=1}^{M}\sum_{j=1}^{N} x_{ij}$$

$$\text{Subject to} \quad \begin{cases} -x_{ij} + \dfrac{y_{ilij}}{\mu_{il}} + \sum_{k \neq i}^{M} \dfrac{y_{klij}}{v_{kli}} \geq 0 \\[2mm] \rho_{ij} x_{ij} - \rho_{ij}\sum_{k=1}^{M}\left(\sum_{l=1}^{N} y_{ijkl}\right) - n_{ij} \geq 0 \\[2mm] \sum_{i=1}^{M} n_{ij} = n_j \\[2mm] x_{ij} \geq 0 \\ y_{ijkl} \geq 0 \\ n_{ij} \geq 0 \end{cases} \tag{19}$$

where $\mu_{il} = \left(\dfrac{SNR_l}{PG}\right)\left(\dfrac{1}{\kappa_l}\right)$ and $v_{kli} = \left(\dfrac{SNR_l}{PG}\right)\left(\dfrac{\rho_{il}}{\rho_{kl}}\right)$.

The solution of the algorithm can be organized as an assignment matrix Z where $z_{ij} = 1$ if $x_{ij} > 0$, $z_{ij} = 0$ if $x_{ij} = 0$. Each column represents a mobile unit where all coefficients in a column except one are zero. Each row represents a base station where the sum of each row is less than or equal to it maximum allowable mobile units.

With this assignment matrix, each mobile unit has been assigned to a single base station where its required transmitted power is minimum. Furthermore, the number of mobile units in each base station is less than or equal to its limit. The global minimization algorithm yields a feasible solution. We are now ready to allocate power to each of these base station-mobile unit pairs using the Local Minimization Algorithm.

6 LOCAL MINIMIZATION ALGORITHM

Once each mobile unit has been assigned to a base station, the system can be described as a matrix where each equation represents the transmitted power required by a base station-mobile unit pair:

$$\alpha_{ij}x_{ij} - \beta_{ij}\sum_{l \neq j}^{N_i} x_{il} - \sum_{k \neq i}^{M}\left(\gamma_{kj}\sum_{l=1}^{N_k} x_{kl}\right) \geq n_j \qquad (20)$$

where $\alpha_{ij} = \left(1 - \dfrac{SNR_j}{PG \cdot \kappa_j}\right)\rho_{ij}; \beta_{ij} = \dfrac{SNR_j}{PG \cdot \kappa_j}\rho_{ij}; \gamma_{kj} = \dfrac{SNR_j}{PG}\rho_{kj}; n_j = SNR_j \sigma_j = \text{constant}.$

The goal of the local minimization algorithm is to allocate the optimal power to each base station-mobile unit pair. This is an optimization problem that can be solved as a linear programming problem of the form:

$$\text{Minimize} \quad \sum_{i=1}^{M}\sum_{j=1}^{N_i} x_{ij}$$

$$\text{Subject to} \quad \begin{cases} \alpha_{ij}x_{ij} - \beta_{ij}\sum_{l \neq j}^{N_i} x_{il} - \sum_{k \neq i}^{M}\left(\gamma_{kj}\sum_{l=1}^{N_k} x_{kl}\right) \geq n_j \\ x_{ij} \geq 0 \end{cases} \qquad (21)$$

The solution of the algorithm can be represented as the power matrix: X. Each column of the matrix represents the transmitted power levels of a base station. If the total transmitted power of each base station is less than or equal to its maximum power, a feasible solution for the local minimization algorithm has been obtained. This solution is also the solution to the power control problem. Hence, if (21) yields a feasible solution, IDPCS has completed its task.

If, however, one or more base station fails to meet its power constraint, an additional algorithm is needed to convert this infeasible solution into a feasible one. This converted algorithm, called SNR Renegotiation, is discussed next.

7 SNR RENEGOTIATION ALGORITHM

If the total transmitted power by a base station exceeds its maximum allowable power limit, the power budget can be reduced by either removing one or more mobile units from the cell or by reducing the required powers of one or more mobile units. Assuming that the policy of the system is to maximize the capacity, IDPCS implements the latter option.

For each base station i whose total transmitted power exceeds its maximum power, its power budget is set at the limit: $\sum_{j=1}^{N_i} x_{ij} = P_{max}$. When the total transmitted power of a base station decreases, the power allocated to each mobile unit in that cell must also decrease. We need a criteria to determine the amount of power that needs to be reduced from each base station-mobile unit pair.

Recall the transmitted power equation of a base station-mobile unit pair: $x_{ij} \geq \frac{SNR_j(I_{ij}^{inter} + I_{ij}^{intra} + \sigma_j)}{\rho_{ij}}$. Since SNR is the only parameter that can be modified in this equation; to reduce the required transmitted power to a mobile unit, its SNR must be reduced. Naturally we would like to minimize the difference between the desired SNR and the provided SNR of each mobile unit. The goal of minimizing the reduction of SNRs must be balanced against the reduction of the power budget. This means the SNR renegotiation algorithm requires a solution of an optimization problem.

Once again, the optimization problem can be formulated as a linear programming problem:

$$\text{Minimize} \quad \sum_{j=1}^{N_i} ((SNR_0)_j - SNR_j)$$

$$\text{Subject to} \quad \begin{cases} \sum_{j=1}^{N_i} x_{ij} = P_{max} \\ SNR_j \geq 0 \\ x_{ij} \geq 0 \\ (SNR_0)_j - SNR_j \geq 0 \end{cases} \quad (22)$$

where $(SNR_0)_j$ is the desired SNR of mobile unit j and SNR_j is the SNR provided to mobile unit j.

Unlike in previous sections, the linear programming problem expressed in (22) cannot be solved readily because the variables in the objective equation, SNR_j, are not the variables in the constraint equation, x_{ij}. To solve the problem, we must first express the SNR of each mobile unit as a function of the transmitted powers of all base station-mobile unit pairs in the cell: $SNR_j = f(x_{ij}, \forall j \in i)$.

(22) can be re-written as:

A global QoS management for wireless network

$$\text{Minimize } \sum_{j=1}^{N_i} SNR_0 - f(x_{ij})$$

$$\text{Subject to } \begin{cases} \sum_{j=1}^{N_i} x_{ij} = P_{max} \\ SNR_j \geq 0 \\ x_{ij} \geq 0 \end{cases} \quad (23)$$

SNR is a nonlinear function of x_{ij}, $\forall j \in i$. Thus, (23) is a nonlinear programming problem of the form:

$$\text{Minimize } f(\underline{x})$$
$$\text{Subject to } h(\underline{x}) = P_{max} \quad (24)$$

Although the solution of the general nonlinear programming problem is difficult to obtain, this problem can be solved easily using the well-known Lagrange multipliers (Luenberger, 1989). The Lagrangian associated with the problem is: $l(\underline{x}, \underline{\lambda}) = f(\underline{x}) + \underline{\lambda}^T h(\underline{x})$ where $\underline{\lambda}$ is the Lagrange multipliers. The solution of the Lagrangian equation is derived by solving a set of equations of the form: $\nabla_{\underline{x}} l(\underline{x}, \underline{\lambda}) = 0$, $\nabla_{\underline{\lambda}} l(\underline{x}, \underline{\lambda}) = 0$ where ∇ is the partial derivative function.

The solution derived from the Lagrangian equation can be arranged as a vector that is comprised of the new SNR of the mobile units in the cell. Each new SNR is the closest value to the desired SNR given the power budget constraint, and hence is the optimal solution.

8 SUMMARY

We have presented a global solution to the cell-assignment and power-control problem in wireless multimedia networks. The system is formulated as a set of linear programming problems, that can be readily solved in a distributed fashion, yet yield global solutions. Experiments with large sets of cells and terminals have demonstrated that the proposed solution is robust and converges rapidly. Current research includes a further simplication of the problem to reduce computational complexity.

9 REFERENCES

[1] Andersen, J.B., Rappaport, T.S., and Yoshida, S.(1995) Propagation Measurements and Models for Wireless Communications Channels, *IEEE Communications Magazine*, vol. 43, no. 4, 42-9.
[2] Bock, F., Ebstein, B. (1964) Assignment of Transmitter Powers by Linear Programming, *IEEE Transactions on Electromagnetic Compatibility*, vol. EMC-6, 36-44.

[3] Foschini, G.J., Miljanic, Z. (1995) Distributed Autonomous Wireless Channel Assignment Algorithm with Power Control, *IEEE Transactions on Vehicular Technology*, vol. 44, no. 3, 420-9.

[4] Gilhousen, K.S., Jacobs, I.M., Padovani, R., Viterbi, A.J. (1991) On the Capacity of a Cellular CDMA System, *IEEE Transactions on Vehicular Technology*, vol. 40, no. 2, 30

[5] Hanly, S.V. (1995) An Algorithm for Combined Cell-Site Selection and Power Control to Maximize Cellular Spread Spectrum Capacity, *IEEE Journal on Selected Areas in Communications*, vol. 13, no. 7, 1332-40.

[6] Luenberger, D.G. (1989) *Linear and Nonlinear Programming, 2nd Edition*; Addison-Wesley, Reading, MA.

[7] Proakis. J.G. (1989) *Digital Communications*, Second Edition, McGraw-Hill, New York.

[8] Sheng, S., Chandrakasan A., and Brodersen, R.W. (1992) A portable multimedia terminal, *IEEE Communications Magazine*, vol. 30, no. 12, 64-7.

[9] Viterbi, A.J. (1995) *CDMA: principles of spread spectrum communication*, Addison-Wesley, Reading, MA.

[10] Yates, R. (1995) A Framework for Uplink Power Control in Cellular Radio Systems, *IEEE Journal of Selected Areas in Communications*, vol. 13, no. 7, 1341-7.

[11] Yun, L., Messerschmitt, D.G. (1995) Variable QOS in CDMA Systems by Statistical Power Control, *ICC'95*.

[12] Zander, J. (1992) Performance of Optimum Transmitter Power Control in Cellular Radio Systems, *IEEE Transactions on Vehicular Technology*, February 1992, vol. 41, no. 1, 57-62.

10 BIOGRAPHY

My Le graduated from University of California at Davis with a BSc in Electrical Engineering in 1983. She had worked as an engineer at National Semiconductor and Hewlett-Packard on FDDI, FDDI-II, IEEE802.6, and ATM projects. My Le has been a graduate student at the University of California at Berkeley since 1990. As a co-leader of the BayBridge project, she oversaw the development of a high speed networking bridge and router between FDDI and SMDS. My Le runs her own consultancy business in the area of computer network design. Her recent research work is in Wireless Networks.

22
Resource Discovery Protocol for Mobile Computing

Charlie Perkins and Harry Harjono
IBM T.J.Watson Research Center
30 Saw Mill River Road (Route 9A), Hawthorne, NY 10532, USA.
Telephone: 914-784-7350.
email: `perk@watson.ibm.com, harjono@cs.columbia.edu`

Abstract
The increasing complexity of modern networks prompts a need for dynamic resource discovery. Mobile clients have the additional need to rediscover the location of local area network resources each time they move to a different LAN. We present a protocol and proposal for the operation of dynamic resource discovery. Our design is simple, extensible, and light weight. We implemented and tested our design with stationary servers, and mobile clients running mobile IP.

Keywords
resource discovery protocol, light weight, service location, directory service, search engine, mobile IP, mobile computing, dynamic configuration, URN, URL, DHCP, RDP, SLP

1 INTRODUCTION

Mobile computing, mobile networking protocols, and the growth of the Internet are combining to make today's mobile computer users feel like global citizens. However, in order to make full use of the Internet, today's citizen need to be able to make numerous configuration choices. Recent efforts, (Droms (1993), Perkins and Tangirala (1995), Perkins (1995)) have begun to chip away at the requirement for reconfiguration of mobile computers as they move from place to place, but there is still much to do.

Access to local computing resources is usually required to sustain the productivity of mobile users, and the network connection to those resources is accomplished with recently developed mobile networking protocols, IETF Mobile-IP Working Group (1995). However, up until now, there hasn't been an easy way for Internet users to find and use the local computing resources without making phone calls, and then reading manuals about how to perform the necessary configuration operations. That often means finding out which of hundreds or thousands of system files need modifications. Worse, as more and more users are faced with the need to perform these administrative functions, it is inevitable that some mistakes will be made. Determining the cause of errors in the course of performing system configuration is often very unpleasant; if network configuration is involved, errors in the configuration of a new system can cause an entire network or subnet to go out of service.

Our resource discovery protocol is designed to alleviate the problem of finding and using computer resources that are located external to a mobile computer. The protocol itself doesn't depend upon mobility, but on the other hand it is especially when a computer becomes mobile that its need for system reconfiguration is drastically increased, compared to computers which are installed in one place and remain there for a long time. Even so, once the resource discovery algorithms become commonplace, we expect that they will become a natural part of the overall organization of most Internet computers.

2 UNIFORM RESOURCE LOCATORS

The recent growth of the World Wide Web has occasioned the even faster growth of the Internet. The utility of the Web is due to the ease with which even beginning computer users can browse untold gigabytes of interesting pictures, stories, games, and other programs and data. These computer resources are delivered to the browser after it finds them by following a *Uniform Resource Locator*, or URL (T. Berners-Lee and L. Masinter and M. McCahill (1994)) for short. URLs are a standardized way to locating and providing access to a large variety of computer resources located on the network. They have the general form <scheme>:<scheme-specific-part>, where a *scheme* is just a string which tells how to understand the *scheme-specific-part*.

URLs are made available to browsers by user selection from stylized menus, which may contain buttons, maps, or other indicators. User interaction, however, is exactly what we would like to minimize or eliminate in the configuration of mobile computers. It is possible to imagine some sort of "mobile computing butler" which interrupts the user upon any indication of resource outage, and presents a menu of newly available resources which can take the place of the resource which is no longer around. Instead, we wanted a way to discover and use Internet resources without any user interaction whatsoever. For instance, if a computer is turned on in unfamiliar surroundings, all necessary network configuration details should be acquired automatically and put into service without user intervention.

The main difference between delivering network resources to Web browsers, and the more automatic way of discovering access to resources just described, is that in the latter case there is no user interaction to specify or "name" the indicated resources. What is needed is not only a URL, but also a URN, or Uniform Resource Name (Paul E. Hoffman and Daniel, Jr., Ron (1995), K. Sollins and L. Masinter (1994)). Then, a computer could acquire the resources it needs to operate by formulating a list of URNs, and then request a URL for each of them. This is exactly the approach we took. However, URNs are not in common use with the Web, and the exact meaning of URNs is still a matter of dispute within the Web technical community. Even so, an evaluation of existing URN proposals, (Mark Madsen (1995)), pointed out that the OCLC scheme, (K. Shafer and E. Miller and V. Tkac and S. Weibel (1995)), is the most promising framework from the point of view of extensibility and future-proofness. Accordingly, we have adopted that scheme in our work.

Aspects of URNs that may be expected if consensus emerges are:

- resource names will be specified with a syntax conforming to URI (T. Berners-Lee (1993)),
- resolution of URNs will exhibit a high degree of location independence,
- URNs will be well suited for identifying and locating particular documents and versions of documents within the global Internet.

For resource discovery, we decided to make use of the possibility that URNs might resolve differently depending upon which agent was doing the resolution. This can be seen as a violation of the spirit of URNs, because we explicitly want different URLs to be associated with the same URN depending upon where the resolution occurs. Given the proposed deployment of URNs as document identifiers, one could well argue that a URN was expected to always resolve to the same document URL. Although we like URNs, our needs for resource discovery have almost nothing to do with document retrieval. Moreover, we optimistically hope that our use of URNs will influence the future direction and standardization of URNs for specifying resource location, since if the proposed Service Location Protocol succeeds in its current form it will be a motivation for the further development of URN technology and protocols.

3 PROTOCOL CHARACTERISTICS

Our intention is to make a resource discovery protocol suitable for automatic operation in sometimes crowded enterprise internetworks, to serve the needs of mobile clients. To do this, we had to adhere to a number of design requirements. The protocol is required to be

- scalable
- self-managing
- distributed, with numerous servers
- compatible with other administrative tools
- compatible with mobile networking protocols

The need for scalability is almost a given in today's Internet. Any protocol which only works well with a few computers on a network will not pass the review within the Internet Engineering Task Force (IETF), so would have no hope for standardization. Since, from our perspective, the point of creating the protocol in the first place is to eliminate the need for user configuration, the protocol must require zero user administration. As a natural consequence of the requirement for scalability, we must also demand that minimal or non-existent configuration be required for even the servers which provide the resource data for the mobile clients. Any burdensome administrative requirement for human intervention or control of the resource discovery servers will be doomed to failure as the Internet continues to provide an ever greater array of possible services to mobile clients.

DHCP (Dynamic Host Configuration Protocol, S. Alexander and R. Droms (1993), Ralph Droms (1993)) servers are one particular administrative tool with which we had to be compatible. In fact, the ability of a DHCP server to configure its client with the address of a Resource Discovery server is perfect for our needs, and shows that the original designers of DHCP looked forward to the day when such protocols as ours would become available.

Besides the above mentioned protocol requirements, we intended to produce a protocol with some additional highly desirable properties:

- lightweight (fast)
- string-based (simple parser)
- easily implemented
- use existing standards where possible
- syntactically flexible

Since our main protocol operation is to supply a pointer to a named resource, we made it one of our main goals to keep the protocol lightweight. We expect that this will go a long way towards enabling the widespread deployment of the protocol for mobile computers. Our definition of lightweight also includes minimal network cost, so that broadcasts and extended negotiations for resources are considered highly undesirable.

The network of the future is likely to be populated with resources and agencies we can only dimly imagine. These resources, although we can't name them now, will certainly have names, and their names could be used as part of a URN. We already had parsers for URLs that work with Web browser software, and those parsers could be made to work well with human readable resource names in the form of URNs. The combination of existing algorithms, flexibility, human readability, and extensibility for the future makes string-based operation quite attractive.

An alternative approach might be to assign numbers for each new resource type, and make the clients request resources by number. However, that approach relies completely on the required registration of new resources with an Internet arbiter such as IANA (Douglas E. Comer (1991)) as well as the timely dissemination of newly registered resource numbers to all interested resource discovery servers. This is fine for resources that are duly registered with IANA, but not so fine for resources that are still experimental, vendor-specific, or site-specific.

Although the string-based approach to naming also requires a conventional agreement between client and server regarding the names of resource, this agreement is more likely with strings in the abovementioned latter cases. For instance, it's a lot easier for the word "printer" to proliferate throughout the administrative and engineering community at a particular site than some arbitrary (numeric) bit string. Thus, we consider strings to be the most obvious candidate for specifying attributes or keywords for selecting among numerous resources of the same general "type", and we settled on the use of strings for naming resources. This was another motivation for our subsequent use of URNs.

Lastly, we explicitly favored ways to re-use existing protocols and language syntax, in the belief that new development is usually better and almost always faster if it uses the hard work of other people. Not only did we have to make fewer decisions about code structure and query format, but we also have been able to avoid all the mistakes that were probably made in the early design of Web protocols. This is, of course, another benefit of aiming our design towards compatibility with the World Wide Web as a collection of resources and resource locators.

4 RESOURCE DISCOVERY PROTOCOL (RDP)

4.1 Introduction

The main objective of RDP is provide a light-weight protocol which a client can use to discover network resources. It is especially targeted for a mobile client whose environment may change often. In RDP, we assume that the client has a means of obtaining the address of the RDP server, either statically via configuration file, or dynamically via DHCP. Of course mobile clients are unlikely to have any static configuration for the address of an RDP server, but stationary enterprise desktops likely would.

Mobile clients are frequently wireless, and wireless stations currently have characteristically poor interactions with TCP when the wireless medium is suffering from a high bit-error rate (BER) (Ramon Caceres and Liviu Iftode (1995)). We wanted to avoid interactions between our protocol and the timeout characteristics of TCP, to keep RDP as light-weight as possible. Thus, we rely on UDP for packet delivery. For simple query/response case, the data can fit into one UDP datagram. When the data is too big to fit into one datagram, it is broken into multiple UDP packets.

The basic operation follows a client-server query-response model. The client queries the server using a URN query; and the server replies with one or more URLs to satisfy the query. The client may then proceed to use the returned URL(s) to access the network resources. In addition to this, RDP also supports dynamic registration and deregistration of network resources which enables the server to manage the resources automatically.

4.2 RDP Database

The database is very simple. It consists of a collection of records, each with the following structure:

```
<resource URL>
        <description1>
        <description2>
        ...
        <descriptionN>
```

The description lines contain descriptions of the resource, which may consist of multiple keywords with optional attribute names. No syntactic structure is imposed on the descriptions. For example:

```
printer://dukprunz.watson.ibm.com/j1j25ps
    name=j1j25ps
    location=j1-j25,j1j25
    queues=j1plain,j1color,j1foils
    os2 postscript personal printer color foils
    access via TCP/IP lpd lpr
    local=129.34.16/24,9.2.46.0/25
```

Notice that the example above uses a non-standard printer URL*. In this paper, for simplicity we do not always append a subnet prefix length specification to relevant IP addresses, although it is done in this example. The details of managing printers and print queues may require enumerating additional parameters in the same way that is indicated here.

4.3 RDP Query and Response

To query the resource location, a client has to construct a valid URN query. The format of the query is:

```
<service>:/[rp]/[na]/<scheme>/<key1>/<key2>/...
```

where:

```
service = n2l or n2c
rp      = resolution path
na      = naming authority
scheme  = URL scheme
keyN    = keywords describing the URL
```

The format of the URN query is borrowed from the *URN Services* (K. Shafer and E. Miller and V. Tkac and S. Weibel (1995)) internet draft with some liberal changes to suit our purposes.

The service field specifies the desired type of resolution. *n2l* maps one URN to one URL: the server will return the first URL it find which satisfies the URN. *n2c* maps one URN to a list of URLs satisfying the URN[†].

The resolution path is optionally specified by the client to direct the query to the desired RDP server. In the absence of a resolution path specification, the query will be sent to the server host returned by DHCP, if the client has requested option 11 from the local DHCP server (S. Alexander and R. Droms (1993)). Note that, in the usual case, the resolution path will just be the IP address of a nearby host computer. Also note that this procedure allows our resolution architecture to scale well.

The naming authority (*NA*) specifies the organizational entity which is authorized to

---

*The printer URL syntax is printer://<lpd-hostname>/<queue-name>.
[†]Other request services have been suggested along the lines of *n2two*, *n2three*, ...

resolve the query, and then by necessity the dictionary which is used to define the relationship between the scheme and the scheme-specific-part of a URL. The IANA naming authority already specifies some universally known schemes, including *printer*, *mailto*, *http*, and *nfs*. Upon receiving a URN query, the RDP server will verify that its naming authority is compatible with that of the query. If it does not, the query will be forwarded to the authorized server. When the naming authority is omitted, the server can skip the verification. This is useful for wandering mobile clients which do not know the naming authority of their local network.

The requested *scheme* is found in the field after the naming authority field. It can be any valid URL scheme recognizable by the server. When the resource database grows large, the resource server may partition the database into disjoint subsets based on the scheme, since the same URL cannot possibly belong to two different schemes under the same naming authority.

The last components are the keywords used to search the database for the matches. A match is found when the scheme and *all* the keywords match. These keywords should form an intuitive description of the desired resources, and could be obtained directly from the end users if necessary. An efficient multi-keyword search algorithm is presented by Sun Wu and Udi Manber (1994). For large databases, the search can be made faster using a two-level indexing scheme described by Sun Wu and Udi Manber (1993) with minimal indexing time and space.

Some examples of valid URN queries are:

```
n2l://ibm/nfs/rdp/src
n2c://ibm/http/research/homepage
n2l:///printer/local/postscript
```

Here, *n2l* means to return only one URL in response to the query, and *n2c* means to return all matching URLs, concatenated in the reply. A high-function server could possibly sort matching URLs in order of decreasing expected usefulness to the client, based possibly on distance. The word *local* has a special meaning; it means that the returned URL should be local to the client (see Section 4.4).

Some examples of valid URL replies are:

```
nfs://slag.watson.ibm.com/src/rdp
http://www.research.ibm.com/
printer://dukprunz.watson.ibm.com/j1j25ps
```

An application using RDP must come equipped with some conventional, built-in vocabulary, in order to be able to send queries to the RDP server. For example, to find a list of local postscript printers, a word processing application should know how to construct the URN query

```
n2c:///printer/local/postscript.
```

This query could be hard-coded into an application as long as it did not change the type of printer needed. The resolution of the query may return different printer URLs depending on which is most beneficial for the mobile client in its current environment.

4.4 Locality

Locality is a tricky problem, which is at the same time intimately tied to user convenience, but also difficult to define in a way that is useful for all contexts. For instance, a user might be interested in *geographical* locality when it is time to select a printer, but locality with respect to *network topology* is more likely to be of interest for connection to resources with which large amounts of data will be transacted. Moreover, in certain situations a sort of administrative locality will be useful. Geographic locality has received some attention, for instance in papers describing progress with active badge systems by Bill Schilit and Marvin Theimer (1994), and by Roy Want and Andy Hopper and Veronica Falcao and Jonathan Gibbons (1992).

Locality will *usually* be defined (at least partially) by the scheme. As just suggested, the printer scheme is more likely to evaluate locality based on the walking distance between the mobile client and the prospective printers. We do not make any attempt to refine the handling of the local keyword in this version of the protocol. However, we note that future versions may include local as a *scope* specifier, and for now only specify locality by describing subnet information. We expect that when printer URLs are registered, they will include keywords that specify which subnets are local. Clearly this handles some high percentage of current needs, while just as clearly there are many refinements to be made.

4.5 Resource Registration and Deregistration

Clients register and deregister network resources using the *reg* and *dereg* requests. The RDP database grows and shrinks accordingly. The format of the reg/dereg requests are:

```
<reg|dereg>:/[rp]/[na]/<url>;<desc1>[;desc2]...
```

where:

```
rp     = resolution path
na     = naming authority
url    = URL to be registered (or deregistered)
descN  = descriptions of the URL
```

Registration can be performed incrementally. The new URL record will be created only if it doesn't exist in the database. If the URL already exists, only the descriptions not in the database will be added. The final URL record contains the URL and the union of all the descriptions.

Similarly, deregistration can be performed incrementally. If no descriptions are specified, the whole URL record is deleted; otherwise, only the matching descriptions are deleted.

An acknowledgement is sent to the client for a successful reg/dereg. If no acknowledgement is received after the timeout, the client may retry the operation. Note that both the reg and dereg are idempotent; this is necessary to ensure the integrity of the database.

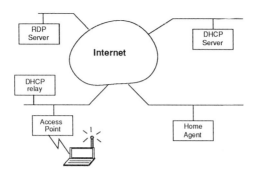

Figure 1 Mobile Client interacting with RDP and other Internet agents

5 MOBILE-IP, DHCP, AND RESOURCE DISCOVERY

Our motivation for investigating Resource Discovery comes from our conviction that it will be a requirement for the convenient operation of mobile computers. We also are convinced that such computers will use the mobile-IP (IETF Mobile-IP Working Group (1995)) protocols being advanced in the IETF. Both protocols are also intimately involved with the use of DHCP. As it turns out, it seems that there there is never conflict in the simultaneous use of the protocols of interest. This results from the fact that resource discovery is used by application level clients, whereas mobile-IP is used in the protocol stack itself, and DHCP is used for system and network configuration parameters. The low level of interaction we have observed validates the notion of modular design of network protocols. However, there is more to be said when a higher level view of the system operation is considered.

Consider the sequence of events when a mobile client first begins operation. In the most demanding case, the client will rely on DHCP for acquisition of both its *own* address (known as a *home address* (Charles Perkins and Jagannadh Tangirala (1995), and Charles E. Perkins (1995)), as well as a care-of address. Note that the RDP server, the DHCP server, the home agent, and the subnet to which the mobile client is attaching via an access point may all be on different subnets. The following might be a typical scenario during the time when a mobile client reboots:

- Mobile queries DHCP for a home address
- Mobile discovers that a mobile-IP registration is needed (if it is not attaching to its home network)
- Mobile queries DHCP for a local IP address to be used as a Care-of Address, including Resource Discovery server option
- Mobile registers the new Care-of Address with home agent
- Mobile queries the Resource Discovery server for needed resources

Note that in this case, DHCP is queried twice, and there is no need to add option 11 for the RDP server[‡] the first time. Moreover, notice that the RDP could reasonably be consulted for all possible resources, on the assumption that things may have changed significantly since the last time the mobile computer has been restarted.

Consider another case. If the mobile node registers with a Care-of Address which is advertised by a nearby foreign agent, there is no immediate need to contact a DHCP server, as long as no network resources were required during the mobile node's stay in the area being served by the new foreign agent.

A system designer could reasonably consider whether it was worthwhile to request new resource pointers for only *local* resources, instead of contacting the RDP server for every resource it might need. It is also likely that the request for new resource locations should be performed on demand instead of upon every cell switch, on the assumption that some resources will not be accessed before another cell switch. There would be no point to resolving new resources when the results might never be used.

We do not offer any conclusion about the best system design for when the Resource Discovery Server should be contacted. So far, our approach has been only to resolve needed resources when the mobile computer reboots. We plan to attack the problems introduced by cell switches after we install some ability for interprocess communications to be triggered by cell switches.

6 IMPLEMENTATION EXPERIENCES

The current implementation of RDP client and server are available in C and C++. We have compiled them for AIX, OS/2, Linux, and SunOS. We expect any BSD socket compliant system should have no problem compiling it. This is a quick implementation, and is not optimal. We just wanted to show the feasibility of the system.

6.1 RDP Client

On startup, the client host gets the address of the RDP server from DHCP, which it then saves in a permanent, conventionally known file. Applications which need to contact the RDP server get the address from this file. This step can be automated by providing a Resource Discovery API (Applications Programming Interface) so that the application does not need to statically configure the filename.

The generic RDP client program simply takes a string from the user or the command line, verifies that it is a valid URN format, and sends it to the RDP server. It then waits and displays the URL reply. The client also keeps a retransmit timer to simulate reliable packet delivery. We use this generic client program in our OS/2 REXX script to communicate with the RDP server.

[‡]Also known as Resource Location Server (RLS) option in the DHCP document (S. Alexander and R. Droms (1993))

6.2 RDP Server

On startup, the RDP server may read an initial list of resources from a configuration file. From then on, the resource database may be changed by registration and deregistration requests.

To serve a URN query, our server verifies the naming authority, and checks the validity of URN format. Then, it does a linear search through its database for matching keywords.

6.3 Mobile-IP testbed

We tested our system with stationary server and mobile clients. The resource database is populated with URL records of local resources such as printers, and NFS mount points. The mobile client obtains the address of the RDP server from DHCP during bootup. In figure 1, the mobile client performs the following steps:

- gets care-of address and address of stationary RDP server from stationary DHCP server via access point and DHCP relay
- registers new care-of address with Home Agent, allowing delivery of packets to the mobile client from anywhere on the Internet.
- contacts a stationary RDP server for each network resource needed

Given the setup depicted in figure 1, we showed that the mobile clients can access local network resources such as printers and NFS filesystems. Since our mobile client runs OS/2, we wrote some REXX scripts to query the RDP server for local resources, and then proceed to access them.

Note that, in the above situation, our mobile client accesses the RDP server at boot time and performs the necessary operations for all possible network resources at that time only. We expect that it will be much more common to access the RDP as the individual resources are needed, possibly after the mobile client has been in operation for quite a while. Moreover, the mobile client will need to contact a (possibly different) RDP server after it moves to a new access point. We haven't tested that operation yet.

Note also, in the following descriptions, that the Internet agents involved usually return the Internet addresses of target hosts, not their human-readable fully-qualified domain names. We use the domain name to make the examples easier to understand. In most cases it is preferable to allow the receiving host to avoid the extra step of having to resolve the domain name into an IP address.

Accessing a local printer is very straightforward (but see Section 4.4). We send a URN query for local postscript printer, and use the returned printer URL in the *lpr* command to send data to be printed. Figure 2 shows the following scenario:

- The wireless mobile client first gets access to a nearby RDP server by querying a DHCP server (usually via a DHCP relay, and physically by way of a wireless access point).
- The DHCP server returns the IP address of the RDP server (muffin.watson.ibm.com), which is on another subnet.
- The mobile client contacts *muffin* to get the address of the local printer service. The URN (namely, n2l:///printer/local/postscript) is delivered to the RDP server (muffin.watson.ibm.com).

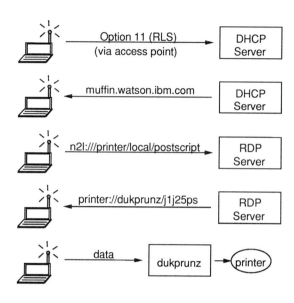

Figure 2 Discovering a Print Service

- *muffin* returns the necessary URL (printer://dukprunz/j1j25ps) to the mobile client.
- Finally, the client is able to access the printer by specifying the correct destination printer address along with the "lpr" command (on our OS/2 and AIX systems).

Notice in this case, the URN is delivered (possibly by use of intervening Internet routers) by unicast to muffin.watson.ibm.com, and no resolution path needs to be specified.

Similarly, we can mount a local NFS filesystem by sending the URN query for a local NFS filesystem, and use the return NFS URL to mount the filesystem. Accessing the NFS filesystem is a bit more work because we have to set some environment variables such as (for OS/2) the PATH, LIBPATH, etc. before the filesystem can be conveniently accessed by common applications. We do this by calling an initialization routine from the mounted filesystem. This is necessary because different filesystems require different initialization. Unfortunately, the environment variable initialization can only affect the newly created shells. Currently, there is no mechanism available to change the environment variable of existing processes – and changing environment variables would be a very tricky and error-prone operation in any case.

There is a little twist in defining the locality of mobile host running mobile IP. Using the local keyword will not work since the mobile host resides in its own virtual network, and is not in the same subnet as the local resources. We assume that the local resources are attached to stationary wired network. The only hint we have about the locality of the

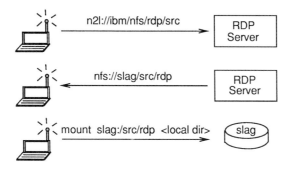

Figure 3 Mounting a Conventionally named NFS Filesystem

mobile host is gleaned from the default routing entry set by the mobile-IP daemon. This entry shows the association of the mobile host to its foreign agent at the wired network. Thus, in order to locate local resources, we have to include a local router (for instance, the care-of address of the foreign agent) in the query keywords.

7 COMPARISON

There has been a lot of work on resource discovery on the Internet. Various tools have been devised for collecting resource information. The tools were created to fill certain needs, and they offer valuable services to the Internet community. In this section, we compare RDP to some existing resource discovery tools. This is not meant to be an exhaustive comparison, but just to give some idea where RDP fits.

7.1 Web Search Engines

Perhaps the most commonly accessed resources in the Web are the search engines (C. Mic Bowman and Peter B. Danzig and Darrenn R. Hardy and Udi Manber and Michael F. Schwartz and Duane P. Wessels (1994), Oliver A. McBryan (1994), Brian Pinkerton (1994)). They collect information from the World Wide Web and condense it into fast searchable indexes. A typical user would enter some keywords and some search criteria into the search engines, and would receive in return a list all the Web resources satisfying the search parameters.

The operation of RDP is similar to some extent, except that RDP server doesn't collect information from the Web. Instead, the client has to explicitly register the URL to the appropriate server as specified by its naming authority. This dynamic registration (and deregistration) is required because RDP will have to deal with a larger, and currently unknown, set of resources, currently including printers and filesystems. For such active resources, the information cannot be collected by traversing the Web. Also, there is a need to update the resource attributes for active resources on demand. Moreover, an RDP server can offer access to resources which are different in kind than what is usually considered to reside on the Web. For instance, a fax machine isn't usually indicated by a button on someone's home page.

For passive resources, the information can be collected manually or using some of the existing collection techniques, and be registered to the RDP database. The RDP database is distributed and partitioned based on the naming authority. The naming authority is usually determined by the organizational entity. Each RDP database is localized, and mainly contains information about its local resources. This distribution is required for RDP scalability.

7.2 Connectionless Lightweight Directory Access Protocol (CLDAP)

CLDAP (A. Young (1995)) is a UDP-based light-weight version of its LDAP and DAP counterparts. The main motivation of CLDAP is to provide access to the X.500 directory service without incurring the full cost of DAP. So, CLDAP functions as a light-weight X.500 directory service front-end for simple applications.

CLDAP is based directly on LDAP, with the differences that CLDAP uses UDP and has a restricted set of operations. CLDAP clients should use a retry mechanism with timeout in order to achieve the desired level of reliability. Only one request may be sent in a single datagram, and only one response may be sent in a single datagram.

While similar to RDP in many ways in its use of UDP, some differences are noted here. CLDAP uses X.500 attribute encoding, whereas RDP uses URN and URL encoding. CLDAP uses X.500 DS, whereas RDP currently uses its own simpler database subsystem.

7.3 Service Location Protocol

The service location protocol (SLP, John Veizades and Scott Kaplan and Erik Guttman (1995)) provides a framework for the discovery and selection of local network services. RDP provides similar facility for discovering resources anywhere on the internet. SLP relies on the multicast support at the network layer; it uses multicast request with unicast response. RDP only relies on UDP, and uses unicast request and response.

Both SLP and RDP allow dynamic registration and deregistration of resources. Currently RDP doesn't have the request message for returning the URL description information equivalent to the attribute request in SLP. This feature will be added in the future to enable browsing the resources.

In terms of data encoding, SLP uses character strings represented as character strings are represented as a type,length,value tuple; and RDP uses character strings conforming to URI (T. Berners-Lee (1994)) specification. For constructing the query, SLP defines its own predicate language which is based on attributes. RDP, on the other hand, uses keywords embedded in the URN query which is far more flexible, and extensible.

Lastly, the SCOPE mechanism used in SLP is meant for use on a single LAN; whereas the naming authority in RDP will scale to a larger network. The use of URN in a RDP query also enable clients to directly specify the resolution path in the query.

We have begun to work with the IETF working group to make modifications to the existing Service Location Protocol. Some of the ideas of RDP have already been assimilated into newer Internet drafts from the working group, having the effect of narrowing the differences between our approaches. We expect the differences between our approaches to eventually disappear.

8 FUTURE WORK

The future surely holds great promise for the development of protocols to automatically locate network resources. In many ways, we have only begun to scratch the surface.

The most immediate direction for further work is to define more resource schemes. Many are possible – for instance, we could use RDP with:

- a white-pages service
- a front end for X.500 queries
- Application-specific libraries
- Local parts databases

Multiple RDP servers should be able to collaborate to provide to their clients the advantage of their combined resource databases. Note that such server-server protocols are difficult at best, and we do not believe that RDP will be an easy case. When RDP servers can cooperate, a query presented to one server might easily be forwarded by that server to another collaborative server and be resolved without undue delay to the client. The manner in which queries are presented to the RDP servers may change whenever a preferred format for URNs is defined within the IETF.

We have not designed any security mechanisms for RDP. Security is not even available with DHCP yet; when the world gets to the point of needing security bad enough with DHCP, we expect to be able to import whatever DHCP has into our work. Access controls may have to be put into place which prevent the discovery of the resources by unauthorized clients. Enabling the secure and perhaps even confidential registration and deregistration of network services is of particular importance. For this purpose, we may reasonably employ recent RFCs for authentication and encryption (Randall Atkinson (1995, 1996)).

Normally, it is expected that access control for most resources will be managed by the agent located by the URL, and not necessarily by the resource discovery server itself.

Applications which rely on environment variables for network configuration options cannot always be expected to work well with Resource Discovery. In order to improve them, we would require operating system support to enable a global change of environment variables for running processes. That seems quite unlikely to happen in the near future.

We expect to specify an applications program interface (API) that would enable applications to easily make use of our protocol. Moreover, this interface should allow applications to do the right thing when their previously obtained URLs for necessary resources are likely to have become invalid.

Last, but not least, we would like to investigate the problems of resource discovery in ad-hoc environments. The convenience of resource discovery may cause mobile nodes (and their applications) to come to rely on resource discovery in enterprise environments rich in computer resources. Applications may be written that rely on the convenience of registering arbitrary and privately named resources at the local Resource Discovery Server. This very powerful feature should then be also available when there isn't any infrastructure, and one of the mobile nodes has to be elected to perform the resource discovery function for the local population of ad-hoc mobile stations. The exact mechanisms by which this might occur will be a fascinating research area.

9 CONCLUSIONS

Obtaining a Resource Discovery Server from DHCP fits naturally within the framework provided by mobile-IP and DHCP. When the mobile node needs to access a network resource in a particular area, it no longer needs to rely on static configuration data that can be valid in only one location. Instead, it contacts the local server to fulfill its needs for the location of nearby resources.

By using keywords, our Resource Discovery Protocol provides a flexible way to get access to dynamic resource information. Since we can only start to categorize and name the kinds of resources that are available on the network, we should avoid any premature attempt to place the resource names into a rigidly controlled system using registered numbers, or sequences of *attribute = value* pairs.

Allowing the dynamic and essentially uncontrolled registration of resources with a local Resource Discovery Protocol server may have big implications for the future extensibility of our protocol. We look forward to experimenting with such facilities, and note here that it would be much more difficult to design such systems if unnecessary structure were placed on the format of the resource specification (and selection) packets.

We feel strongly that URL strings should be used as the resource discovery medium for communicating resource location information, since that is their precise purpose. We believe that it is important to take advantage of existing work contributing to the dominant success of the World-Wide Web. Thus we claim that it is also important to take whatever lessons we can take from existing work on naming resources within the Web, and consequently our design of the resource query was modeled after URNs.

It is also important to attend to details like minimizing the number of broadcast packets, scalability, protocol complexity, and implementability. We feel that our approach is a successful attempt to balance the needs and engineering tradeoffs required for the protocol, and our ability to produce a working system within three months after starting is a solid indication that our protocol can indeed be successfully implemented and made available for use within the Internet.

10 ACKNOWLEDGEMENT

Thanks to Erik Guttman for stimulating conversations about RDP and the IETF protocol, and for his willingness to incorporate some of our ideas into the IETF draft specification for the Service Location Protocol.

REFERENCES

S. Alexander and R. Droms. DHCP Options and BOOTP Vendor Extensions. RFC 1533, October 1993.

T. Berners-Lee. Universal Resource Identifiers in WWW. RFC 1630, November 1994.

T. Berners-Lee, L. Masinter, and M. McCahill. Uniform Resource Locators (URL). RFC 1738, December 1994.

C. Mic Bowman, Peter B. Danzig, Darrenn R. Hardy, Udi Manber, Michael F. Schwartz, and Duane P. Wessels. Harvest: A Scalable, Customizable Discovery and Access System. Technical report, University of Colorado, Computer Science Department, University of Colorado, Boulder, August 1994. CU-CS-732-94, revised March 1995.

Douglas E. Comer. *Principles, Protocols, and Architecture*, volume 1 of *Internetworking with TCP/IP*. Prentice Hall, Englewood Cliffs, N.J., second edition, 1991.

R. Droms. Dynamic Host Configuration Protocol. RFC 1541, October 1993.

IETF Mobile-IP Working Group. IPv4 Mobility Support. ietf-draft-mobileip-protocol-12.txt - work in progress, September 1995.

Paul E. Hoffman and Jr Ron Daniel. Generic URN Syntax. draft-ietf-uri-urn-syntax-00.txt, April 1995.

Mark Madsen. A Critique of Existing URN Proposals. draft-ietf-uri-urn-madsen-critique-00.txt, July 1995.

Udi Manber and Sun Wu. GLIMPSE: A Tool to Search Through Entire File Systems. Technical report, University of Arizona, Computer Science Department, Tucson, Arizona, October 1993. TR 93-34.

Oliver A. McBryan. GENVL and WWWW: Tools for Taming the Web. In *Proceedings of the First International World Wide Web Conference*, CERN, Geneva, May 1994.

Charles Perkins and Jagannadh Tangirala. DHCP for Mobile Networking with TCP/IP. In *Proceedings of IEEE International Symposium on Systems and Communications*, pages 255–261, June 1995.

Charles E. Perkins. DHCP Home Address Option. draft-perkins-homeaddr-dhcpopt-01.txt - work in progress, October 1995.

Brian Pinkerton. Finding What People Want: Experiences with the WebCrawler. In *Proceedings of the Second International WWW Conference '94: Mosaic and the Web*, Chicago, October 1994.

K. Shafer, E. Miller, V. Tkac, and S. Weibel. URN Services. draft-shafer-uri-urn-resolution-00.txt - work in progress, June 1995.

K. Sollins and L. Masinter. Functional Requirements for Uniform Resource Names. RFC 1737, December 1994.

John Veizades, Scott Kaplan, and Erik Guttman. Service Location Protocol. draft-ietf-svrloc-protocol-06.txt - work in progress, July 1995.

Sun Wu and Udi Manber. A Fast Algorithm for Multi-pattern Searching. Technical report, University of Arizona, May 1994. TR 94-17.

A. Young. Connection-less Lightweight Directory Access Protocol. RFC 1798, June 1995.

11 BIOGRAPHY

Harry Harjono (*harjono@cs.columbia.edu*) is a PhD student at the Columbia University Computer Science Department where he works as a graduate research assistant at the Mobile Computing Lab under the supervision of Dr. Dan Duchamp, his academic and research advisor. His research interests include dynamic resource discovery, mobile computing, and distributed operating systems. He received an M.S. in Computer Science from Columbia University in 1994.

Charles Perkins (*perk@watson.ibm.com*) is a research staff member at IBM T.J. Watson Research, investigating mobile and ad-hoc networking, resource discovery, and automatic configuration for mobile computers. He is serving as the document editor for the mobile-IP working group of the Internet Engineering Task Force (IETF), and serves on the editorial boards for ACM/IEEE Transactions on Networking, ACM Wireless Networks, and IEEE Personal Communications. Charles holds a B.A. in mathematics and a M.E.E. degree from Rice University, and a M.A. in mathematics from Columbia University.

TRACK 2: TRUSTING IN TECHNOLOGY, AUTHENTICATION, SECURITY

23
The Future of Smart Cards : Technology and Application

Vincent Cordonnier
Professor - University of Lille - France
RD2P - CHR CALMETTE
59037 LILLE Cedex
Chairman of the IFIP Special Group on Smart Cards

1 INTRODUCTION

For the twenty last years, the smart card market has grown from nothing to approximately a billion cards a year for 1996. Many applications have been identified as typical areas of that technology, mostly electronic money and identification of individuals. However, for many service providers and application designers, the smart card domain is still hazardous and not perfectly well identified in terms of technology and capabilities. The ratio of project failures is extremely high and many existing applications do not fit with their initial definition. This remark leads to give the greatest importance to the identification of that technology and its existing and potential characteristics and market. Then, it becomes easier to look at the future as the convergence between technological improvements and innovative applications requirements.

2 AN OVERVIEW OF THE SMART CARD STATUS

A smart card can be roughly identified as a portable, secure and intelligent memory. As a portable device it is almost always related to an individual and gives this individual access facilities to various services when moving around. Security is provided by both temper resistance and encryption of stored and exchanged data. Intelligence appears in both an efficient management of stored data and the capability to adapt the activity to the identified reader or terminal or external partner. Whatever can be the embarked application, a card usually behaves as a passive component. It just has to react to questions, commands or access requirements which are delivered by the terminal.

From the technological point of view, a card is composed of a data memory, a control unit which can either be a logical circuit or a microprocessor. In that case, the program to be executed is permanently stored in a ROM memory and cannot be altered. Two other components, the input-output module and the physical security module must be added to realise a complete device.

Typical characteristics are 1 to 8 kilobytes for the data memory, 1 to 16 kilobytes for the ROM memory, 9,600 bits per second for the serial communication link and a 8 bit microprocessor with a 4 MHz clock provided by the terminal. These limits will not grow faster than the average silicon improvement for the five next years.

From the application point of view, the success has led to an impressive number of services covering the international level (credit cards) and the local level (cities, companies, universities). In every case, a card can be seen as a replacement tool for various portable devices such as identity papers, purses, portable data files. It is more reliable, more secure and can support automatic exchanges with the service provider.

3 THE FUTURE OF TECHNOLOGY

There are many possible tracks for improving the flexibility and intelligence of smart cards. They will appear at the component level, the software level and the functional level. Then many innovative applications will be made possible. Actually, it seems more logical to first look at the requirements of these possible new applications and then to design smart cards of the future. As for many other technologies, the final result will probably by issued from a convergence of these two complementary approaches.

3.1 The component level

As the surface of the chip is restricted to 25 mm2, most of the improvements will not result from drastically increasing the number of gates. For the microprocessor itself, a RISC approach will probably offer a better ratio between performances, complexity and electrical consumption. Some manufacturers prepare a 32 bits approach (Cascade project). The resulting power will serve security for high level algorithms, protocols and biometrics. Another important issue is the communication link which will require higher rates and propose possibilities for contactless accesses. Eventually, it seems interesting to partly merge the ROM and the EPROM memories into a unique memory space. Dedicated hardware for fast encryption-decryption algorithms is another interresting issue.

3.2 The software level

So far smart cards manufacturers and application issuers have decided to write in a single ROM code, the application itself and the basic functions of the operating system. Such a software architecture is interesting because of security and memory space saving. As many people now think of multi application cards, it will become necessary to clearly separate these to layers.

Operating systems for smart cards will become a major issue for the next generation. Obviously, this operating system will be characterised by a few properties which are typical of the smart card working environment:

- A fixed and dedicated memory,
- A single communication link,
- A unique and simple file and data management,

- A high level of security against any possible fraud, and, for the most advanced
- Capability for multi application, multi services and multi users.

3.3 The functional level

As previously mentioned, smart cards are mostly considered as memories and the microprocessor is mostly used to bring the most secure and intelligent management of that memory including communication with external local or remote devices. The ultimate example of this memory management approach is the CQL card which works as a relational data base with a great flexibility. There are many ways to give a card more capabilities and to use the microprocessor and its operating system for other purposes.

Code importation looks like a promising track. Assuming that a limited memory space does not allow to store many programs, it becomes interesting to make the card capable of directly obtaining the program it must execute from the terminal. Code importation is the most promising approach for the universal card.

Active cards represent another innovative possibility. By processing an external request and its stored data, the card can produce a new information to be presented to the terminal. Such a function is an excellent mean to save privacy. Eventually, the active card will execute more processing to provide the convenient response regarding to its content, the status of the partner and the security level of the transaction.

Biometry, including expert system, will take place in smart cards. Because the presently used PIN number is known as not secure enough and easy to fraud, a better identification technology become mandatory. Prototypes which analyse the voice, recognise fingerprints or the shape of the hand already exist.

4 THE FUTURE OF APPLICATIONS

For many years, application designers hesitated between a centralised solution using a network to access a server and a decentralised one using distributed cards. Now, everybody agrees that networks and cards are totally complementary. Internet will certainly needs smart cards to authentify secured accesses as, for example to manage money transfers or to protect private communications; pay-tv channels needs smart cards to personalise an access. But, potentially, the most powerful network is the mobile telephone. Smart cards have been closely associated to GSM to support identification and auxiliary services. Smart card also use the GSM facilities to access many other applications. A key to the future is probably to look at the smart card as the natural extremity of any secured communication network.

5 CONCLUSION

The estimated production before the end of the century is more than four billion smart cards a year. The second generation is still to be designed and validated. Manufacturers and major application issuers will promote new technologies if and only if they bring a possibility to extend the market and to collect more and more transactions. Research on smart cards will propose a lot of possible improvements. They will only be accepted if they offer a good balance between quality of services, flexibility and security.

Industrial State-of-the-Art Report

24
Use of smart cards for security applications by Deutsche Telekom

Bernd Kowalski
Deutsche Telekom AG, PZ Telesec
Untere Industriestr. 20, 57250 Netphen, Germany,
telephone: +49 271 708-1600, fax: +49 271 708-1625
X.400email: kowalski@04.sgn.telekom400.dbp.de

Abstract

Smart cards will be a key technology for secure electronic commerce and electronic payment applications on the Internet. They offer the unique advantage to keep cryptographic mechanisms securely in a tamper-proof equipment. Smart cards will be used for access control instead of passwords, for the generation of digital signatures, for encryption/decryption, as an electronic purse and as a repository of any confidential information.

Several years ago smart cards have mostly been used at card telephones and for banking applications and there usage was restricted to dedicated terminals like public telephones and ATMs. With the introduction of the health insurance card in Germany the first multi-application smart card terminals for PCs appeared on the market. Since 1993 prices for PC smart card interfaces dropped dramatically and made the card terminal a standard device for PC-hardware at least in Germany. In the following months the number of card applications for PCs and networks grows rapidly.

Already some time ago Deutsche Telekom recognised that Information Security will become a key issue for telecommunication companies. Telekom also decided that a big company`s security should be independant from specific software or hardware suppliers. This strategy, however, required compatibility of security devices and applications between different products. This proved to be very difficult because of the lack of security standards and appropriate products on the market.

A Security Platform has been developped consisting of a number of security components like a smart card, a smart card terminal, a Trusted Third Party for keymanagement, a high speed PCMCIA encryption card, various authentication and encryption terminals and a security management system.

The smart card plays a key role in Telekom`s information security strategy. The card includes a public key calculation unit for RSA, symmetric cryptography and a session key generator.

In 1995 Telekom started to issue smart cards for all its employees. Today more than 100.000 of them have access to information processing and all of them will use the card for digital signature, encryption and access control shortly.

Besides information security Telekom`s „Company Card" is also used for physical access to buildings and rooms, flexitime, electronic purse and electronic railway tickets. This makes the smart card very cost-efficient. On the other hand the card solves the problem of managing user passwords accross different applications in a distributed and interconnected environment.

Several products will be offered to Telekom`s customers with the smart card`s security features. One important project is a country-wide intranet backbone. The network services are protected by a user smart card that provides for authentication and encryption between client and server. Applications use the smart card for signing electronic documents or to encrypt confidential messages. Keymanagement and card personalisation are performed by a Trusted Third Party and the network`s directory service.

Deutsche Telekom has also developped a smart card solution to protect an ordinary analogue local loop. Also the operation of the management system of the digital exchanges in the telephone network (TMN) is protected by smart cards for operators.

Keywords

Access control, authentication, company card, cryptography, digital signature, encryption, public key cryptography, smart cards, Trusted Third Parties

PART ONE

Protocols for Authentication, Secure Communication and Payment

25

An Authentication and Security Protocol for Mobile Computing

Yuliang Zheng
Monash University
McMahons Road, Frankston, Melbourne, VIC 3199, Australia
Phone: +61 3 9904 4196, Fax: +61 3 9904 4124
Email: yzheng@fcit.monash.edu.au

Abstract
The main contributions of this paper are: (1) to analyze an authentication and key distribution protocol for mobile computing proposed by Beller, Chang and Yacobi in 1993, and reveal two problems associated with their protocol. (2) to propose a new authentication and key distribution protocol that utilizes a broadcast channel in a mobile network. A particularly interesting feature of the new proposal is that it allows the authentication of a base station by a mobile user to be conducted "at the background", which yields a very compact protocol whose total number of moves of information between a mobile user and a base station is only 1.5 !

Keywords
Authentication, Cryptography, Key Distribution, Mobile Computing, Security

1 SECURITY ISSUES IN WIRELESS NETWORKS

Recent years have seen an explosive growth of interest in wireless (information) networks that support the mobility of users (and terminals). These networks serve as a foundation of future universal, mobile and ubiquitous personal communications systems.

Emerging wireless networks share many common characteristics with traditional wire-line networks such as public switched telephone/data networks, and hence many security issues with wire-line networks also apply to the wireless environment. Nevertheless, the mobility of users, the transmission of signals through open-air and the requirement of low power consumption by a mobile user bring to a wireless network with a large number of features distinctively different from those seen in a wire-line network. Especially, security and privacy becomes more eminent with wireless networks. To this end, we will be primarily concerned with security issues related to or caused by the mobility of users/terminals, open-air transmission of signals and low power supply of a mobile user.

250   *Part One   Authentication, Secure Communication and Payment*

When examining security in a wireless network, a range of issues have to be taken into account. These issues include:

1. identification of a mobile user
2. anonymity of a mobile user (protection of identity)
3. authentication of a base station
4. security of information flowing between a mobile user and a base station
5. prevention of attacks from within a base station
6. hand-over of authentication information
7. the communication cost of establishing a session key between a mobile user and a base station, which is indicated primarily by the total number and length of messages to be exchanged
8. the cost of communications between a mobile user's home domain and a foreign domain where he is currently located, as well as security requirements on the communication links between the two domains
9. the computational complexity of achieving authenticity and security
10. the complexity of computations to be carried out by a mobile user's terminal which is in general much less powerful than a base station

Some issues contradict one another. For instance, to prevent mobile network resources from being abused by a fraudulent user, a network relies on the identification/authentication of a mobile user, which generally requires the user to reveal his or her identity. On the other hand, however, a user who wishes to make anonymous communications may be unwilling to reveal his or her identity. Two recent articles (Brown 1995, Wilkes 1995) survey in details many issues related to security and privacy in mobile networks.

In this extended summary, we assume that the reader is familiar with basic concepts in cryptography, including digital signature, public-key and private-key encryption systems.

2   PREVIOUS PROPOSALS

A concise summary of the authentication and security protocol employed by the global system for mobile telecommunication or GSM (Rahnema 1993) can be found in (Brown 1995). A description of the proposed security and privacy mechanism used in the cellular digital packet data (CDPD) in the US is provided in (Frankel, Herzberg, Karger, Krawczyk, Kunzinger & Yung 1995), where potential threats and attacks to the mechanism, together with possible solutions, are also discussed. Other notable works include (Beller, Chang & Yacobi 1993), (Aziz & Diffie 1994), (Molva, Samfat & Tsudik 1994), and more recently, (Herzberg, Krawczyk & Tsudik 1994), (Asokan 1994) and (Samfat, Molva & Asokan 1995). In the full version of this paper, an outline of each of these protocols, together with a comprehensive comparison of various aspects of these protocols, will be described.

2.1   Beller-Chang-Yacobi Protocol

Among the protocols mentioned above, the one presented in (Beller et al. 1993) deserves special attention, as it represents one of the earliest solutions employing a combination of both private-key and public-key encryption algorithms. The protocol is based on two computationally infeasible problems: factorization and discrete logarithm. (For this reason, Beller, Chang and Yacobi call their proposal *MSR+DH protocol.*)

As Beller-Chang-Yacobi protocol is partially based on the Diffie-Hellman public key distribution scheme (Diffie & Hellman 1976), it uses a large prime N and a generator α for the multiplicative group $GF(N)^*$. (Note: in their exposition, Beller, Chang and Yacobi also consider a more general case where N can be the product of two large primes.) Both N and α are public. A mobile user m has a pair of public-secret keys (P_m, S_m), where $P_m \equiv \alpha^{S_m} \pmod{N}$. Similarly a base station b too has a pair of public-secret keys (P_b, S_b), where $P_b \equiv \alpha^{S_b} \pmod{N}$.

As in many security solutions, the protocol further requires the existence of a trusted certification authority ca who issues a certificate to each mobile user as well as each base station to certify their public keys. (See for instance (Chokhani 1994) for discussions on certification services.) The core part of a certificate issued by the certification authority ca "o the mobile user m is a digital signature defined by

$$sig_{ca,m} \equiv \sqrt{h(m, P_m)} \pmod{N_{ca}}$$

and similarly, the core part of a certificate issued to the base station b is defined by

$$sig_{ca,b} \equiv \sqrt{h(b, N_b, P_b)} \pmod{N_{ca}}$$

where h is a one-way hash function known to the public, P_m and P_b are the public keys of the mobile user m and the base station b respectively, N_b is the product of two large primes associated with the base station b, and similarly, N_{ca} is the product of two large primes associated with the certification authority ca. While N_b and N_{ca} are made public, their prime factors must be kept secret by their respective owners.

Finally a private key encryption algorithm is used in the protocol. One may choose a private key encryption algorithm from a large set of potential candidates, including DES (National Bureau of Standards 1977), IDEA (Lai 1992), RC5 (Lai 1992) and SPEED (Zheng 1996). In the following discussions, we assume that a private key encryption algorithm has been selected, and denote by $Encrypt_K(M)$ the encryption of a message M under a key K, and by $Decrypt_K(C)$ the decryption of a ciphertext C under K.

As indicated below, Beller-Chang-Yacobi protocol consists of five (5) moves (or steps) of information between a mobile user and a base station.

Beller-Chang-Yacobi Protocol

1. Mobile User $m \Longrightarrow$ Base Station b
   a message indicating a request for services.
2. Mobile User $m \Longleftarrow$ Base Station b
   Upon receiving the request from m, b sends to m four numbers:

   $$(b, N_b, P_b, sig_{ca,b})$$

   where $P_b \equiv \alpha^{S_b} \pmod{N}$ and $sig_{ca,b} \equiv \sqrt{h(b, N_b, P_b)} \pmod{N_{ca}}$ is the digital signature issued to b by the trusted certification authority ca. h, α, N and N_{ca} are all public parameters.
3. Mobile User $m \Longrightarrow$ Base Station b
   Upon receiving $(b, N_b, P_b, sig_{ca,b})$ from b, m checks if

   $$h(b, N_b, P_b) \equiv sig_{ca,b}^2 \pmod{N_{ca}}$$

   The protocol is aborted if the numbers fail to pass the check.

Otherwise, m sends to b two numbers e_2 and e_3 defined by

$$e_2 \equiv x^2 \pmod{N_b}$$
$$e_3 = Encrypt_x(m, P_m, sig_{ca,m})$$

where x is a number chosen uniformly at random from between 1 and $N_b - 1$, $P_m \equiv \alpha^{S_m} \pmod{N}$ is the public key of m and $sig_{ca,m} \equiv \sqrt{h(m, P_m)} \pmod{N_{ca}}$ is the signature issued to m by the certification authority ca.

Upon receiving (e_2, e_3) from the mobile user m, the base station b extracts x from e_2:

$$x \equiv \sqrt{e_2} \pmod{N_b}$$

by the use of the two secret prime factors of N_b. b then uses x to decrypt e_3:

$$(m, P_m, sig_{ca,m}) = Decrypt_x(e_3)$$

and checks whether

$$h(m, P_m) \equiv sig_{ca,m}^2 \pmod{N_{ca}}$$

The protocol is aborted if m, P_m and $sig_{ca,m}$ fail to pass the test.

4.&5. Mobile User $m \iff$ Base Station b

Now the mobile user m can calculate $\eta \equiv P_b^{S_m} \pmod{N}$ and $sk = Encrypt_\eta(x)$. Symmetrically the base station b can calculate $\eta \equiv P_m^{S_b} \pmod{N}$ and $sk = Encrypt_\eta(x)$.

To confirm that they have the same session key sk, m and b exchange two known messages which are encrypted under sk. For instance, m can send $Encrypt_{sk}(m)$ to b in exchange of $Encrypt_{sk}(b)$ from b. If the messages are decrypted correctly, sk becomes an authentic session key between m and b.

Note that in practice, a mobile user's computations are all carried out by his personal smart card and/or mobile terminal.

2.2 Problems with Beller-Chang-Yacobi Protocol

After a close examination, we have identified two problems with Beller-Chang-Yacobi protocol. The first problem is related to the inefficiency of the protocol, while the second is concerned with replay-attacks that can be mounted against the protocol.

As the protocol is based on public-key cryptography, an attacker can obtain the public key, as well as its associated digital signature, of a mobile user m, namely $(m, P_m, sig_{ca,m})$. The attacker can then impersonate the mobile user and initiate the protocol with a base station b. The attacker will have no problems in successfully carrying out all the operations involved in the first three (3) moves of the protocol, including passing the test by the base station b in the third move of the protocol. Therefore, the fourth and fifth moves of the protocol must be executed in order for a base station and a genuine mobile user to confirm the consistency of their session keys. Such a 5-move protocol may be inefficient for applications in mobile computing.

The next problem has more serious consequences. Consider an attacker who is malicious towards a mobile user m. The attacker may record communications, including the five moves in Beller-Chang-Yacobi protocol, between the mobile user and a base station b. Some time after m and b complete their communication session, the attacker can initiate a communication with the base station b by replaying messages previously sent to b by m in Beller-Chang-Yacobi protocol. Clearly the messages will pass all the tests by the base station b, which results in the attacker being successful in impersonating the mobile user m. Now the attacker may be able to transmit, using the name of the mobile user m, an arbitrarily long, but perhaps random and meaningless, message to an arbitrary third user, even though it is computationally infeasible for the attacker to find out the session key sk. This could result in the mobile user m being charged a large amount of money for a communication he never conducted !

3 A NEW PROPOSAL

This section proposes an authentication and key distribution protocol based on a broadcast channel in a mobile network. This protocol is remarkably simple: it consists of only 1.5 moves.

3.1 Certification Authority

As in the case of Beller-Chang-Yacobi protocol, we assume that a mobile network involves a trusted certification authority ca which provides participants of the network, including mobile users and base stations, with public key certification services.

We further assume that the certification authority employs DSS or Digital Signature Standard (National Institute of Standards and Technology 1994). An equally good candidate is a digital signature scheme by Schnorr (Schnorr 1991). The two signature schemes are closely related to each other, and both are based on discrete logarithm over a finite field.

DSS involves three public parameters (p, q, g), where

1. p is a large prime.
2. q is a (large) prime factor of $p-1$.
3. $g \equiv h^{(p-1)/q} \pmod{p}$ with h being an integer satisfying $1 < h < p-1$ and $h^{(p-1)/q} \mod p > 1$. Note that g is also said to have order $q \mod p$.

These three parameters are known to all network participants. In addition, DSS requires each user to have a pair of public-secret keys. In particular, the pair of public-secret keys of the certification authority ca is (y_{ca}, x_{ca}), where $y_{ca} \equiv g^{x_{ca}} \pmod{p}$ and x_{ca} is a secret number chosen randomly from $[1, q-1]$.

The pair of public-secret keys of a mobile user m, denoted by (y_m, x_m), and that of a base station b, denoted by (y_b, x_b), are defined in a similar way.

Now the certification authority ca can use DSS to create a certificate for a participant, say a base station b, by digitally signing on a message M using x_{ca}, where M may contain such information as certificate serial number, validity period, the ID of b, the public key of b, the ID of ca, the public key of ca, etc. (See (Chokhani 1994) and (ITU 1993) for a proposed standard format of a certificate.) The digital signature of ca on M is composed of two numbers r and s which are defined as

$$r \equiv (g^k \mod p) \pmod{q}$$
$$s \equiv (h(M) + x_{ca} \cdot r)/k \pmod{q}$$

where k is a random number chosen from $[1, q-1]$, and h is a one-way hash function. NIST specifies SHS (National Institute of Standards and Technology 1995) as the one-way hash function used in DSS.

Given (M^*, r^*, s^*), one can verify whether (r^*, s^*) is indeed a genuine signature of the certification authority on M^* by the following steps:

1. calculates $v \equiv (g^{h(M^*)/s^*} \cdot y_{ca}^{r^*/s^*} \mod p) \pmod{q}$.
2. accepts (r^*, s^*) as a genuine signature of ca on M^* only if $v = r^*$.

3.2 Making Use of a Broadcast Channel

Typically a mobile network uses a broadcast channel to continuously propagate from a base station to mobile users various types of control information such as synchronization parameters, available services, network time data, base station ID etc. The authentication protocol to be proposed in the following uses part of the capacity of the broadcast channel for a base station to propagate to mobile users the certificate associated with its public key. For simplicity, we assume that the certificate takes the form of

$$cert_{ca,b} = (b, y_b, sig_{ca,b})$$

where y_b is the base station's public key, and $sig_{ca,b}$ is the certification authority's digital signature on (b, y_b) created using the DSS scheme. (More information on the format of a certificate can be found in (Chokhani 1994) and (ITU 1993).)

To keep himself abreast of the various types of network information such as synchronization data, types of services, current network time, and the public key and certificate of a base station, a mobile user (through his mobile terminal) continuously monitors the broadcast channel. In ¨oing so, it will be able to check whether $sig_{ca,b}$ is indeed a genuine signature of the certification authority ca on the base station's public information b and y_b, and hence to authenticate the base station "at the background". For this reason, we say that this process contributes 0.5 move to the protocol.

3.3 Key Distribution and Authentication of a Mobile User

As discussed above, a mobile user can authenticate a base station "at the background". In particular, this process can be completed immediately after his mobile terminal is switched on, or he roams into a new cell, without being noticed by the mobile user.

Now we assume that the mobile user m is in the cell covered by a base station b, has successfully authenticated the base station "at the background", and wishes to initiate a communication session. The mobile user m sends two data items c_1 and c_2 to the base station. Here c_1 and c_2 are constructed in the following way:

$$c_1 \equiv g^x \pmod{p}$$
$$c_2 = G(y_b^x \mod p) \oplus (K, T, cert_{ca,m}, tag)$$

where $tag = h(K, T, cert_{ca,m}, y_b^{x_m + x} \mod p)$.

The meanings of other symbols used in c_1 and c_2 are as follows: x is a random number from $[1, p-1]$, $\oplus$ denote bit-wise exclusive-or, K is a random session key, both chosen by the mobile user m, x_m is the secret key, $cert_{ca,m} = (m, y_m, sig_{ca,m})$ is the certificate and y_m is the public key of m, while y_b is the public key of the base station, T is the current network time stamp taken from the base station's broadcast channel, G is a cryptographically strong pseudo-random number generator, and finally h is a

Authentication and security protocol for mobile computing

one-way hashing function such as SHS or HAVAL (Zheng, Pieprzyk & Seberry 1993). As is the case for $cert_{ca,b}$, more information may be included in the certificate $cert_{ca,m}$.

Note that the involvement of T, the current network time stamp taken from the base station's broadcast channel, is to ensure the freshness of the message. Also note that the main ideas behind the formation of (c_1, c_2) are from (Zheng & Seberry 1993), where three practical public key cryptosystems have been designed to resist against chosen ciphertext attacks.

Upon receiving (c_1^*, c_2^*) which may differ from (c_1, c_2), the base station calculates $G((c_1^*)^{x_b} \bmod p) \oplus c_2^*$, and splits the result into four parts

$$K^*, T^*, cert_{ca,m}^*, tag^*$$

Here $cert_{ca,m}^*$ consists of $(m^*, y_m^*, sig_{ca,m}^*)$.

The base station b then verifies the certificate $cert_{ca,m}^*$ and also checks the freshness of T^*. It aborts the protocol if either $cert_{ca,m}^*$ is invalid or T^* deviates too far from the current network time. Otherwise, if both $cert_{ca,m}^*$ and T^* are OK, the base station performs the hashing operation

$$d = h(K^*, T^*, cert_{ca,m}^*, (y_m^* \cdot c_1)^{x_b} \bmod p)$$

The base station is convinced of the identity of the mobile user and accepts K^* as a valid common key shared with the mobile user only if $tag^* = d$.

The new proposal is summarized in the following:

A 1.5 Move Protocol

1. Mobile User $m \Longleftarrow$ Base Station b
   The base station b broadcasts to mobile users

   $$cert_{ca,b} = (b, y_b, sig_{ca,b})$$

   where y_b is the base station's public key, and $sig_{ca,b}$ is the certification authority's digital signature on (b, y_b).
   The mobile user m monitors the broadcast channel and verifies, "at the background", the authenticity of the certificate and hence of the base station.

2. Mobile User $m \Longrightarrow$ Base Station b
   When the mobile user m wishes to initiate a communication session, he sends to the base station two data items c_1 and c_2 constructed by

   $$c_1 \equiv g^x \pmod{p}$$
   $$c_2 = G(y_b^x \bmod p) \oplus (K, T, cert_{ca,m}, tag)$$

   where $tag = h(K, T, cert_{ca,m}, y_b^{x_m + x} \bmod p)$. The meanings of other symbols used in c_1 and c_2 are: x is a random number in $[1, p-1]$, K is a random session key, both chosen by the mobile user m, x_m is the secret key, $cert_{ca,m} = (m, y_m, sig_{ca,m})$ is the certificate and y_m is the public key of m, while y_b is the public key of the base station, T is the current network time stamp taken from the base station's broadcast channel, G is a cryptographically strong pseudo-random number generator, and finally h is a one-way hashing function.
   Upon receiving (c_1^*, c_2^*) which may differ from (c_1, c_2), the base station b calculates $w = G((c_1^*)^{x_b} \bmod p) \oplus c_2^*$, and then splits w into four parts

   $$K^*, T^*, cert_{ca,m}^*, tag^*$$

where
$$cert^*_{ca,m} = (m^*, y^*_m, sig^*_{ca,m})$$
The base station then verifies the certificate $cert^*_{ca,m}$ and also checks the freshness of T^*. It aborts the protocol if either $cert^*_{ca,m}$ is invalid or T^* deviates too far from the current network time. Otherwise, if both $cert^*_{ca,m}$ and T^* are OK, the base station checks whether
$$tag^* = h(K^*, T^*, cert^*_{ca,m}, (y^*_m \cdot c_1)^{x_b} \bmod p)$$
The base station is convinced of the identity of the mobile user and accepts K^* (which should be identical to K) as a valid common key shared with the mobile user only if the above equation is satisfied.

Once a shared session key is established between the mobile user and the base station, they can use the session key together with a private-key (block or stream) cryptosystem to conduct secure communications.

4 REMARKS

The following observations on the new proposal can be made:

1. Due to the participation of the mobile user's secret key x_m in the formation of (c_1, c_2), the chance for an attacker to make a valid pair (c^*_1, c^*_2) is negligibly small, even if the attacker has the full knowledge of y_m and $cert_{ca,m}$. Hence successful completion of the protocol guarantees that $K^* = K$, namely, the mobile user and the base station have an identical shared key. Consequently, unlike Beller-Chang-Yacobi protocol, there is no need to confirm the consistency of the keys through the exchange of known messages.
2. As (c_1, c_2) employs network time information, replay attacks, such as the one applicable to Beller-Chang-Yacobi protocol, can be effectively thwarted by limiting the valid life span of (c_1, c_2), say, to a fraction of a second.
3. The proposed new protocol consists of 1.5 moves of information: 0.5 move for the authentication of a base station by a mobile user, and a single move for the authentication of the mobile user by the base station and the establishment of a session key.
4. The protocol provides anonymity of the mobile user with respect to an onlooker: as all messages exchanged between a mobile user and a base station, including the identity of the mobile user which is part of his certificate $cert_{ca,m}$, are transported in their encrypted form, an outsider or onlooker cannot figure out which mobile user is communicating with the base station.
5. The protocol prevents the impersonation of a mobile user by a fraudulent base station: the messages sent from a mobile user to a base station contains enough information for the base station to authenticate the mobile user, but not enough for a fraudulent base station to masquerade the mobile user. In fact the only entity who can create a correct pair of (c_1, c_2) is the mobile user who knows his secret key x_m. The reader is directed to (Zheng & Seberry 1993) for more information on the unforgability of (c_1, c_2).
6. The creation of (c_1, c_2) can be partially pre-computed before the mobile user wishes to start a communication session. This further shortens the time required to establish a session.
7. The protocol can also be applied to networks and distributed computing where broadcast channels present.

Currently we are in the process of conducting a detailed analysis of the proposed new protocol, covering time complexity of the protocol, strategies for pre-computation by a mobile user, roaming, the procedure for a visited base station to contact the "home network" of a mobile user and other issues.

REFERENCES

Asokan, N. (1994), Anonymity in a mobile computing environment, *in* 'Proceedings of 1994 IEEE Workshop on Mobile Computing Systems and Applications'.

Aziz, A. & Diffie, W. (1994), 'Privacy and authentication for wireless local area networks', *IEEE Personal Communications* **1**(1), 25–31.

Beller, M., Chang, L.-F. & Yacobi, Y. (1993), 'Privacy and authentication on a portable communications system', *IEEE Journal on Selected Areas in Communications* **11**(6), 821–829.

Brown, D. (1995), 'Techniques for privacy and authentication in personal communications systems', *IEEE Personal Communications* **2**(4), 6–10.

Chokhani, S. (1994), 'Toward a national public key infrastructure', *IEEE Communications Magazine* pp. 70–74.

Diffie, W. & Hellman, M. (1976), 'New directions in cryptography', *IEEE Transactions on Information Theory* **IT-22**(6), 472–492.

Frankel, Y., Herzberg, A., Karger, P., Krawczyk, H., Kunzinger, C. & Yung, M. (1995), 'Security issues in a CDPD wireless network', *IEEE Personal Communications* **2**(4), 16–27.

Herzberg, A., Krawczyk, H. & Tsudik, G. (1994), On travelling *incognito*, *in* 'Proceedings of 1994 IEEE Workshop on Mobile Computing Systems and Applications'.

ITU (1993), Information technology — open systems interconnection — the directory: Authentication framework, Recommendation X.509, International Telecommunications Union.

Lai, X. (1992), *On the Design and Security of Block Ciphers*, ETH Series in Information Processing, Hartung-Gorre Verlag Konstanz, Zürich.

Molva, R., Samfat, D. & Tsudik, G. (1994), 'Authentication of mobile users', *IEEE Network*.

National Bureau of Standards (1977), Data encryption standard, Federal Information Processing Standards Publication FIPS PUB 46, U.S. Department of Commerce.

National Institute of Standards and Technology (1994), Digital signature standard (DSS), Federal Information Processing Standards Publication FIPS PUB 186, U.S. Department of Commerce.

National Institute of Standards and Technology (1995), Secure hash standard, Federal Information Processing Standards Publication FIPS PUB 180-1, U.S. Department of Commerce.

Rahnema, M. (1993), 'Overview of the GSM system and protocol architecture', *IEEE Communications Magazine* pp. 92–100.

Samfat, D., Molva, R. & Asokan, N. (1995), Untraceability in mobile networks, *in* 'Proceedings of MobiCom'95'.

Schnorr, C. P. (1991), 'Efficient signature generation by smart cards', *Journal of Cryptology* **4**(3), 161–174.

Wilkes, J. (1995), 'Privacy and authentication needs of PCS', *IEEE Personal Communications* **2**(4), 11–15.

Zheng, Y. (1996), 'The SPEED cipher'. Submitted for publication.

Zheng, Y., Pieprzyk, J. & Seberry, J. (1993), HAVAL — a one-way hashing algorithm with variable length of output, *in* J. Seberry & Y. Zheng, eds, 'Advances in Cryptology — AUSCRYPT'92', Vol. 718 of *Lecture Notes in Computer Science*, Springer-Verlag, Berlin, New York, Tokyo, pp. 83–104.

Zheng, Y. & Seberry, J. (1993), 'Immunizing public key cryptosystems against chosen ciphertext attacks', *IEEE Journal on Selected Areas in Communications* **11**(5), 715–724.

26
Design of Secure End-to-End Protocols for Mobile Systems

V. Varadharajan and Y. Mu
Department of Computing, University of Western Sydney, Nepean,
PO Box 10, Kingswood, NSW 2747, Australia

Telephone: +61 47 360192, Fax: +61 47 360 800
Email: vijay@st.nepean.uws.edu.au

Abstract

Use of mobile personal computers in open networked environment is revolutionalising the way we use computers. Mobile networked computing is raising important security and privacy issues. This paper is concerned with the design of authentication protocols for a mobile networked computing environment. We propose secure end-to-end protocols between mobile users using a combination of public key and symmetric key based systems. These protocols enable mutual authentication and establish a shared secret key between mobile users. They also provide a certain degree of anonymity of the communicating users to other system users.

Keywords

Authentication Protocols, Mobile Security, Hybrid Approach, Anonymity

1 Introduction

Information and communication technology is on the threshold of new style of computing (Cox, 90). First, the telecommunications industry is witnessing the development of Personal Communication Systems that are "person-specific" with person to person logical connections. Such systems rely more and more on wireless communications, both in the fields of voice and data communications between mobile personal computers and computer systems. Second, the computer industry is in the phase of practical implementation of distributed systems concept. In particular, the notion of open systems is a major driving force. Whereas today's first generation notebook computers and personal digital assistants are self-contained, networked mobile computers are part of a greater computing infrastructure. This raises several issues with regard to information security and privacy, system dependability and availability (Varadharajan, 95; Molva, 94).

The paper is organised as follows. We begin in Section 2 by outlining the mobile computing environment. Section 3 gives the security requirements that need to be addressed in the design of the protocols. Section 4 proposes both intra and inter domain end-to-end protocols which can be used to provide authentication of mobile users. We use a hybrid approach involving both symmetric key and public key based systems. The design of protocols using only the symmetric key approach has been considered in (Varadharajan, 96). Finally, Section 5 discusses the important characteristics of the proposed protocol.

2 Mobile Environment

A simple mobile computing environment is shown in Figure 1. Mobile Computing Stations (MS) access the mobile network via a mobile network system. For instance, the network system may consist of Base Stations, Location Register, and Mobile Switching Component. The Location Register contains information related to the location and the subscription of the users in its domain. We will assume that an Authentication Server is present in every domain. This is logically a distinct entity; in practice, it may be co-located with the Location Register. The Authentication Servers store confidential information such as keys and are assumed to be physically protected. The mobile stations can move from one place to another, either within its domain (referred to as the "home" domain or move outside its home domain to a "visiting" domain. We will collectively refer to the authorities in the home domain as H and the authorities in the visiting domain as V.

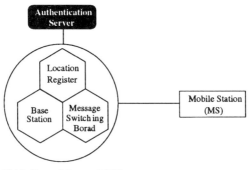

Figure 1: Mobile Networked Computing Environment.

We assume that when accessing the network in the home domain, the mobile user is authenticated with a server-based authentication mechanism. Users of every network domain are registered with that domain's authentication server. The authentication server of a domain can be replicated or partitioned within the domain but the set of all partitioned and duplicated authentication servers represent a single domain-level authority.

We will assume that users share a long term secret key with their home domain, i.e., A in domain H shares a secret key K_{AH} with H and B in domain V shares a secret key

K_{BV} with V. Authentication between domains is achieved using public key cryptography. That is, H and V have public key and private key pairs. When A travels to the visiting domain V, a shared session key between A and V must be established. In this paper, the user and the mobile computing station are regarded as an intact part.

3 Security Requirements

- *Authentication* : The authentication service should provide to the communicating parties the confidence that at the time of request, an entity is not attempting to masquerade as another and is not mounting a reply attack. In a secure mobile system, this implies that the end parties A and B should be confident that they are in fact communicating with each other. In order to achieve this, they rely on the guarantee provided by the Authentication Servers in their respective domains. That is, the mobile stations and the users of a domain trust both the competency and the honesty of their Authentication Server. For the Authentication Servers to be able to offer this guarantee, they need first to be able to reliably verify the identity of the communicating parties. Authentication between domain servers is achieved using a public key system.

- *Secure Communication* : The communication should not be vulnerable to attacks from other users and eavesdroppers. This is achieved by establishing a communication session key at the end of the authentication process. This session key is used to secure the communication between the end parties. We will be using symmetric key based cryptosystems for securing communications between mobile stations. This choice is due to less computational time required for performing symmetric key based computations compared to the public key ones.

- *User Identity Confidentiality* : In practice, there may be several reasons why the users might wish to keep their identities secret from other users. There can be different degrees of anonymity. For instance, a mobile user A may wish to be known only to the network authorities (e.g. the Authentication Servers) and to the other communicating party B, while remaining anonymous to other network users. We will refer to this form of anonymity as the first degree anonymity. At a higher level, a user A may wish in addition to remain anonymous even with respect to the visiting domain's Authentication Server. We will refer to this as the second degree anonymity. In principle, there is no need for the visiting authority to know the real identity of a user from another domain. What it needs is only a proof of the solvency of the entity accessing the service and enough information to bill the user's home authority correctly.

We address this issue of anonymity by introducing the notion of a subliminal identity (a form of alias), written as ID_s. Each user is issued a subliminal identity by the home domain H at the time of initial registration. The subliminal ID is composed of a number (e.g. a sequence number) along with a timestamp. This will allow H to perform efficient search of the database when required to locate a specific subliminal ID. Only H knows the mapping between this subliminal ID and the real user ID.

The use of subliminal IDs help to conceal the real user IDs to other network users. A user's subliminal identity can be updated at the end of each authentication session as part of the protocol.

- Non-repudiation of Service: For the service provider, it is desirable that a mobile user subscriber cannot deny the bill for the service he requested. At the same time, the subscriber should not be wrongly charged due to any billing error or security faulty on the network. Theoretically, both goals can be achieved through the use of digital signatures. Given the practical constraint that a mobile unit is not able to perform computationally intensive public key functions, only a limited form of non-repudiation service can be achieved.

- Protocol Design Principles : Domain specific secret information such as a user's long term secret key should not be propagated from the home domain to the foreign visiting domains. Furthermore, it is important to minimize the number of exchanges in the protocol between the home domain and the foreign domain in the setup phase, given that the distance between domains may be large.

4 Security Protocols

4.1 Notation

- $[\ ]_K$: Encryption under key K.
- $h(\ )$: Strong one-way hash function.
- $[h(\ )]_K$: Encryption of hashed digest under K.
- PK_X: Public key of X.
- SK_X: Private key of X.
- K_s: Session key shared between A and B.
- H: Home Authentication Server.
- V: Visiting Authentication Server.

4.2 Intra-Domain Authentication and Secure Communication

We now describe the end-to-end authentication protocol between A and B in their home domain (see Figure (2)).

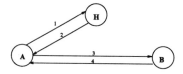

Figure 2: Secure End-to-End Protocol : Intra Domain

4.2.1 Assumptions

- Mobile Station User A (or B):

  Belongs to domain H. Has subliminal identity A_s issued by H and a secret symmetric key K_{AH} (or K_{BH}) shared between A (B) and H.

- Home Server H:

  Has the mapping between the subliminal identity A_s (and B_s) to real identity A (and B). Has public key - private key pair PK_H and SK_H. Has secret symmetric keys K_{AH} and K_{BH}.

4.2.2 Protocol

1: $A \to H$: $A_s, H, n_A, [A_s, B]_{K_{AH}}, [h(A_s, B, H, n_A)]_{K_{AH}}$
2: $H \to A$: $H, A_s, n_H, n_A, [B, B_s, K_s]_{K_{AH}}, [A'_s]_{K_{AH}}, [h(H, A_s, B, B_s, A'_s, K_s, n_A)]_{K_{AH}},$
$[A, A_s, K_s]_{K_{BH}}, [h(H, A, A_s, B_s, K_s, n_H)]_{K_{BH}}$
3: $A \to B$: $A_s, B_s, H, n_H, n'_A, [A, A_s, K_s]_{K_{BH}}, [h(H, A, A_s, B_s, K_s, n_H)]_{K_{BH}},$
$[h(A_s, B_s, H, n_H, n'_A)]_{K_s}$
4: $B \to A$: $B_s, A_s, n'_A, [h(B_s, A_s, n'_A)]_{K_s}$

In Step 1, mobile computer station A sends a message to its home server H requesting a secure communication with B. In Step 2, H returns a response message including a session key K_s encrypted under the shared key K_{AH} and a new subliminal identity A'_s. The message also includes the K_s encrypted under K_{BV}, which will be passed to B in Step 3. Step 4 completes the authentication and the key establishment process.

Note that the use of the subliminal identity helps to conceal the real identity of the initiator to other system users. In our protocol, we have carefully separated the information which needs to be signed (for integrity and authentication) from that which needs to be encrypted (for confidentiality). It is particularly important to adhere to this principle in the design of protocols; mixing these two aspects leads to lack of clarity in protocol design which is often an important source for protocol flaws. Furthermore this separation is useful when it comes to obtaining export licenses where it is necessary to justify to the authorities the functionality of the various cryptographic interfaces and their use.

4.3 Inter-Domain Protocol and Secure Communication

User A travels to a foreign domain V. When A requests a service to securely communicate with B in V, V needs to ensure that A is a legitimate user from the domain H before providing the service. This authentication process relies on the mutual trust between H and V. Following the authentication process, a secret key to protect communications between A and B can be established. Regarding anonymity, as we mentioned earlier, the real identity of A may need to be hidden from both the eavesdroppers as well as V. There should also be a mechanism for H to issue a new subliminal identity to A. This may be optional.

In the following, we consider a security protocol for the situation where A from H travels to a domain V and requests for secure communication with B in V (see Figure (3)).

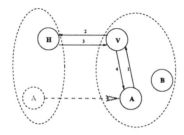

Figure 3: Secure End-to-End Protocol : Inter Domain

4.3.1 Assumptions

- Mobile Station Users A and B:

    Belong to H and V respectively.
    A has subliminal identity A_s and a secret symmetric key K_{AH}, both issued by H.
    B has subliminal identity B_s and a secret symmetric key K_{BV}, both issued by V.

- Home Server H:

    Has the mapping from the subliminal identity to real identity for A.
    Has secret symmetric key K_{AH}.
    Has public key - private key pair PK_H and SK_H as well as the public key of V, PK_V.

- Home Server V:

    Has the mapping from the subliminal identity to real identity for B.
    Has secret symmetric key K_{BV}.
    Has public key - private key pair PK_V and SK_V as well as the public key of H, PK_H.

4.3.2 Protocol

1: $A \to V$: $A_s, H, n_A, [A_s, B]_{K_{AV}}, Token_{AHV}, [h(A_s, H, n_A)]_{K_{AV}}$,
   where $K_{AV} = f(K_{AH}, A_s, V)$ and $Token_{AHV} = [A, H, V, n_A]_{K_{AH}}$
2: $V \to H$: $V, H, n_V, A_s, Token_{AHV}, [h(V, H, n_V, A_s, Token_{AHV})]_{SK_V}$
3: $H \to V$: $H, V, n_V, [K_{AV}, A_s]_{PK_V}, [h(H, V, K_{AV}, A_s, n_V)]_{SK_H}, [h(H, n_A)]_{K_{AH}}$
4: $V \to A$: $V, A_s, n'_V, n_A, [B, B_s, K_s]_{K_{AV}}, [h(V, A_s, B, B_s, K_s, n_A, n'_V)]_{K_{AV}}$,
   $[A_s, K_s]_{K_{BV}}, [h(V, A_s, B, n'_V, K_s)]_{K_{BV}}, [h(H, n_A)]_{K_{AH}}$
5: $A \to B$: $A_s, B_s, V, n'_A, n'_V, [A_s, K_s]_{K_{BV}}, [h(V, A_s, B, n'_V, K_s)]_{K_{BV}}$,
   $[A, A_s]_{K_s}, [h(A_s, A, B_s, V, n'_A, n'_V)]_{K_s}$
6: $B \to A$: $B_s, A_s, n'_A, [h(B_s, A_s, n'_A)]_{K_s}$

In Step 1, A sends V his subliminal identity A_s, communication request $[A_s, B]_{K_{AV}}$, a token $Token_{AHV}$, a nonce n_A, and the signed hash value. $Token_{AHV}$ is encrypted under K_{AH} and it needs to be passed to H by V in Step 2. $Token_{AHV}$ conveys to H that A at present in V wishes to use a service in V. At this stage, V cannot understand the communication request and cannot verify the authenticity of the hash value as it does not have K_{AV}. K_{AV} is generated with a publicly known strong one-way function f. Knowing V and A_s, H is able to calculate K_{AV}, because it shares K_{AH} with A. The mapping of A_s to A is maintained at H.

In Step 2, V sends H the $Token_{AHV}$. The communication is signed by V using its private key (of the public key pair). At the end of this step, H is able to verify that A is making a request to V at the present time.

In Step 3, H sends V the key K_{AV} (encrypted under public key of V), the subliminal identity A_s (encrypted under public key of V). It also sends response information (the freshness component n_A) encrypted under K_{AH} for V to pass to A.

Upon the receipt of H's message, V retrieves K_{AV}, and uses it to verify the hash value that it initially received from A in Step 1. It then generates a session key K_s for use between A_s and B. The K_s is encrypted under K_{AV} for A, and is also encrypted under K_{BV} for B. All of this is then sent to A in Step 4. The information $[h(H, n_A)]_{K_{AH}}$ that V received from H is also passed to A.

In Step 5, A passes to B the session key that it received from V. Finally Step 6 completes the authentication and the key establishment process.

As an option, H may issue a new subliminal identity A'_s for A as part of this protocol. This can be done by including the following information in Step 3: $[A'_s]_{K_{AH}}$, $[h(H, A'_s, n_A)]_{K_{AH}}$ (instead of $[h(H, n_A)]_{K_{AH}}$). This portion of the message will then be passed by V to A in Step 4.

5 Discussion

It is worth discussing the following characteristics of the above protocol.

- The real identity of the mobile station user A is not revealed to system users other than B. That is, the first degree anonymity is achieved. Furthermore, the real identity of A is not revealed to V. Similarly the identity of B is not revealed to H. Hence the second degree anonymity is also achieved. A useful discussion of anonymity can be seen in Ref. (Asokan, 94).

- The protocol provides an option for changing the subliminal identity during every session if required. On the other hand, for the period of time that A is going to be within the domain V and using its services, it may be sufficient to keep the same subliminal identity. In this case, V and A can share K_{AV} for this period of time resulting in the optimization of subsequent protocol exchanges.

- It is possible to include the nonce n_A as part of the calculation of K_{AV}. In this case, K_{AV} can be changed every request even if the A_s parameter remains constant for a period of time.

- Given that K_{AV} can only be calculated by H and A, and that H is trusted by both V and A not to generate $[h(A_s, H, n_A)]_{K_{AV}}$ illegally, the possession of communication request encrypted under K_{AV} by V indicates that it has been generated and sent by A sometime earlier. At the same time, given that A trusts H and that V (or any one else) cannot generate the request under K_{AV}, one can ensure that A is not charged wrongly.

- Session key K_s that is used for communication is changed every request. Hence even if one K_s is compromised, future communications can be protected.

- There are several options as to who generates the key K_s. In the protocol proposed, we have assumed that V generates K_s and distributes it to A and B. This seems natural when both A and B are in V's domain. When A and B are in two different domains, either H or V can generate the session key K_s.

- There are various options for distributing the session key K_s to B. We have chosen the option whereby V passes K_s to B via A. Other alternatives include V passing the session key directly to B (which may lead to timing problems) or asking B to pass the session key to A.

- At the end of Step 4, A has received from V the information $[h(H, n_A)]_{K_{AH}}$ sent by H in Step 3. This helps A to believe that its request has been passed to H and H has provided the right information to V.

- Note the asymmetricity in the use of A_s and B_s in the protocol exchanges. This is based on the principle that H only knows the true identity of A (and not B), and V only knows the true identity of B (and not A).

- Public key system has been only used between V and H and not between H and A or V and A. This reduces the complexity of the computations.

REFERENCES:

Asokan, N. (1994), Anonymity in a mobile computing environment, In *Proceedings of*

1994 IEEE Workshop on Mibile Computing Systems and Applications.

Cox, D. C. (1990), *IEEE Communications Magazine* **27**.

Molva, R., Samfat, D., and Tsudik, G. (1994), *IEEE Network* , 26–34.

Varadharajian, V. (1995), Security for Personal Mobile Networked Computing, In *Proceedings of the International Conference on Mibile and personal Communications Systems.*

Varadharajian, V. and Mu, Y. (1996), Design of Security Protocols for Mobile Systems : Symmetric Key Approach, In *Proceedings of the rth International Conference on Wireless Communications,* (to appear).

27
Yet Another Simple Internet Electronic Payment System

J. Zhao, C. Dong and E. Koch
Fraunhofer Institute for Computer Graphics
64283 Darmstadt, Germany
Phone: +49 (06151) 155-412 (or 147)
Fax: +49 (06151) 155-444
Email: {zhao, dong, ekoch}@igd.fhg.de

Abstract

The potential service providers, customers, as well as financial institutions are constructing geographically the largest global business net in the business history over the Internet. The payment mechanism needed for this new business medium is a very important issue. This paper presents a simple debit-based electronic payment protocol QIPP (Qudro-way Internet Payment Protocol) and its implementation in an electronic commerce system on the World Wide Web. The aim of the QIPP is to provide sufficient security means and least changes to conventional business transaction and payment model for all participants involved in the global electronic commerce.

Keywords

Electronic Payment, Electronic Commerce, Security, Internet, World Wide Web (WWW)

1 INTRODUCTION

The Internet, especially the World Wide Web, is moving from a free, academic domain to a profitable commercial world. There is no need to emphasize the importance of electronic payment for the merging electronic market on the Internet and the unavoidability of such an Internet market. The decentralization of the Internet leads to less personal contact, less trust of participants in this gigantic market. Easiness of eavesdropping and forgery on the Internet incurs a variety of attacks from many sides. Security concerns of the potential customers, merchants and acquirers should be solved not only in technical aspects, but also in social aspects.

An electronic payment system should meet the following requirements: (1) sufficient security means for all the participants in the electronic commerce; (2) similar running scenario as the traditional business when ever possible to ease the doubts of the public and encourage them to participate; (3) less changes on the current financial system to avoid tremendous costs when electronic commerce is introduced.

In this paper we will present a simple, secure Internet electronic payment protocol QIPP (Qudro-way Internet Payment Protocol), which is based on an online debit model. In the following, first the state-of-the-art electronic payment systems and proposals are presented. Second the QIPP protocol is described. Finally we will discuss briefly the implementation of the QIPP on the WWW through a flexible Web-development-independent security architecture.

2 RELATED WORK

Nowadays, commercial companies start up their Web shops on the Internet. In many cases, service order requires pre-registration of user. Where online payment is supported, credit card data or account PINs will be sent over Internet with more or less security protection in many cases. There are many approaches which target the electronic payment system to solve the problems.

IBM has proposed the iKP (Bellare et al., 1995), which is an online payment system applying CA-based security (i.e. the one-sided security scheme). The iKP can be implemented in different level, just as its name indicated: the 1KP, 2KP and 3KP. Different level of iKP offers different security level: The 1KP does not provide non-repudiation; The 2KP provides only non-repudiation of messages produced by merchant; The 3KP achieves non-repudiation for all messages and parties involved. In the iKP, the authorization of payment is based on the credit card number and associated PIN. The PIN will be encrypted with the public key of the acquirer, so that the merchant will have no chance to abuse the credit card of the customer. The iKP assumes that the PIN is not of necessity in these circumstances, since the signature of the customer already offers protection for the account of the customer.

The NetBill (Sirbu & Tygar, 1995) proposed by Carnegie Mellon University also relies on the public key security system (Kerberos). Micropayments are the major target of this online payment system. Both debit model and credit model are supported in NetBill. NetBill does not describe how to prevent malicious users from re-spending preloaded electronic fund in debit model. In the credit model, the risk of non-payment exists. Moreover, the PIN of a customer is clear to the bank site, so a malicious bank employee is a potential threat to the customer's account. On the other hand, the decryption key is known to the NetBill server (bank) and logged there, anyone who has access to the NetBill server database can intercept the electronic products sent to customer and then use the decryption key to decrypt them.

HP proposal (Mao, 1995) of Hewlett Parkard is another system for online payment. According to the disclaimer by the system designer W. Mao, the system does not utilize public key certification infrastructure, although some observers do not agree with him. The HP proposal applies the one-way hashed values of passwords which are centrally stored in financial institutions. The central point of this proposal is that no decryption algorithm is used for secret retrieval. This aims at the avoidance of the export and import restrictions set by USA and other governments, according to HP. The payment requirement in this system is sent by the merchant to the bank without the signature of the merchant. This could be vulnerable, since anyone can reuse the message in this step and send it to the bank without the real identity being found.

The DigiCash (Chaum, 1992) is an online electronic cash system. The DigiCash system aims to provide the privacy of customers, based on blind signature. When the customer consumes digital cash, the DigiCash multiplies the note number by a random factor and sends it to the bank for signing. Thus, the bank knows nothing about what it is signing except that it carries customer's digital signature. After receiving the blinded note signed by the bank, the customer divides out the blind factor and uses the note as before. The blinded note numbers are unconditionally untraceable. That is, even if the shop and the bank collude, they cannot determine who spent which notes. Because the bank has no idea of the blinding factor, it has no way of linking the note numbers that merchant deposits with customer's withdraws. The anonymity of blinded notes is limited only by the unpredictability of customer's random numbers.

The AT&T has proposed an Anonymous Credit Card (ACC) system (Low et al., 1994). The ACC utilizes an intermediary to separate the information among the parties involved. To achieve the privacy of the customer, a double locked box (DLB) is used to let message initiator send the message without knowing the destination and vice versa. The major difficulty in implementing this proposal is that the conventional financial system suffers great changes. Social acceptability and legal concerns are other obstacles to this proposal: anonymous credit card requires anonymous account at bank, which may cause problems in some countries where opening anonymous account at bank is not allowed.

CAFE (Conditional Access for Europe) is a project in the European Community's program ESPRIT (Boly et al. 1994). In general, the CAFE system is a prepaid offline payment system and utilizes electronic cash (like the DigiCash). The basic devices of the CAFE system are tamper-resistant electronic wallet, and POS (Point-Of-Sale) terminals which accept the payment from the electronic wallet. The core technique to prevent the same digital cash from multiple uses is to place a guardian chip with a crypto processor inside a wallet. No transaction will be possible without guardian's cooperation: no payment is accepted unless the guardian says OK, and for each unit of electronic money, the guardian gives its OK only once. This is somewhat like a signature by the guardian, but it is a blind signature in the CAFE scenario to protect the privacy of the customer.

Recently, MasterCard and Visa have announced the SET (Secure Electronic Transaction) draft (MasterCard & Visa, 1996) for public comments. This draft also images electronic commerce built on the CA-based security. The SET includes a payment section, which is able to deal with different credit cards. The SET applies acquirer payment gateway which is able to authorize using the existing bankcard networks. In the authorization request sent by merchant to acquirer, the purchase instruction of customer enables the acquirer to verify that the merchant and customer agree as to what was purchased and how much the authorization is for.

Many other electronic payment systems have been developed and more and more are being proposed, for example, the payment system in the European R&D Semper (Secure Electronic Marketplace for Europe) project (Semper, 1995), Kerberos-based NetCash (Medvinsky & Neuman, 1993) and NetCheque systems developed at the Information Sciences Institute of University of Southern California.

3 QIPP – A SIMPLE, SECURE INTERNET ELECTRONIC PAYMENT

QIPP (Qudro-way Internet Payment Protocol) to be described in this section aims to provide a simple yet secure electronic payment for the electronic market on the World Wide Web. Unlike most other payment systems discussed above, the payment will be initiated by the customer, and the merchant is not directly involved in the payment process. Thus, the merchant is virtually excluded from the attacks towards account of customer at the bank, i.e., re-charge or multi-charge is impossible for the merchant. The QIPP simulates the conventional payment in the following three steps: Customer gets the bill from the shop serviceman, pays it by the cashier, and finally shows the receipt to the shop and to get the goods. The interests of the partners in this payment system are well protected, in particular the interests of the customer. In QIPP, customer's bank accounts are double-protected. QIPP introduces minimal changes to current financial systems and to the customer's shopping habits, and naturally does not intend to supplant any other payment systems.

QIPP Overview

An architecture of Internet electronic market is illustrated in Figure 1. As shown in this figure, there are four kinds of players: Customers, Service Providers (or called merchants), Financial Institution presented by acquirer gateway linking to the conventional financial clearing network, and Trusted Third Party (TTP) or Certificate Authority (CA) responsible for all kinds of certification and certificate archiving (Schneier, 1994). The main process of the QIPP protocol consists of four dataflows between the customer and the acquirer. They will be described in detail in the following.

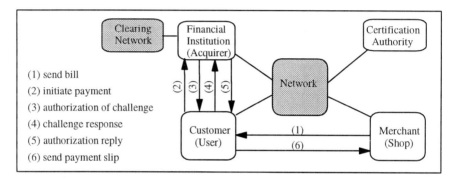

Figure 1 QIPP in an Internet market architecture

There are two assumptions in this protocol. First, the acquirer has widely distributed its public key to the public, so that both customer and merchant has this public key at hand. Second, the public key of both the customer and the merchant shall be exchanged when the business transaction begins. In addition, the widely applied Certificate Authority plays an important role in issuing the certificates to all participants of the electronic business.

Before proceeding, we explain the following notations which will be used to describe the protocol in detail:

ACQ: The acquirer, which is a gateway to the conventional financial network. We also call the acquirer the bank in this paper.
MER: The merchant (or called the service provider).
CUS: The customer (or the user).
PAYMENT: The information segment indicating payment requirement.
RESPONSE: The information segment indicating challenge response.
ACCEPT: Positive response to the prior request.
REJECT: Negative response to the prior request.
PK_X: Public key of party X, e.g. PK_CUS is the public key of customer.
TID: Transaction ID, which must be a unique number produced for each transaction.
SIG_X: Digital signature of party X using X's private key, e.g. SIG_MER is the signature of the merchant.
ENC_X: Encryption operation using the public key of party X.
H(*data*): One way hash operation on *data*, e.g. md5 (Schneier, 1994).

RAN_X: Random number created by party X.
DATE_X: System date and time obtained by party X during the transaction.
DESC: Description of the articles provided by the service provider, including article name, article price, amount of article the customer wants to purchase, clearing type (debit or clearing while purchasing), payment type (pay before or after delivery), credit card type, delivery type (on- or off-line), purchase date, delivery date, delivery address (physical or electronic). Besides the essential information described below, the merchant can also extend this structure.
ID_X: The bank account ID of party X.
MON_X: The amount of payment requested by party X, e.g. MON_MER is the payment amount that merchant wants customer to pay for the GOODS.
OT: One-time-key for online delivered goods used by merchant for encryption and by customer for decryption.
INVOICE: It contains H(DESC), ACCEPT, RAN_MER, DATE_MER, TID, ID_MER, MON_MER. It may optionally include ENC_ACQ(ENC_CUS(OT)).
GOODS: The article in the transaction, including digital articles (pictures, videos, etc.) as well as non-digital physical articles, which must be delivered offline.

Double protection

Security based on the public key certification authority utilizes a pair of complimentary encryption keys. The public/private key pair serves for secure communication between partners. The key pair is usually physically stored in the system on which the secure application of user runs. Therefore, it relies on the security ability of the system to protect the key pair against potential abuse. If the private key is stolen under certain circumstances, then the security based on this key pair will be a dangerous false friend for the user if the security of all the secret of the user relies on this security key pair, especially when user pays against his account over Internet. This leads us to the idea of double protection, i.e., another protection layer in addition to normal complimentary key pair. The credit card number and its associated PIN, which may be known to the merchant or to the bank, are not suitable for this purpose since a malicious merchant or bank employee can virtually do every thing with this credit account what the user can do with.

To prevent the bank from knowing customer's PINs, the QIPP stores a hashed value of the customer account's PIN in the bank. The following hash function given in (Schneier, 1994) is used in the QIPP:

$$g^x \bmod p \tag{1}$$

if g is a generator modular p and g is less than the prime number p, there will be some x where $g^x \equiv n \pmod{p}$ for each n from 0 to p–1. When p is a large prime (Mao, 1995), say, p=2q+1, where q is a prime. Then, the user can choose his/her PIN, u, in the interval [3, q–1] and [q+1, p–2] with the conditions that $u \neq 1$, $u \neq 2$, $u \neq q$, and $u \neq 2q=p-1$. The user gives the result of $g^u \bmod p$ (denoted by $\mathcal{G}^u$) to the bank, stored as a hashed PIN in the bank database.

The following equation is used for challenge verification:

$$(g^x \bmod p)^n \bmod p = (g^n \bmod p)^x \bmod p. \tag{2}$$

During the payment, the bank sends a challenge $g^n \bmod p$, denoted as $\mathcal{G}^n$. Only the holder of the PIN can answer this challenge correctly: The holder of the PIN replies the challenge with

$(\mathcal{G}^n)^u \bmod p$ to the bank. The bank computes $(\mathcal{G}^u)^n \bmod p$ and then compares it with $(\mathcal{G}^n)^u \bmod p$ – if they are equivalent the challenge is answered correctly.

QIPP protocol descriptions

(1) Send bill:

    ENC_CUS(DESC,
                SIG_MER(H(DESC), ACCEPT, RAN_MER, DATE_MER,
                      TID, ID_MER, MON_MER,
                      [ENC_ACQ(ENC_CUS(OT))]   # this item is an option
)
) or

    ENC_CUS(DESC, SIG_MER(INVOICE))

The item RAN_MER and DATE_MER prevent replay and provide transaction record. The bank account ID_MER is given to the user for money transfer. The ENC_ACQ(ENC_CUS(OT)) is only for the case of online delivery.

(2) Initiate payment:

    ENC_ACQ(PK_CUS,
                SIG_CUS(H(DESC), PAYMENT, RAN_CUS, DATE_CUS,
                      TID, ID_CUS, MON_CUS
                  ),
                H(SIG_MER(INVOICE))
)

The MON_CUS is given by the customer to the bank, which should be the same as MON_MER, otherwise, the merchant will not deliver the goods afterwards. The H(SIG_MER(INVOICE)) is kept by the bank for the legal evidence in case of discrepancy.

(3) Authorization of challenge:

    SIG_ACQ(RAN_ACQ, DATE_ACQ, $\mathcal{G}^n$,
                TID, H(INVOICE), ID_CUS
)

The nonce to generate $\mathcal{G}^n$ is a large random number, so that each payment challenge will have different $\mathcal{G}^n$. The random number RAN_ACQ is used to prevent replay.

(4) Challenge response:

    SIG_CUS(RESPONSE, ID_CUS, RAN_CUS, DATE_CUS,
                TID, H(INVOICE), $(\mathcal{G}^n)^u$
)

The $(\mathcal{G}^n)^u$ is only correct when the customer gives right PIN, i.e., the u.

(5) Authorization reply:

    SIG_ACQ(ACCEPT/REJECT, MON_ACQ,
                TID, H(INVOICE), ID_MER, ID_CUS, DATE_ACQ, RAN_ACQ
)

The MON_ACQ indicates that the bank has the payment requirement in amount of MON_ACQ.

The invoice sent by the merchant to the customer is signed by the merchant to provide non-repudiation. In case of shopping electronic information goods, an one-time key (OT-key) encrypted with customer's public key and then re-encrypted using the acquirer's public key is also included in the invoice. The OT-key will be used in the online delivery for encryption of information goods in order to protect the interests of both the customer and merchant, which is realized in the subsequent delivery protocol.

During the payment process, the bank sends a challenge to the customer, only the customer who owns the clear PIN can reply with the correct answer. The payment request sent by the customer is encrypted with acquirer's public key, so that eavesdropper cannot get any information about the payment. The payment amount and other contractual information included in the payment request are signed by the customer to provide authenticity and non-repudiation. Finally, the payment response from the bank is also signed by the acquirer so that both the customer and the merchant can keep acquirer's signature as legal evidence.

4 IMPLEMENTATION

The QIPP has been implemented as one of the three components in an electronic commerce system ELITE (Secure Electronic Commerce on the Internet) developed at the Fraunhofer Institute for Computer Graphics in Darmstadt. The other two protocols in ELITE are for electronic negotiation and for online and offline goods delivery. ELITE has been implemented on the World Web Web without any specific prerequisites (on security or technology). Figure 2 shows the implementation architecture of the ELITE.

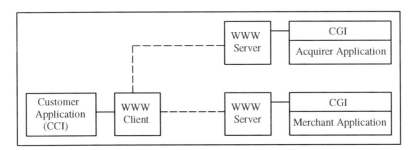

Figure 2 Implementation architecture

Each customer in this architecture is equipped with a WWW browser (NCSA Mosaic 2.7) and a CCI (Common Client Interface) application. The CCI-application, which interacts with the browser according to the CCI protocol (Malgiery & Sanderson, 1995) and communicates with the WWW-server via the browser, is responsible for fulfilling the security functions on the client's side. Each merchant or each acquirer is provided with three components: a WWW-server, a merchant or acquirer application, and a CGI (Common Gateway Interface) program connecting them. The merchant/acquirer application runs as a background daemon to securely perform customer's requests. A set of specific MIME (Multipurpose Internet Mail Extension) types are defined between the CCI applications and the CGI program for re-routing messages from the customer's browser to the CCI applications.

The security packages used in the CCI application and in the merchant/acquirer application are based on an implementation of the GSS API (Generic Security Service) (Lim, 1994), which in turn is based on a secure development security toolkit SeCUDE (Schneider, 1995).

5 CONCLUSION

The QIPP is a simple, flexible protocol for Internet electronic payment system. In sum, it provides the following features: (1) Proof of the payment acceptance for both customer and merchant, (2) Exclusion of unauthorized payment by double protection on customer's account, (3) Protection against eavesdropper, (4) Customer's privacy (but no full anonymity), (5) Minimized changes to current financial system, (6) Minimized attack possibility towards customer's bank account, (7) Unchanged shopping habit for the customer.

The QIPP has been implemented on the World Wide Web and is completely compatible to the Web application in the sense that neither the Web browser nor the Web server need to be changed in order to apply the QIPP. Moreover, the approach to the implementation of the QIPP allows the participants in an electronic commerce to use security mechanisms that are independent of the Web development.

6 REFERENCES

Bellare, M. et al. (1995) ikP – A family of Secure Electronic Payment protocols, in *Proc. of the First UNENIX Workshop on Electronic Commerce (New York, July 1995)*, also available in *http://www.zurich.ibm.com/Technology/Security/extern/ecommerce*.

Boly, J. P. et al. (1994) The ESPRIT Project CAFE – High Security Digital Payment Systems, in *EXORIC94 (Third European Symposium on Research in Computer Security)*, LNCS 875, Springer-Verlag, Berlin, 1994, pp.217-230.

Chaum, D.; Fiat, A.; Naor, M. (1988) Untraceable Electronic Cash, in *Advances in Cryptology CRYPTO'88*, S. Goldwawwer (Ed.), Springer-Verlag, pp. 319-327.

Lim, J. (1994) Generic Security Service Application Program Interface, IETF RFC 1508, available as *ftp://ftp.internic.net/rfc/rfc1508.txt*.

Low; S. H.; Maxemchuk, N. F.; Paul, S. (1994) Anonymous Credit Cards, in *Proc. of 2nd ACM Conference on Computer and Communication Security (Nov. 2-4, 1994)*.

Mao, W. (1995) A Secure, Cheap, Scalable and Exportable/Importable Method for Internet Electronic Payment, available as *http://www.hp.co.uk/projects/vishnu/main.html*, May 19, 1995.

Medvinsky, G. and Neuman, B. C. (1993) NetCash: A design for practical electronic currency on the Internet, in *Proc. of the 1st ACM Conference on Computer and Communication Security (Nov. 1993)*.

Malgiery, T. and Sanderson, B. (1995) Writing Web Software through Client APIs (Mosaic CCI, Netscape), *Tutorial of the Forth International World Wide Web Conference (Boston, Dec. 11-14, 1995)*, also available as *http://www.w3.org/pub/Conferences/WWW4/Tutorial_Abstracts.html#D*.

MasterCard & Visa (1996) Secure Electronic Transaction (SET) Specification, available as *http://www.visa.com/cgi-bin/vee/sf/set/settech.html*, Feb. 23, 1996.

Schneider et al. (1995) SeCUDE 4.4 Overview, available as *http://www.darmstadt.gmd.de/TKT/security/secude/overview.html*, GMD-TKT, Darmstadt, Germany, 1995.

Schneier, B. (1994) *Applied Cryptography: Protocols, Algorithms, and Source Code in C*, John Wiley & Son, Inc., New York et al., 1994.

Semper (1995) SEMPER – Project AC026 – ACTS Programme, in *http://semper.zurich.ibm.com*. March 1995.

Sirbu, M. and Tygar, J. D. (1995) NetBill: An Electronic Commerce System Optimized for Network Delivered Services, in *IEEE Personal Commun.*, August 1995, pp. 34-39.

PART TWO

General Security Aspects in Mobile Communication

28
Difficulties in Achieving Security in Mobile Communications

I. Nurkic
Security Analyst
16/501 Wilson St. Chippendale. N.S.W. 2008, Australia, tel. 61-2-99025725, fax. 61-2-9902-5141, medan@ee.su.oz.au

Abstract

This work investigates specific challenges in implementing secure mobile digital communications. The objective was to look at some aspects of the problem which may contribute to an exchange of new ideas. Use of an asymmetric key protocol relationship for key management and subscriber authentication, in combination with a chosen symmetric algorithm to protect mobile sessions, is suggested. Dependence on the security of the chip card (smartcard, Integrated Circuit (IC) card), for the mobile unit, is addressed. Consideration of the emerging security protocols and key management methods for open networks (Internet) is included. Ideas are presented generically for a mobile digital system and some comments on the existing Global System for Mobile Communications (GSM) security implementation are included.

Keywords
Security, mobile, communications, GSM, cryptography, asymmetric, smartcard, key

1 BRIEF SUMMARY OF SECURITY REQUIREMENTS

Three overall network security requirements, confidentiality, data integrity and availability (Ford, 1994), apply to mobile voice communications as well. Specifically for mobile communications, security requirements may be expressed as:

- Privacy of information: Prevent the eavesdropping on the radio path. Although the same requirement may be also applied to the traditional telephony (fixed networks), the nature of mobile communications (availability of signals for reception) makes the requirement for privacy stronger in wireless telephony.

- Privacy of movements of mobile users: Prevention of signal tracking, i.e. "confidentiality of signaling information" (Unknown, 1995).

- User authentication: In order to protect both the business and services of the mobile service provider and its customers, service should be available only to the legitimate

(authenticated) users. In other words, service stealing either by an impersonation or other sophisticated means, should be prevented.
- Subscriber anonymity: Avoid the transfer of genuine user identification information over the network.

If taking into account that there are some security issues related to the charging service, for example how to prevent subscribers repudiating used services, non-repudiation may also be a relevant security requirement in future systems, although it is not specifically addressed in this paper.

2 CRYPTOGRAPHY IN MOBILE COMMUNICATIONS

Security relies on cryptography to achieve data privacy and user authentication. While it is beyond the scope of this paper to present/ analyse cryptographic algorithms, it is important to say that there are some concepts which are common to the cryptographic algorithms currently used in the commercial sector, namely they use a secret parameter - encryption key - to achieve the needed 'unpredictability' of the encryption process. Then, depending on whether an algorithm requires exactly the same encryption key value both to encrypt and decrypt data, or there are two mathematically correlated values used, where one is required for encryption and the other for decryption, algorithms are classified in two families: symmetric and asymmetric algorithms, respectively. The best known algorithms of both classes are publicly documented, some of them standardised either on the international or on a national level, and most of them are licensed.

If looking at some implemented solutions in the mobile environment, 'standard' cryptographic algorithms did not seem applicable in the mobile telephony. There seem to be three reasons for this:
1. Publicly known symmetric key algorithms, like DES (Data Encryption Standard), at the time when considered, could not satisfy mobile environment's specific security (key management) requirements. (Also, some characteristics of older symmetric algorithms - namely key lengths, already were not good enough for security of emerging mobile systems.)
2. Asymmetric key algorithms, like RSA (Rivest-Shamir-Adleman), were not considered for the implementation in the design of first secure mobile systems either (see more in Section 5).
3. Standard GSM security has adopted a specific security design which employs very specific algorithm(s).

Identified specific implementation difficulties, when implementing security for mobile networks, are:
- Difficulties in synchronising cryptographic key material (keys, initial vectors, seeds) between the mobile unit and the centre. There are only few options available in achieving cryptographic key synchronisation requirement:
  - Broadcast a seed value which is an input parameter in the key generation process. Any solution with broadcasting actually relies on the secrecy of either the encryption algorithm (method) used or some unique value (e.g. user authentication code) in the mobile unit.

- Rely on sets of pre-generated keys which are stored in the mobile unit. However, it would be impossible to store a large number of random encryption keys (without any functional relationship among the keys), presumably required to last for the lifetime of the mobile telephone. It is not acceptable to use permanent session keys because of the risks of exposure over an extended lifetime. Also, it is not acceptable to store non-random keys because of the risk of discovering the key relationship.
- Broadcast new keys encrypted under the initially set, 'encrypting' (master) keys. For the schemes using symmetric cryptographic algorithms only, this would result in an unacceptable burden of managing (generating, loading into the handset, storing in the service provider's centre) large number of symmetric master keys (similar as if using symmetric algorithms for secure communications in large networks with a key centre, in general).
- Broadcast new keys using asymmetric keys (see more in Section 5).

- Air-emitted information is accessible to anyone with a receiver without any additional effort required (see more in Section 4).
- Performance factor: In voice communications, due to the speed required to complete all cryptographic operations, the complete family of current asymmetric algorithms is still unusable, at least in achieving data privacy. However, these algorithms may have a practical value in user authentication, if implemented in a more sophisticated chip card technology, with an integrated modulo arithmetic co-processor.
- Commercial factor (cost) up to date was an obstacle in getting more processing power in the chip card. However situation in this area changes rapidly since the cost is actually dictated by the amount of silicon used - therefore for new micro chips it is no longer true that an enhanced chip design would automatically cost more.

3 END-TO-END SECURITY: UTOPIA?

There are two 'credos' in the data communications security framework. Firstly, complete security relies on an implementation of a standard (universal) method over the complete path. Secondly, a security design is as good (secure), as its weakest component.

An end-to-end security design is still a desired aim of the modern telephony. Namely, the traditional voice communications network provide neither data privacy (confidentiality) nor user authentication using sophisticated cryptographic techniques. For that part of the communication path data information (voice and signaling information) in most cases will not be protected in any existing scenario.

Also, the current mobile digital communications (GSM) carriers base the security on the corresponding European Technology Standards Institute (ETSI) recommendations, which protect information only within the radio domain of that carrier. Note also that there exist weaknesses in the GSM security management, i.e. that the security information generated in the Authentication Centre (AUC) and required in the remote locations of HLR (Home Location Register) and VLR (Visitor Location Register), are distributed over the fixed network. Securing of this sensitive information (triplets used for the challenge-response authentication method) is left to the discretion of the network provider (e.g. link encryptors).

Even further, ETSI recommendations (ETSI GSM 02.09., 1992), (ETSI GSM 03.20., 1992) are broad enough to allow significant differences in the level of security (i.e. quality of implementation) among the implemented systems, which all formally comply to the referenced standards.

For example, the referenced ETSI recommendation allows the use of different authentication and key generation algorithms (corresponding algorithms A3 and A8, respectively, are not specified). It is not clear if the algorithms used by the network providers had to pass some verification process before implementing them. It is believed that, as long as there are no issued guidelines/ standardised requirements on these algorithms, quite simple reversible functions may have been implemented. Preferably, these algorithms should have the features of an one-way hash function. (Due to the explicit compatibility requirements, the third algorithm used in GSM, the encryption algorithm A5, is a specified, although kept confidential, stream cipher.)

Also, a quality of random number generators used is not mandated in ETSI standards. There may exist a large discrepancy in the quality of implemented generators. Some of them may still use ordinary seeds (e.g. date and time) where all samples can easily be collected in a database used for a 'dictionary attack'. Desirably, implemented generators should use more unpredictable sources of random information and more complex functions for deriving a final 'random' value (namely, RAND used as a challenge in the challenge-response scheme used in GSM). Once again, if there are some guidelines available to follow, this rapidly developing industry would follow them.

4 SECRET ENCRYPTION ALGORITHMS: ARE THEY REALLY A NECESSITY?

Today, when achieving communications security in the majority of all industries/ areas involved in electronic communications, beyond the military/ state security sector, public cryptographic algorithms (e.g. DES; RSA) are used. In such public algorithms the encryption method is known and available for an endless process of crypto-analysis therefore their resistance to crypto-analysis may be proven. Security of those systems relies on the secrecy of encryption keys. (A number of security measures is involved but almost all fit into 'key management', i.e. mechanisms and controls, including physically secure, tamper-resistant devices, around cryptographic keys.)

Current mobile security implementations (GSM) are arguably based on the secrecy of encryption algorithm(s). Reasons behind this phenomenon seem to be the following:
1. Air-emitted information is accessible to anyone with a receiver without any additional effort required i.e. it is easy to collect unlimited amounts of 'crypto-material'. As a consequence, it is considered that the risks of attack are much higher than in the conventional communications media.

    (Note that the aims of attack in mobile communications are quite different. Specific nature of the voice communications is such that it is still too complex to change live messages. The prime aim is either a passive attack - to hear personal information - or a special version of user impersonation - to find out subscriber specific information to use mobile services with no charge, etc. However, in generic data communications, an

attacker may be more interested in changing sensitive messages, e.g. value transactions.)
2. Difficulties in synchronising keys between the subscriber's mobile telephone unit and the mobile service centre (as stated in the Section 2 above).
3. Mobile networks are seen as a phenomenon of new technologies, but the concept of wireless communications was already mature in the military domain. Therefore, it seems that security for mobile telephony just followed some data confidentiality principles established in the military scientific circles. In addition, international standards which would dictate a more common implementation approach and which could be based on a public standard algorithm and corresponding key generation process instructions, are not yet available.
4. It seems that the politics (export restrictions on encryption technology) could be responsible for the current restricted access to the security algorithms used in GSM. According to the (Dmargrav, 1995), cellular telephone manufacturers must agree to non-disclosure and obtain special licenses from the British government, since the algorithms were developed in Britain.

The current situation may be overcome with the promises of new technologies, e.g. the IC cards technology improves daily its processing and storage power versus cost. It seems that the advantage of asymmetric key (public key) cryptography, which eliminates the need for key distribution, in combination with a chosen symmetric encryption algorithm, may provide some very good security tools in the area of mobile networks. Also, there is one set of algorithms, so-called zero-knowledge techniques, which may be potentially attractive for future mobile digital systems (Unknown, 1995), but they are not discussed in this paper. The suggested scenario of using asymmetric algorithms is presented in the next section.

5 USAGE OF ASYMMETRIC KEY CRYPTOGRAPHY

Asymmetric key cryptography has three merits very important to resolve issues in network security and, specifically, in mobile communications:
1. Public key, of the asymmetric key pair, of any entity participating in communication may be given to all potential users freely[*] without any need to hide it and to distribute it via secret channels.
2. Decryption of the original message or of a hash of the message by the sender's own secret key (digital signature), before transfer, represents on its own three guarantees (see also merit no. 3):
   - that the message has not been changed in transport;
   - that it has been originated by the entity which 'signed' the message;
   - that the sender cannot deny that it sent the message (non-repudiation).

---

[*] The key should be properly certified, i.e. there should not be any suspicion on who stands behind it. Certificates are data formats containing 'signed' public key and some other specified control information like the expiry date, granted by a 'Certification Authority', which is a trusted party whose own public key is too well known and trusted to initiate the chain of trust. Therefore public keys should always be provided together with their certificates to be trusted.

3. There is no need that the entity's secret key ever leave the location where key pair is generated. If it is stored securely (i.e. this location is a tamper resistant entity), this contributes to the security value of the digital signature as an ultimate method to achieve entity authentication and data integrity.

These three features of asymmetric key algorithms can resolve all major key management problems in setting up security for a mobile network system. Actually, there exist all prerequisites to successfully implement security with asymmetric cryptography, within the domain of the mobile digital provider:

- The scheme of the establishment of trust is simplified since the service provider itself may act as a trusted party (Certification Authority) for its customers.
- Each handset contains a piece of tamper-resistant hardware (SIM chip) where both the service provider's public key and the asymmetric key generation software are pre-loaded.

The proposed scenario of setting up security with asymmetric algorithms is the following:

- Mobile service provider's public key is pre-loaded into all mobiles serviced. The centre's secret key is stored securely, probably in a Secure Cryptographic Device (SCD) which has actually generated the centre's key pair and which would perform all other cryptographic services for the central system.
- Subscriber's asymmetric key pair is generated in the mobile's SIM (Subscriber's Identity Module) card. The generated public key is input into the subscriber's database at the time of the customer's registration, probably centrally or as a part of a manually controlled procedure. Corresponding secret key never goes out from an inaccessible IC location.
- At the time of registration, which has to be on-line in this scenario, the public key in the handset has to be certified (signed by the service provider's secret key). This certification function is performed in the SCD's secure environment.
- The resulting User Public Key Certificate has to be loaded into the SIM card. This must be an automatic process, since these certificates are, for required key lengths (more that 512 bits), long unintelligible strings of digits, and it would be both impractical and error prone if they have to be manually entered. (Another, less secure, option is that this step of entering the user's public key into the subscribers database (registration) plays in itself a role of 'public key certification'.)
- The established public key relationship then enables a basis for a safe broadcast of any other keys, which may be organised in various simple or more complex hierarchies, depending on the specific key management design.

In choosing algorithms, their security qualities, speed of cryptographic operations and other factors, are researched. In a commercial environment it is much more attractive then to use a certified public algorithm than a secret algorithm of unknown qualities. This also applies to the choice of data encryption algorithms.

In addition, the quality of key generators is very important. Any fixed pattern in generating subsequent keys should be prohibited.

6 CHIP CARD TECHNOLOGY

Being a separate discipline in modern data security, chip card security is obviously relevant for security of mobile communications. Ideally, mobile devices should implement IC cards which represent the best possible security design in the chip card technology. As indicated in Section 2, cost should not be any longer a reason to employ older chip technology and also not to use chips specialised for calculations used in asymmetric algorithms (i.e. with arithmetic co-processors).

In recent years, a question on how secure are, after all, chip cards, imposes on all industries which use the chip card technology and whose services rely on security of this media. While for some threats to mobile service providers industry, e.g. 'device cloning', security of chip design is irrelevant (since it is much cheaper to replace the original chip), a risk of successful intentional compromise of the complete system by a successful probing into the SIM in the handset, is a viable threat.

To prevent this happening, it may be necessary to mandate a set of minimum security requirements for the Integrated Chips used in mobile communications (hardware, software, initialisation process, etc.).

For example, recent experiences show that the access to the test mode is the most vulnerable feature in chip design. Assuming that an organised crime is ready to invest in equipment (and expertise) required, this vulnerability of the vast majority of all chips manufactured at present, is an open door into the 'hidden secrets' of chip internals.

Assuming that it is possible to access the chip internals by exploring the mentioned weakness in the chip design, another requirement becomes very relevant, namely the internal logical design (access control features of the application and of the underlying operating system).

7 NEW TECHNIQUES - DYNAMIC SETUP OF CRYPTOGRAPHIC KEYS

This Section may also be titled 'What can be learnt from new Internet security developments?'. Actually, only one aspect present in the recent Internet communications security models is relevant to this Section and it is a dynamic setup of cryptographic key relationship required for user authentication and communications privacy.

Current intensive development of security techniques and of underlying key management methods for totally open networks with massive access (Internet), may help to derive ideas on new methods applicable in mobile communications; since if a dynamic key distribution method is viable in the Internet environment, it may be viable for mobile communications as well.

If the new security concepts for emerging Internet applications prove to be adequately secure (for example using a specialised language, like Java, which is designed such that it prevents many harmful activities, like writing into external data space, etc.), features they could support like a dynamic downloading of key generation code, may revolutionise mobile security as well. In particular, such 'dynamic' systems may provide solutions for the third generation systems (beyond GSM) which may require distributed databases.

CES

...94) Computer Communications Security. *Principles, standard protocols ...ques*. PTR Prentice Hall, New Jersey.
ETSI GSM 02.09., (1992) Recommendation 02.09. *Security Aspects*.
ETSI GSM 03.20., (1992) Recommendation 03.20. *Security-related Network Functions*.
Dmargrav, (1995) GSM Security and Encryption. //www.utw.com/~dmargrav/paper
Unknown author, (1995) Report on security mechanisms in mobile telecommunications system, Trondheim, Norway.

9 BIOGRAPHY

Ilhana has a B.Sc. in Mathematics from University of Sarajevo, Bosnia and Herzegovina and a Masters in Electrical Engineering from the University of Technology in Sydney, Australia. She started to work on communications security issues in eighties in Sarajevo, when she also participated in the CEC COST 11ter security project. Currently she works as a bank's security consultant and for the software company 'Microhit Australia'. She is involved in the work of the corresponding security committees in Standards Australia.

GSM Digital Cellular Telephone System
A Case Study of Encryption Algorithms

Tony Smith
Deputy Director
Centre for Telecommunications Information Networking,
University of Adelaide, 33 Queen Street, Thebarton, South Australia,
5031.   Tel: +61-8-3033222   Fax: +61-8-3034405
Email: tsmith@ctin.adelaide.edu.au

Centre for Telecommunications Information Networking

Abstract:

There have been many reports in the newspapers of eavesdropping and cloning of telephones in the analogue network. However is the digital network free from these attacks? Security was seen as a high priority and was addressed in the development of GSM, this paper looks at fraud in GSM and shows how it has not been eliminated, in fact due to the flexibility and the new services it introduced GSM has created new forms of fraud. GSM includes three algorithms for the purposes of Authentication, Air Interface Encryption and Key Delivery (A3, A5 and A8). The paper looks at the history of the algorithms and the different versions.

Keywords
Global system for Mobiles, GSM, A3, A5, A8, Encryption, Fraud,

Part Two   General Security Aspects in Mobile Communication

)DUCTION

voice networks are being deployed extensively throughout the world with 140 or ks now in existence. It must be noted no matter what the GSM community would like the reality is GSM will not become the exclusive mobile network technology but will share the world market with many other technologies. GSM should not be considered as a competitor but as a platform to understand the issues that will confront a mobile data network. The important underlying issue is the move from fixed to mobile communications, be it voice or data and the problems this brings. Interception of the radio link without knowledge is the first issue followed by cloning and lack of knowledge of the calling parties location. Billing and charging problems increase and is similar to the credit card industry where authentication of a person's identity is required. In a fixed network the location of the wall socket and terminal equipment bears a direct relationship to the creditor. GSM has also had extensive experience due to its deployment in many of countries, handsets are being sold extensively into the consumer market whereas many mobile data products are still penetrating the lower risk corporate market.

In designing GSM an assessment of risk was used to assist in the work of the security expert group. The first objective was to produce a robust authentication mechanism, the second to make the radio path as secure as the fixed line telephone network (PSTN) to preserve the caller's anonymity and privacy. The third objective was to prevent one operator from compromising the security of another network either by accident or due to pressures from competition. All of this must be met with penalty using a cost effective solution.

2   EXISTING ANALOGUE SYSTEMS

Analogue mobile telephone networks are known for a third party to eavesdrop on a conversation, with many notable examples worldwide, another problem is handset cloning for the purposes of committing fraud. Eavesdropping is common, all that is required is a cheap scanner, that along with some spare time, a tape recorder and you are well on your way to obtaining your 15 minutes of fame. Cloning is not as simple but is achieved by capturing mobile handset codes which are sent on an AMPS and TACS system in clear mode every time a handset registers or makes a call. Devices readily available will capture these and then put these into a modified handsets for use. One technique called 'Tumbling' uses each cloned code once before the next set are installed automatically into the handset. This method builds on the knowledge that a typical mobile customer doesn't check their account on a call by call basis and so one extra call goes unnoticed. Even if it is noticed the problem is how to prove to the network operator you didn't make that call and then who is responsible for fraudulent calls.

3 THE ROLE OF STANDARDS IN GSM DEVELOPMENT

GSM has had a somewhat unique development life cycle as it is one of the success stories of a system developed under a standards environment. Typically standards bodies are perceived as being slow and conservative, some steps behind the industry leaders who develop and market proprietary solutions. GSM provides a different model to that typically seen in the IT industry as it is cooperative approach to systems development. GSM began in the early 80's with the realisation of the need for a pan European mobile telephone system, it built on the experience of the NMT analogue cellular system which covered the Scandinavian countries. A French-German agreement saw the standards organisation CEPT form the Groupe Speciale Mobile in 1982. Digital was thought to be the solution with a data rate of 16kbps or something close to it and to resolve the principles of operation detailed technical simulations on spectrum efficiency, delays and queuing were performed from 1984 to 1986. During 1985 France and Germany joined to build prototypes based on four alternatives and in 1987 the official decision was made on the preferred solution. A harmonised Europe issued an EC Directive establishing the European Telecommunications Standards Institute (ETSI) in 1988 with the work of CEPT being transferred to ETSI in 1989. ETSI Special Mobile Group (SMG) is the committee which ratifies work from its nine technical committees (SMG1 to SMG9), these are supported by a full time project team (PT12) which manages the ongoing development.

The important lesson from GSM is the how the participants with different goals and business priorities took a strategic position that although they may have preferred their own solutions they were prepared to put the alternatives to a committee vote and adopt the result. They saw the benefits stemming from commonality and standardisation as being far more important than owning the Intellectual Property. The benefit coming rather than proprietary solution is in the profits derived from mass production for a world market.

GSM developed in the same time frame as the rising demand for mobile telephone services, solving the problem of roaming across Europe has provided a solution for roaming on a 'global' basis. Today as the system is reaching what many think is development maturity in Europe it is going through extensive enhancement particularly with the introduction of enhanced data services under the title of General Packet Radio Service (GPRS) which when implemented will provide a connectionless packet data service to mobiles. Equally the migration of the technology into the US market has provided stimulation for the rapid introduction of enhancements using improved handset processing capacity. The US has seen the introduction of a new voice codec which provides improved voice quality, the timing of the release of new spectrum (1900Mhz) has motivated GSM to finalise the enhanced features under Phase 2 of the standard.

4   SIM CARD & ISO STANDARDS

Role of the SIM Card in GSM
The Subscriber Identity Module (SIM) contains ALL the information relevant to the customer and so the SIM can be swapped from one handset to another for the purpose of handset maintenance. One initial concept was to be able to issue a SIM card to a person for them to use in any vehicle, private or company, taxi etc. This would allow people to remove the card from their low power pocket portable and insert it into their higher power car mounted mobile. It is interesting to see how an original concept failed to materialise as networks were being built for high traffic levels resulting in relatively small cell sizes which work very well with pocket portable units. This eliminated the need for high power mobiles, where as vehicle adaptor kits providing battery charging and handsfree operation are popular.

The SIM was a marketing feature of GSM but is neither provides a benefit or problem for customers, what it has become is the system component that provides the network operator - customer relationship. Customers can roam onto a GSM network elsewhere in the world and the costs are billed by their home operator, the home operator provides all the authentication services as they are held responsible for the intercarrier charges. Also GSM does not divulge to the other network operators secret information about their customers.

SIM Design
The SIM is built to ISO standard 7816 for contact type smartcards with the data elements and communication protocols under the ISO framework. To achieve this a GSM directory was established separate from the established Telecom directory, this contains information about preferred networks, prohibited networks, active services etc. It also provides the mechanism to communicate the challenge (RAND) and response (SRES) to and from the card for authentication.

SIM Card Manufacture and Personalisation
SIMs are manufactured as blank cards although the operator determines the algorithms which are built into the mask of the SIM memory. Each operator can load the customer specific data 'personalise the SIM', however the most common method is to outsource this to the SIM card manufacturer. A third and more expensive method is to personalise the cards at the point of sale by using a system which communicates securely with the operator's Authentication Centre.

5   FRAUD

5.1   Impacts of Fraud

Fraud in cellular is being committed as a organised business, one European operator published figures that showed the number of fraud events being constant as their network grew from 50,000 to 150,000 customers. They didn't see the fraud rate increase in proportion to their customers base and the conclusion is that vast majority of customers are honest and methods to limit fraud will impact on all trustworthy customers.

Fraud has a major impact on all sides of the industry it can't be ignored as being small to insignificant. It WILL occur and its a matter of managing it to keep it to an acceptable level. The credit card industry manages it at about 1.5% of turnover and are working on ways to reduce it further. It will impact customers, the operators revenue and network performance as well as the industry by reducing growth.

The question an operator should ask themselves is; why should I bother focussing on fraud? Why bother committing scarce human resources to fraud control and why spend money on fraud control?

Impact On Customers

Customers who end up with mystery calls on their accounts become dissatisfied. In the majority of cases these calls go unnoticed as cloning normally only places one additional call on each account. As most people don't check their accounts down to the transaction level these extra costs go unnoticed.

However if the additional call is noticed how do you convince the operator that the call wasn't made by yourself. This leads to dissatisfaction with the service being provided, it usually involves a number of people at the customer service centre and it's common to see the customer arguing with the operator.

Another aspect of fraud is that being an organised business it occurs in specific locations in large quantities, this leads to increased network traffic and from the customers perspective is seen as poor service due to network blocking.

Impact On The Operator

Financially, fraud can exceed 2% of turnover with 3% being easily achieved before it is noticed. Due to the large network capital investment a fraud level of 1-2% of turnover can easily equate to 5-10% of profit. With future competition resulting in lower margins and with the average revenue per customer reducing as sales move from the corporate to the consumer market the effect of fraud will become more significant in reducing profit margins.

Fraud also increases the operational costs of customer service centres. Every account enquiry takes a significant amount of time to resolve and often require multiple calls are required to resolve a dispute.

Customer dissatisfaction not only leads to additional churn, which is already excessively high but it also effects sales as personal recommendation is a significant part of the sales effort, we also have found bad publicity takes some 2 years to be negated.

Network loading is often affected by fraudulent traffic and is often see on a localised basis. Localised blocking not only affects customers but also sees the operator responds by installing more base stations, the associated costs which will never be recovered. A UK operator temporarily blocked international call forwarding from part of their London network and found that the majority of the traffic disappeared, this load was seen to be fraudulent.

Industry Impact

Fraud reduces growth of the industry, i.e. it affects people who are potential customers. The consumer market is particularly sensitive to price and have little margin to absorb fraudulent charges, therefore if they have the perception they will incur fraudulent charges they refrain from becoming customers.

In the US analogue networks where methods have been introduced to limit fraud they have introduced barriers to the use of services. e.g. to make a long distance call the customer calls an operator and uses a credit card. Due to the inconvenience, these measures inhibit the use of the mobile phones, but they do provide an opportunity for new networks to compete for customers without price being the sole differentiator.

Law enforcement agencies have been known to influence the industry particularly when it is associated with organised crime. In Australia the launch of GSM was delayed due to the issues relating to call interception and in Pakistan the GSM network was temporarily shut down soon after launch due to concerns about the lack of the law enforcement agencies ability to intercept calls.

5.2 Fraud Types

Fraudsters have no intention for paying for the service they are either using themselves or reselling for their own profit. They arrange to have the airtime charged to a false account, to someone else's existing account or even onto a non existent account.

One method of classifying fraud is Technical, Subscription and Internal.
- **Technical fraud** is about exploiting weakness of the technology or the network. A simple example, network fraud could be achieved by entering the network by a modem connected to the Mobile Swithing Centre (MSC) for remote maintenance. More common forms use diversion, cloning of identity codes etc.
- **Subscription fraud** is the most common method of fraud and is the classic fraud committed within the fixed telephone network. It is also the same fraud experienced by the Credit Card industry and so operators can use a number of existing strategies to reduce this form of attack.
- **Internal fraud** is less obvious but can account for a significant proportion of fraud losses particularly as other forms are reduced.

5.3 Fraud Examples

Subscription Fraud

Subscription fraud relies on providing false identity, this is easier to achieve in a mobile network where as in a fixed network the service requires a wall socket and terminal equipment which is related to a physical location or building and therefore a person.

It can be significantly reduced by asking for proof of identity at the point of sale. This needs to be managed carefully as these documents can be stolen or altered. If you accept electricity or gas accounts for proof then it must be understood that these are easily stolen and are not reliable. There is an industry built around producing false documents and any fraud management team should learn how to obtain these documents so they know how to identify them. It is best to only use documents that theft can be verified in real time e.g. Cheques, Credit Cards are good examples.

Once the persons identity is known, credit checking can be added as these agencies have already established multiple sources of information.

The third method is to limit the risk by building a relationship with the customer. Letting the customer build a credit history is a good method as few fraudsters will pay accounts for 6 to 12 months and then commit fraud for a one or two month period. Credit limits can be built into the network as an Intelligent Network (IN) function as this will block outgoing calls until the account is paid. Deposits, or money in advance also deter fraudsters. In Recent times payment in advance systems are being introduced, these can be network based, a variation of the credit limit IN function, or terminal based, e.g. a prepaid SIM Card. Limiting risk also implies limiting access to high cost services such as International Dialling, International Diversion, Roaming.

Diversion Fraud

One of GSM's billing principles has the caller paying the standard charge irrespective of the handset location with the mobile owner paying the additional international charges when roaming internationally.

The fraud method uses a SIM, it is set for diversion to an overseas number so that the account owner pays the international charge. Although authentication back to the home operator occurs for each call the costs are charged back to the home operator each month. This method requires to be conducted in another country and it exploits the one month delay in billing, SIMs are obtained with a false identity or are stolen but still unreported.

The MSC provides all the switching function and when diversion is set the mobile number is released once the call is established. A new Diversion Number can be established while the other call(s) are in progress and so provides a mechanism for a call reselling 'business'.

A typical sequence of events are:-
1) The handset sets a diversion to an international number (See Figure 1)
2) A call is made to the handset, the caller paying local call fee only, the handset owner pays for the cost of the diverted call, i.e. the international call costs. Nothing is paid if the call is made from a cloned analogue phone
3) Note the mobile side of the network is not involved and so this form of fraud is not detected by network congestion etc
4) The mobiles number is released once the call is established
5) While the first call is taking place the mobile sets the next diversion to another international number.
6) The next call is made to the mobile number. (See Figure 2)

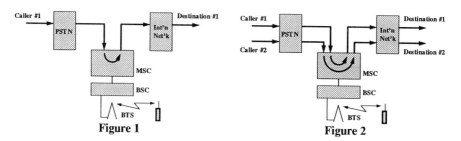

Figure 1 **Figure 2**

Diversion fraud can be detected by monitoring for simultaneous calls, long duration calls, call diversions being set/reset on a continuous basis. However reacting to this has the potential to upset legitimate customers. It is important to know how to handle customer liaison, if handled correctly it can be a bonus.

As most customers stay within their home country few need international roaming BUT even less need International Call Forward. Therefore to prevent this method the operator should separately block and release International Roaming from International Call Forward.

Roaming - Fraud

This is based on obtaining a SIM which is either stolen or obtained using subscription fraud. The SIM is obtained in one country and used in another; or used on another network within the same country if intercarrier roaming is available. It relies on the long bill reconciliation time (30+ days) even if "Home" operator has credit limiting in place.

It can be detected by speed up the loop between calls being made and charging information being passed back to the home operator. GSM is reducing the occurrence by:-

- Daily processing rather than end of month processing of charging records. The MoU now has set a maximum of 72 hours for the forwarding of charging records and this will be reduced to 36 hours by Sept '96.
- Daily processing of accounts above a threshold, Fax or EDI notification to home operator. MoU recommend this to be approximately $150.

- Real Time processing rather than end of day/month batch processing. Many operators now will only set up a roaming agreement if online EDI reconciliation or twice a day EDI is available.
- Use an online Clearing House. The Clearing House only implements the notification, reconciliation as specified by the operator. It is up to the operator to establish a roaming agreement with all the necessary contractual requirements and then specify these to the Clearing House.
- To be effective operators need to move beyond a series of bilateral agreements.
- Build a relationship with other operators and develop a single and consistent agreement for charging and fraud management.

Internal Fraud

It can be caused by sales staff who are paid a commission for each new customer. They can invent customers and receive their sales commission. These 'customers' are easily detected as they don't make calls. One European operator reported that this accounted for 26% of all fraud and accounted for 7% of fraud losses.

Another method of internal fraud relates to activating a customer on the network by adding the customer directly to the network database, the same customer doesn't appear on the billing system. This can be prevented by activating all customers via the Billing systems including all test SIMs. In addition a comparison should be regularly run between the network databases and billing system to audit the procedure.

Cloning Prevention

Cloning is not a problem in GSM. To date there isn't evidence although there have been a number of claims. The industry investigates each very carefully and has found other causes rather than cloning. However if you don't look you won't see it should it happen!

If someone should clone a SIM by copying then they have cloned one card, they cannot guess any other cards IMSI and Ki combination from the knowledge gained.

Cloning in the analogue networks is detected by looking for multiple calls at one time or by Velocity Checking, here the handset cannot travel the distance between the two call locations with the time between calls. Artificial Intelligence systems are being introduced to detect the random calling nature of cloned handsets and finally systems are now going into service which monitor the unique RF 'fingerprint' of each mobile which determines if the handset is genuine or a clone.

Resale of Subsidised Handsets

When handsets are sold under a high subsidy environment, e.g. a purchase price of $100 or less, there is the temptation to purchase the handset, pay the 12 months contract charge, make no calls and resell the handset. Due to the handset subsidy the handset is still valued in excess of the purchase price plus contract fee. Often the handset is purchased at a highly subsidised price and resold at wholesale price to another dealer, who may even claim subsidy again from the same operator! Network operators relying on usage to repay the subsidy. Is this illegal as the contract conditions have been met?

5.4 Conclusion on Fraud

Although there are no reports of GSM been compromised, it is the increased functionality of the system that opens it to new forms of fraud. Fraud in a mobile environment also provides a new set of opportunities which do not occur in a fixed network. Roaming and the delays in inter operator charging have been exploited as one of the biggest opportunities for fraud.

Remember fraud will continue unless their profit margins can be reduced so it isn't a worthwhile business, alternatively fraud will decrease by reducing the convenience of their service. However it is important to balance this against the impact and inconvenience on genuine customers.

6 HANDSET SECURITY

As all customer information is contained in the SIM there potentially is no method to remotely to identify a handset. Stealing a handset and removing the SIM no longer identifies the handset in terms of the customer and the mobile number, therefore GSM has provided a handset electronic serial number, the International Mobile Equipment Identifier (IMEI).

During 1994, the UK industry reported 12,000 were being stolen each month! Stolen handsets are an international trade and SIM sales without some check on the handset origin will only encourage this market. Handsets prices are usually subsidised by the network operator, in Australia we have recently seen handsets being offered for $9. Replacing stolen and subsidised handsets creates a problem for the operator; Who replaces the handset? Can the customer afford a full price replacement handset? Is the replacement also subsidised? The dilemma is that the replacements usually aren't subsidised but the subsidies exist to stimulate the market by overcome the entry barrier of handset cost. To overcome this the UK operator 'Orange' includes handset insurance in the monthly service fee.

Customer service centres must also provide fast deactivation once a customer reports a handset stolen. Some network operators require a PIN code before a customer care centre will make any changes, i.e. caller authentication, and without the PIN code the service won't be deactivated. There has been the case where a customer rings the customer service centre to report a handset stolen but can't remember their PIN code. "I'm sorry we can't deactivate your service without your code", comes the response. It should be remembered that people forget codes and this needs to be built into the system, one method that can help is to not only record the customers pin code but also a prompt that will help them remember the code. The operator then says the prompt in the form of a question and the customer replies with the PIN code.

To manage the IMEI information each network can optionally establish an Equipment Identity Register (EIR) which has three potential categories. The White list is all satisfactory handsets which meet Type Approval, the Grey list which contains doubtful handsets that should be tested and repaired as they exceed the technical specifications and the Black list contains the IMEI of handsets reported stolen.

The global EIR is planned to be established in Ireland, each operator optionally can run their own EIR with an data maintenance procedure providing the appropriate updates. If an operator runs an EIR its another database which needs maintenance, it incurs costs and it is questionable whether it returns a benefit. There have been cases where handsets on the Black list are not blocked but the customer details are passed to the Police for investigation. Thailand has an EIR, although the details are vague it is understood it only allows service to those handsets sold by the operator, i.e. if an operator makes money from handset sales, as opposed to a subsidy, then they want to provide an incentive for customers to purchase handsets from the operator!

7 ALGORITHMS

Three encryption algorithms are used within GSM, these are known as the A3, A5 and A8 algorithms.

A3 algorithm is used to Authenticate the SIM and therefore the customer.
A8 is used to deliver the key for use of the Air Interface algorithm.
A5 is used to encrypt the Air Interface.

The A3 and A8 algorithm are contained within the SIM card and are therefore developed and distributed at the discretion of the network operator. The A5 algorithm is installed in the handset and the Base Transceiver Station (BTS).

7.1 Call establishment process

Each customer is identified by their International Mobile Subscriber Identity (IMSI) number which is contained in their SIM card. When a call is attempted the handset forwards the IMSI to the home operator's Authentication Centre, the IMSI contains both the home operator's identity plus the customers identity. See Figure 3.

The home operator's Authentication Centre queries its data base to obtain that customer's Authentication Algorithm Key (Ki). It then generates a Random Number (RAND) which is sent to the handset transparently via the home operator's Mobile Switching Centre (MSC), the MSC of the host network, the Base Station Controller (BSC), Base Transceiver Station (BTS) and the Mobile Equipment (ME) (the handset), which in tern forwards it to the SIM card. It is the SIM which computes the Signed Response (SRES) which it sent back as far as the roaming network's MSC.

Meanwhile the home network's Authentication Centre has been computing the Signed Response for the handset using the same Ki. This is forwarded to the host operator's MSC to check against the mobiles SRES. If they are equal the call establishment process proceeds, if not the call is terminated.

If successful the voice traffic will be encrypted using the A5 algorithm, the key for this must be delivered to both the handset and the BTS. Delivery of this A5 key (Kc) is performed by the A8 algorithm which is installed in the SIM. Kc is also calculated by the home network and delivered to the host network for use in the host network's BTS. The full key for the A5 algorithm is a combination of the Key Kc and the frame number of the radio link frame (n).

The home operator on receipt of request from a host network, simultaneously forwards RAND, SRES and Kc, which is known as a triplet. As the home operator and the host network may be some distance apart and due to transmission delays, the home network forwards three triplets to allow subsequent calls to be authenticated rapidly with the forwarding of replacement triplets being not as time critical.

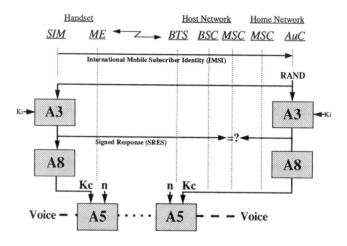

Figure 3   GSM Encryption Algorithms

7.2   A3, A8 Algorithms

Cloning and authentication is far more sensitive than eavesdropping and so although the A5 algorithm is shared the management of a single authentication algorithm was seen as being too difficult. If a single algorithm was used the consequences of a broken algorithm would allow all mobiles to be cloned as has happened in AMPS. The technique employed in GSM allows each operator to choose an algorithm, if one should be broken then other networks haven't been compromised. It also allows replacement algorithms to be implemented without requiring network changes.

RAND is 128 bits long, SRES must be 32 bits and Ki is any format and length as is the choice of A3/A8 at the discretion of the network operator.

The complexity of the A3 and A8 algorithms are limited by the memory capacity and processing speed of the Smart Card and so compromises are made on the level of security it provides. This when added to the limited distribution, ability to change at the desire of the operator allows GSM to continue to be a secure network. Each operator can select their own version of the A3 and A8 algorithm as they issue the SIM card to a customer.

Further to protect the anonymity of the mobile user once they first register with a system (by using their IMSI) they are issued a Temporary Mobile Subscriber Identity (TMSI) which replaces the IMSI.

7.3 A5 Algorithm

The algorithm is owned by the GSM MoU and is tightly copyright protected. Its interface requirements however are public and it works by taking a 22 bit Frame Number (n) from the air interface and a 64 bit key (Kc) to produce a 114 bit sequence which is then Exclusively Or with the voice data stream. Two 114 bits streams are produced one to decode in the incoming voice and the other to encode the outgoing voice streams.

Versions A5(1) & A5(2)

Two different versions of the A5 algorithm exist, the original A5(1) and a later version A5(2) was developed to meet the export licensing conditions of some countries. The need arose for two versions as GSM was originally designed for use in Western Europe and licensing did not allow for export outside this region. The rules for use are: countries who are full members of CEPT use A5(1) and all others use A5(2). Future handsets built to Phase 2 specifications will contain both A5(1) and A5(2) algorithms and will select the appropriate algorithm used in each country to allow for free circulation of handsets throughout the world. Interim use of A5(1) is allowed in some countries until the handsets are available which use A5(2) otherwise the air interface can work in clear mode.

The GSM MoU has a policy to establish a single algorithm for use worldwide.

Activation process

To allow for multiple air interface encryption algorithms and also where regulation prohibits the use of encryption the air interface starts in the clear mode and moves to the cipher mode in a managed way. This allows for handsets with incompatible A5 algorithms to make calls in the clear mode when roaming.

To move from clear mode to cipher mode is a three step process with the handset and BTS starting the communication process in clear mode. The BTS instructs the mobile to start sending on the uplink in encrypted mode, if it cannot decipher the stream it instructs the mobile to move back to the clear mode. When it has established communication on the uplink (mobile to BTS) in the ciphered mode the BTS notifies the mobile it is about to move the down link to the ciphered mode. The mobile reports back to the BTS if it can successfully decode the cipher stream otherwise the BTS reverts the downlink back to clear mode.

The cipher mode is established well before voice or data traffic starts and it occurs part way through the call set up process when only signalling information is being sent. To save attempting this process for all incompatible handsets each handset first sends a list of its capabilities via its Mobile Station Classmark code which includes details on revision level, power capability, A5 algorithm type, etc.

8 FUTURE DEVELOPMENTS

As detection systems become more sophisticated and eliminate external fraud attacks internal fraud will rise as a proportion of total fraud and so we will see a focus on internal fraud.

The development of Dual Mode handsets, e.g. AMPS/GSM mainly to allow wider roaming using a single handset will open GSM users to the weaknesses of the other system, e.g. GSM users will be exposed to AMPS cloning fraud.

The biggest unknown is the potential for Value Added Services (VAS) to provide opportunities and weaknesses for exploitation. History has shown the more sophisticated the service the more sophisticated the form of fraud and the harder it is to detect. With so many VAS being considered and the ability to build and implement a VAS rapidly there is a big temptation to implement a VAS in response to a competitor without the due design process which includes a fraud risk assessment.

Whatever direction the technology goes, there is a bright future for people working in the area of security and risk assessment.

PART THREE

New Security Algorithms and Methods

30
A new algorithm for smart cards

*Ch. Marco, P. Morillo**
Dep. de Matemática Aplicada y Telemática, E.T.S.E.T.B.
Universidad Politécnica de Cataluña
C/ Gran Capitán s/n, módulo C3
Barcelona (España)
Tel. (34) (3) 401 65 38
E-mail: cmarco@mat.upc.es
**Tel. (34) (3) 401 60 08*
**E-mail: paz@mat.upc.es*
Fax (34) (3) 401 59 81

Abstract
In this paper we present a new cryptographic scheme based on the discrete logarithm problem which allows to perform encryption, authentication and digital signatures. Due to its versatility and efficiency it can be appropriated for the identification and exchange of information between smart cards and between smart cards and terminals.

Keywords
Cryptography, discrete logarithm, smart card

1 INTRODUCTION

At the present, a lot of security services, many of them based on cryptographic systems, are invading our every day life. The limitations of conventional criptography have been already overpassed, since public key cryptography provided the solution to the two basic problems of the secret key cryptography: the secure exchange of secret keys and the possibility to generate electronic signatures.

We can say that, since 1976, when the concept of public key cryptography was presented by Diffie and Hellman in (Diffie-Hellman, 1976), public key cryptosystems has shown to be ideal for an open communications world. Typical applications of public key cryptography, such as identification, access control, exchange and/or storage of enciphered data, software protection, etc... are commonly used in our days, and their use in the future seems to be unlimited.

According to this criteria, it is not very intrepid to think that smart cards can play a very important role in the world of security. In fact, it seems that smart cards will become the basic security device in the future due to three fundamental reasons:

1. As it is well known, a smart card is able to store and process information and, therefore, to perform cryptographic algorithms. We can say that a smart card is a true microcomputer.

2. Smart cards are very easy to use (the user only must remember his own secret key to authenticate himself against the card) and, moreover, they are roughly of the size of common credit cards.

3. Smart cards offer both logical and physical security.

Nevertheless, in spite of the continuous evolution of the technology, the limitations of the cards in terms of memory capacity and computational power must be taken into account when designing cryptosystems for smart cards. According to this idea, Schnorr in (Schnorr, 1991) proposed an algorithm which minimizes the length of the keys and Fiat and Shamir in (Fiat-Shamir, 1987) proposed an algorithm which minimizes the number of operations.

In this paper we propose a new cryptographic protocol which is based on the ElGamal encryption algorithm (ElGamal, 1985), so it is based on the discrete logarithm problem. Our algorithm includes identification, encryption with message authentication, and signature generation. The most important feature of our scheme is its flexibility since it provides different security functions. Due to its characteristics the algorithm can be considered very suitable for interactions between smart cards and terminals. As we will show, it can be used to authenticate the terminal or the cards, to exchange confidential and authenticated messages and to generate and verify digital signatures. We will also show that the new scheme can be a good alternative when choosing a cryptographic algorithm for smart cards.

2 THE NEW SCHEME

We assume the existence of a trusted authority (a government, a credit card company, a computer center, a military headquarters, etc.) which issues the smart cards to the users.

The cards contain an identification number generated by the KAC (Key Authentication Center), I, and a signature for the pair (I, v), where v is the user public key. The pair (I, v) allows to identify the card. The identification number and the public key of the card have to be verified by the terminal (or other card) reading I and v from a public file or checking the KAC's signature using the KAC's public key. So, any interaction between a card and a terminal or between two cards which requires the use of their respective public keys must contain a previous verification of such keys.

On the other hand, we have just mentioned that the new scheme is based on the discrete logarithm problem in a subgroup of Z_p^* (McCurley, 1990) (Odlyzko, 1985). So, a modulus prime p, a prime q which divides p-1, and an integer b, are required. The private and public keys are the integers s modulo q and $v=b^{-s}$ mod p, respectively. Also a one-way hash function, h, is used.

2.1 General scheme: Encryption with authentication and signature generation

The algorithm follows the next steps:
1. A picks a random integer $0<r<q$.
2. A calculates $z = b^r$ mod p.
3. A calculates $x = m \cdot v_B^{r \cdot e \bmod q}$ mod p, where the message, m, is an integer, $0<m<p$, which includes at a minimum:

• The identity of A and B, to show that the message was generated by A and intended for B.

• The information, m_1.

• A timestamp, t_A, including the generation and expiration time to prevent delayed delivery of messages.

• $m_2 = h(m_1, v_B^{r \cdot e \bmod q})$. This value must be unique within the expiration time of the message. Then, the user B can store it until the message expires and reject messages with the same value m_2 which are received during that period.

The value e is calculated as $e = h(m_1, z)$.

4. A calculates y such that $s_A \cdot y = -(r \cdot e + 1)$ mod q.
5. A sends to B: (x, y, z).

Message reception and signature verification:

6. B deciphers calculating $m = x \cdot v_B \cdot v_A^{y \cdot sB \bmod q}$ mod p.
7. B authenticates the message checking that $m_2 = h(m_1, x/m \bmod p) = h(m_1, v_B^{r \cdot e \bmod q}$ mod p).
8. The signature is verified checking that $(z^{-e} \cdot v_A^y)$ mod $p = b$.

One special and important feature of the proposed scheme is that encryption includes authentication. This is a basic difference with respect to the general encryption algorithms since it makes possible to authenticate a message independently of the signature generation, and simultaneously to the encryption.

The receiver doesn't have to check the signature to verify the authenticity of the message. Such verification is performed very easily after deciphering (the redundant information included in the message, m, allows the verification of the authenticity of the message). Signature verifications (if the message is signed) would be performed only if later authentications were necessary.

Finally, the new scheme not only allows to encrypt and authenticate messages and/or generate electronic signatures. With slight variations the algorithm can be also used as an identification protocol. The identification can be performed with two different procedures which are based on the encryption and authentication scheme and on the signature scheme, respectively. Both options, denoted as one-way and two-way identification schemes, are presented.

2.2 One-way identification

1. A picks a random integer r, $0<r<q$.
2. A builds a messsage, m, such as it has been explained before, but using a random value m_1.
3. A calculates $x = m \cdot v_B^{r \cdot e \mod q} \mod p$, where e can be a random integer in $\{1,..., 2^t-1\}$ or simply $e = 1$.
4. A calculates y such that $s_A \cdot y = -(r \cdot e + 1) \mod q$.
5. A sends to B: (x, y).
6. B calculates $m = x \cdot v_B \cdot v_A^{y \cdot sB \mod q} \mod p$ and checks that $m_2 = h(m_1, x/m)$.

2.3 Two-way identification

1. B sends a random integer, k, $0<k<q$ (or with shorter length).
2. A picks a random integer $0<r<q$.
3. A calculates $z = b^r \mod p$.
4. A calculates y such that $s_A \cdot y = -(r \cdot e + 1) \mod q$, with $e = h(k, z)$.
5. A sends to B: (z, y).
6. B calculates $e' = h(k, z)$.
7. B checks that $(z^{-e'} \cdot v_A^y) \mod p = b$.

3 SECURITY

The security of the new protocol lies on the difficulty to calculate discrete logarithms. The identification and signature protocols and the encryption algorithm, only can be broken if the attacker gets the secret keys.

So, let's see how the algorithm can be attacked.

Including $m_2 = h(m_1, v_B^{r \cdot e \mod q} \mod p)$ we avoid known plaintext attacks since m_2, which contains random information, is never made public. An attacker can't obtain a pair plaintext-ciphertext.

To supplant A, a forger can't use an old one-way identification message of A since its contents would reveal to B that it is a false message. Moreover, without the secret key of B, an attacker is unable to understand a message from A and, therefore, he can't obtain any information.

To falsify a signature of A (or supplant A in the two-way identification procedure), a forger could fix y, and then try to find z such that $z^{-e} \cdot v_A^y \mod p = b$, with $e = h(m_1, z)$ (or $h(k,z)$), or he could fix z, and try to calculate y. In both cases the problem is very hard.

Both the authentication and signature schemes, apparently, don't reveal any information about the user secret key, since all values are calculated using random variables. Therefore, the enciphered messages always depend on a random number.

So, we can conclude indicating that we have not found any efficient attack to the new scheme.

4   COMPLEXITY

In order to show the validity of the new scheme it is interesting to compare it with two schemes which are considered suitable for smart cards. These schemes have been already mentioned and they are the Schnorr and Fiat-Shamir algorithms. The parameters considered are the number of operations and interchanged bits (see Table 1) and the length of the keys. These parameters are directly related to the limitations of the cards in terms of memory space, computational power and slow serial I/O.

Table 1 Complexity of the new scheme

	Average number of multiplications (t=80)		Number of interchanged bits
	Transmission	Reception	
Encryption	240	240	672
Signature	240	260	672
General algorithm	480	240+260	1184

Some observations must be done:

1. The number of operations can be reduced by optimization.

2. Part of the calculations in transmission can be performed in a pre-processing stage and/or using an algorithm to simulate random exponentiations, since the exponentiation is independent of the message. So, the most-time consuming operations can be precomputed in the case of the signature generation and in the encryption scheme (or in the authenticacion procedures). In the case of the general scheme, including encryption, authentication and electronic signature, one of the two exponentiations can also be precomputed.

3. Verifying the signature requires to check that $(z^{-e \mod q} \cdot v_A^y) \mod p = b$. It can be computed such as Schnorr shows in (Schnorr, 1991):

$$\text{Initiation: } e = \sum_{i=0}^{79} e^i \cdot 2^i, \; y = \sum_{j=0}^{159} y^j \cdot 2^j, \text{ with } e_i, y_j \in \{0, 1\}.$$

1. We calculate and store $(z^{-1} \cdot v_A)$.
2. x=1.
3. $x = x^2 \cdot z^{ei} \cdot (v_A)^{yi}$ for i=159 to 0.
4. 1=x.

Finally, we must indicate that the secret and public keys are 160 and 512 bits long, respectively.

4.1 Comparison

The scheme presented by Schnorr is based on the discrete logarithm problem. Therefore, the components of the Schnorr algorithm and the components of the new scheme are the same. On the other hand, the protocol proposed by Fiat and Shamir is based on the difficulty of extracting modular square roots if the factorization of the modulus, n, is unknown. The modulus, $n \geq 2^{512}$, is the product of two secret primes p and q. Public and secret keys are integers v_j and s_j (smallest square root of v_j^{-1}), respectively. A hash function, h, in $\{0,..., 2^{wk}-1\}$, is used.

The complexity of the algorithms is shown in the Table 2.

Table 2 Complexity

	Schnorr (t=80)	Fiat-Shamir (k=9, w=8)
Length of the keys (bits)		
Secret key	160	4.608
Public key	512	4.608
Authentication scheme		
Average number of operations	240+260	44+44
Number of interchanged bits	752	8.201
Signature scheme		
Average number of operations	240+260	44+44
Number of interchanged bits	240	8.201

In order to compare the schemes we must take into account that the number of communication bits of the signature scheme presented in the Table 2 doesn't include the length of the signed message. Moreover, in the case of the algorithm of Schnorr, the amount of communcation for authentication can be reduced from 752 to 320 bits by using a hash function, and , on the other hand, part of the computational effort can be done in a preprocessing stage.

So, we can see that the Fiat-Shamir scheme has a very significative low number of operations, but it uses very long keys and messages.

The new cryptographic protocol and the Schnorr scheme are very similar in terms of complexity. The length of the keys is the same (quite short), and also the number of transmitted bits and the number of operations are approximately the same. Therefore, the main difference is their functionality. The new scheme, due to its versatility, can be used in many applications of security services.

5 CONCLUSIONS

First of all, we can say that such as we have shown, the complexity of the new scheme is nearly equal than the complexity of the protocol proposed by Schnorr. The scheme proposed by Fiat and Shamir needs a lower number of operations but uses extremely long keys, and the total number of interchanged bits is enormous.

With the proposed scheme, the identification can be done with only a one-way transmission. This suposes that A doesn't have to wait for any message from B to generate its authentication message.

We would like to emphasize that with our scheme, the same algorithm allows to encrypt, authenticate and sign messages. With the Schnorr scheme it is possible to sign messages but, if we want to encipher this messages then another cryptosystem must be used. So, we should combine two different algorithms. With the new scheme, we are adding new features that can be very useful.

So, we have proposed a new cryptographic scheme which is efficient and flexible, and which can be used to encrypt (with simultaneous authentication), to authenticate the origin (identification), and to sign messages (totally compatible to the encryption and authentication scheme). Moreover, we have shown that the new squeme compares well, in terms of complexity, with algorithms which are considered specially suitable for smart cards. The basic features of the new scheme are:

- Flexibility.
- Most of the computational work is performed by the receiver. Moreover, part of the calculations in transmission can be done by the card in a precomputation stage, which is independent of the message.
- The length of the keys and the number of communication bits are pretty short.

Due to its characteristics the proposed algorithm can be considered particularly suited for interactions between smart cards and terminals.

Finally, we can mention that the new scheme can also be implemented using elliptic curves (Koblitz, 1987). In this case, shorter keys would be used and lower number of bits would be exchanged. Nevertheless, the number of calculations would be increased. We consider that it would be worth studying such implementation.

6 REFERENCES

Diffie, W. and Hellman, R. (1976) New directions in cryptography. *IEEE Trans. on Inform. Theory*, **IT-22**, 644-654.

Schnorr, C.P. (1991) Efficient signature generation by smart cards. *Journal of Cryptology*, **4**, 161-174.

Fiat, A. and Shamir, A. (1987) How To Prove Yourself: Practical Solutions of Identifications and Signature Problems. Advances in Cryptology - Crypto'86, *Lect. Notes in Comp. Science*, **263**, 186-194.

ElGamal, T. (1985) A Public Key Cryptosystem and a Signature Scheme based on Discrete Logarithm. *IEEE Trans. on Inform. Theory*, **IT-31**, 469-472.

McCurley, K. (1990) The discrete logarithm problem. Cryptol. Computat. Number Theory, *Proc. Symp. Appl. Math.*, **42**, pp. 49-74.

Odlyzko, A. (1985) Discrete logarithms and their cryptographic significance. Adv. in Cryptology, Eurocrypt´84, *Lect. Notes in Comp. Science*, **209**, 224-314.

Koblitz, N. (1987) *A Course in Number Theory and Cryptography*. Springer-Verlag, New York.

7 BIOGRAPHY

Christian Marco received the Engineer of Telecommunication degree from the Universidad Politécnica de Cataluña, Spain, in 1993. Actually he is doing his doctoral work at the Universidad Politécnica de Cataluña under Paz Morillo.

He began his academic career at the Centro Politécnico Superior de Zaragoza, Spain, as Assistan Professor in 1993. Since 1995 he is Assistant Professor at the Escuela Técnica Superior de Ingenieros de Telecomunicación de Barcelona, Spain. He has been working in the field of telematics with particular emphasis on security services in communications. His main research interest lie in the area of public key cryptography and its applications.

Paz Morillo studied mathematics at the Facultad de Matemáticas de la Universidad Autónoma de Barcelona, Spain. In 1987, she received the Ph. D. in computer science at the Facultad de Informática de Barcelona, Spain.

In 1985, she joined the Escuela Técnica Superior de Ingenieros de Telecomunicación de Barcelona, Spain, where she became Full Professor in 1987. She has been working in the field of applied mathematics, specially in graph theory and number theory. Her current research interest is the application of number theory in cryptology. In particular, she works in the areas of public key cryptography, secret sharing schemes and primality testing.

31
A New Approach to Integrity of Digital Images

D. Storck
Institute for Computer Graphics, Fraunhofer Gesellschaft
Wilhelminenstr. 7, 64283 Darmstadt, Germany,
+49 6151 155 418 (444 fax),
email: schoko@igd.fhg.de, URL: http://www.igd.fhg.de/~schoko

Abstract

This paper describes a new approach to integrity for digital images. Different to the most mechanisms, this approach describes a method which is resistant against a loss of integrity by slight modifications like compression or conversion. The suggestion to archive this goal is a transformation to the frequency-domain similar to the jpeg-compression.

Keywords

Integrity to digital images, frequency-domain

1 INTRODUCTION

Ever since the invention of the photograph, people have believed in what they saw in pictures. The same goes for newspapers and, to a much higher degree, for TV news coverage. Supported by existing techniques and more and more affordable hardware to perform these on, creating photo-realistic scenes which have never happened or manipulating real pictures becomes more and more easy. This may be a great opportunity for entertainment. Think about a movie with Tom Hanks and Marilyn Monroe. Two people who've never acted together in the same movie star in a perfect illusion (as has already been shown in „Forrest Gump" or, even before that, in Woody Allen's „Zelig"). On the other hand, the door for misuse is ajar. Pictures of famous people can be modified to provide a different information. *„Our lack of clarity produces both overly optimistic (trusting) and overly pessimistic perceptions. In the world of digital information, the tools and mechanisms of ensuring integrity are complex and exotic, and our unfamiliarity with these tools leads us to distrust their efficacy. Thus we regard the digital information environment as basically lacking integrity. These feelings are supported by, for example, the arresting examples of image manipulation in the digital domain, which run counter to our assumptions about photographic images as recorded vision*

(Mitchell 1992)." (Clifford 1994). To avoid such forgery and fraud, certain mechanisms must be used to check the integrity and origin of pictures. Common techniques used for textual data may also be used for images, but there are several disadvantages involved in this approach. The traditional techniques used to provide integrity detect every modification in the image data to some extent. This means a scaled or gamma-corrected image will lose its integrity, which may not be the intention of the sender or originator of the picture. Integrity should only be corrupted if the semantic information of the picture is modified.

The department Security Technology for Graphics and Communication Systems of the Fraunhofer Institute of Computer Graphics is developing a method to detect manipulations of images which modify the semantic information of the images. This work is part of the PLASMA-project, an object-oriented security-platform for multi-media data. Modifications which affect the quality (like gamma-correction, scaling or compression) should not necessarily trigger a detection mechanism. The new approach makes it necessary to transform the image data into the frequency domain. This is done by a Discrete Cosine Transformation (DCT) of 8x8 pixel blocks. The coefficients resulting from this transformation have different influence on the picture. The first value (DC coefficient) and the next few AC coefficients are the most important factors in the appearance of the picture. The higher AC values have only little perceptual influence on the picture and are often modified by compression standards to get good compression rates. Therefore these values of the higher frequencies are irrelevant for the integrity check. Severe manipulation of the picture always affects the lower frequencies and the DC value, therefore the protection of integrity is based on these coefficients.

2 RELATED WORK

A very good overview to the new problems of integrity of digital information is given in (Clifford 1994). Many parts of the next two sections are taken from this paper.

One of the key advantages of digital information is that each copy is of equal quality to the original. Indeed the copies are identical to and indistinguishable from the original digital object, whether the object represents text, images audio video, or executable computer software. But anyone who is involved with computers quickly learns that it is simple to alter bits or characters in a file and pass it to others as it would be original. The amount of tools allowing to change files in a sophisticated manner has increased significantly over the last ten years.

To avoid such fraud, a number of „digest" algorithms have been developed which compute a relatively brief „hash" or digest of a given digital file, typically considerably less than 1000 bytes in size. Most famous algorithms were developed by Rivest. They are known by the identifier MD4 and MD5 (Rivest 1992).

The characteristics of these algorithms are such that it is extremely improbable for any two files to generate the same digest, particularly if the two files are „close" in content. It is possible to approximate mathematically the likelihood that two different files would produce the same digest; this probability is very small. The problem with most of these algorithms is that they only work if one assumes that the integrity is maintained only if there is a bit-for-bit equivalence between the copy and the original data. Unfortunately, this definition is often too limited to be of much use. With the continued proliferation of storage and interchange standards for various types of digital objects, and the development and deployment of increasingly intelligent systems to convert them from one format to another, bit-by-bit equivalence will lose much of its value as a definition of „identical" objects in the future (Clifford 1994). As Clifford describes in his paper: „*We have very few tools to help us deal with digital objects at the level of abstraction of intellectual content. Our understanding of digest algorithms, for example does not extend beyond bit-level equivalence to deeper notions of document content. The development of digest algorithms at the intellectual content level (or even heuristic digest algorithms that measure similarity rather than precise equivalence, and that perhaps only work on limited classes of material, or only work on some rather than all instances of such objects) would represent a substantial breakthrough. In my view, the development of such methods, as well as a general theoretical framework within which to categorize them and to analyze their performance, represents a challenging but important area for ongoing research.*"

Surely, a realization of the approach described in this paper will not be the perfect solution for all these problems. But it may be a first step towards an algorithm for integrity of digital images which is robust against several modifications. In the following section some requirements are spelled out. They describes modifications to pictures which destroy the bit-to-bit equivalence between copy and original image, but do not destroy the semantic information of the image.

3 REQUIREMENTS TO AN ALGORITHM FOR THE INTEGRITY OF DIGITAL IMAGES

As described above, in the world of digital information algorithms are often applied to digital images to translate them from one form to another. These algorithms change the bytes in an digital image but do not touch the semantic information. They can be roughly categorized into two classes: lossless and lossy. In the case of lossless algorithms, no information is lost, and normally the original data can be reconstructed from the converted file. In the case of lossy conversion some information of the image data gets lost. But if it is only a loss of the coefficients from high frequencies, the difference between the original and the altered copy of the image can only be detected by the most discerning eye. In the most cases not even by any human being.

For example the popular JPEG compression algorithm supports some lossy compression options which discard information. In fact, typically JPEG compressors let the user manipulate settings that control the degree of loss permitted in compression, with higher settings for permissible loss producing a greater degree of compression. Whether information is actually lost in a specific case depends on the settings used in JPEG compression and the content of the original image. But it may be difficult or impossible for the user receiving a JPEG-compressed image to be aware of or to detect this loss of information.

According to these given parameters, the integrity of an image is not destroyed by this methods and an algorithm for the integrity of digital images should be robust against conversion and lossy compression (especially the JPEG compression).

Due to the differences of the most computer monitors it may be necessary to correct the brightness of a received image. There are many tools available to adjust the brightness of an image to a specific CRT. These gamma-corrections do not destroy the semantic information and therefore should not affect the integrity of the image.

If an image is scaled or enhanced by interpolation to a higher pixel density there is no loss of information (at least no loss of semantic information, only if the image is scaled down to an icon) if one knows how the interpolation is done, since it should be possible to reconstruct the original. Due to this it should be possible, that the integrity is not lost by (at least) scaling an image.

3.1 Requirements - Summary

An algorithm that provides integrity checks for digital images should detect any manipulation that destroys or changes the semantic information of the image. The following manipulations do not effect the semantic information and there should not be a loss of integrity if they are used.
- Format-conversion
- Compression (especially JPEG)
- Gamma-Correction
- Scaling

4 SPECIFICATION

4.1 The frequency domain and the JPEG-compression

As described at the beginning, with the JPEG-compression the pixel information of an image is transformed into the frequency domain. The resulting coefficients affect the appearance of the image in different ways of which the DC-value and the coefficients of the low frequencies are the primary influences. The coefficients of the medium and the high frequencies have only little effect to the information of the image (Hung 1993).

These different influences are important for a first approach to integrity of digital images. It seems to be sufficient to use only the DC-value and perhaps the coefficients of the lowest frequencies for the integrity (see figure 1). The final solution will be determined by future tests and fine-tuning. In this first approach (only the DC-value is used) the robustness against compression on the high frequencies or other low-pass filtering should be achieved. Further tests have to been made to check if this approach is also robust against manipulations. To avoid that color modification like gamma correction destroy the integrity it is necessary to transform the RGB-values after the Discrete Cosine Transformation into YUV-components. The Y-values are not affected by changes in the color space, so they are used to build the message which is protected by a common integrity mechanism.

If already slight modifications to the image changes the DC-value of the Y-components it becomes necessary to encode this value in a special way. A possible solution for this encoding is shown in the next section.

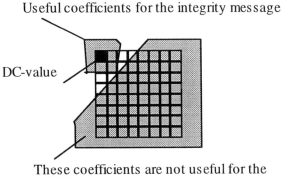

Figure 1 Not all coefficients are useful to built the integrity check message. The grid shows the DC-value and the 63 AC-values. Only the DC-value and few of the lowest frequency coefficients are used in the approach described.

4.2 Building the integrity check message

To create the integrity check message the common techniques are used (Schneier 1994). But different to the common approaches not the complete data is used to build the message. As shown in figure 2, only some data is extracted from the image and used with the classic algorithms to create a signature which allows to check the integrity of the image.

Encoding of the coefficients
If the described approach is not robust enough against slight modifications by using the DC-values directly it becomes necessary to encode the coefficients. Once this has been done, the hash-value (using MD4 or MD5) is not built directly from the DC-values, but rather built from the result of the *integer*-function of the difference between two following DC-values. Even if slight modifications of the image affect the difference between two DC-values, it is not to expected that the sign of the difference between the DC-values of two following DCT-blocks is changed. This robustness assumes that the modification is done to the complete image. If the modification is confined to a region of the image or to manipulation of some objects from the image, it can be expected that some differences change completely and the integrity will be lost.

In this way, it is possible to protect an image against the loss of integrity if there is a manipulation of the semantic information of the image (This assumes that the semantic information is related to the shape and the position of the object and not to their color.). Figure 3 illustrates the encoding of the DC-values.

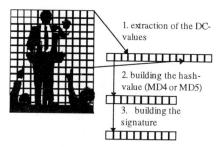

Figure 2 Creation of the signature for protection of integrity. First the DC-values are extracted. These values are the input for a MD4 or MD5 hashing function. The hash value can be signed with any common public-key algorithm like the RSA.

Integrity and scaling
An integrity protected image with the techniques described above, would clearly lose its integrity by scaling. This can be avoided by adding the original size as part of the signed information if the integrity of a picture should be checked. It will become necessary to restore the original image size before checking the integrity. The following example shows the possible structure of an integrity check message for images.

```
typedef struct {
  int width, height;
  BOOL difference_flag;
  Signature SizeSig;
  Signature ImageSig;
}ImageSignature;
```

The *difference_flag* will be set to TRUE if the image signature is calculated from the results of the *integer*-function from two following DC-values. If the signature is calculated from the DC-values directly the flag is set to FALSE.

Integration of the image signature
The signature of the image should not appear as a separate file and has to been integrated directly into the image file. The different file formats offer several ways to integrate additional data. In GIF-files the signature can be put in the **application extension block** and in JFIF-files (the common exchange format for JPEG-compressed images) the signature can be integrated after a so called **APP0-marker**.

Sadly, some imaging tools do not write back these values if an image is stored. So the integrity of an image may be lost because the signature was not stored by an application even if there were no changes to the image. In this case the signature in a separate file may be helpful.

DC-values

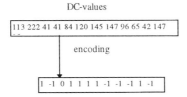

result of the *integer*-function of the differences

Figure 3 The encoding of the DC-values may become necessary if already slight modifications produce a loss of integrity.

5 CONCLUSION

This new approach may be a step towards a new understanding of integrity for digital information. Even if the first tests produced the expected (promising) results, a lot of work still has to be done. The first realization works fine with simple pictures as shown in figure 4. The different possibilities to create the integrity check message (selection of the coefficients to calculate the signature, calculate the differences between two values, etc...) allow to tune the mechanisms for special requirements.
Future test will show if it is possible to create completely different pictures which produce the same signature.

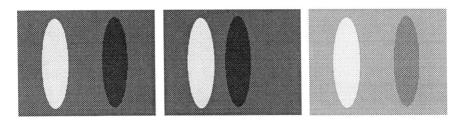

Figure 4 The new approach already produces the expected results with simple images. Part a) shows the original image. Part b) shows the same image, but the distance between the two objects was reduced. Part c) is a brightened version of the original image. The tests show that the integrity is still correct in image c). The integrity check for image b) fails because 20% of the calculated message differ from the message produced by the original image. With common mechanisms like the MD5 hashing function and the RSA algorithm the integrity check would fail for both images.

6 REFERENCES

Clifford, A.L. (1994) The Integrity of Digital Information: Mechanics and Definitional Issues. *Journal of the American Society for Information Science* 737-744, 45(10).

Hung, A..C. (1993) PVRG-JPEG Codec 1.1. Portable Video Research Group. Stanford University

Mitchell, W.J. (1992) The reconfigured eye: Visual truth in the post photographic era. Cambridge, MA: MIT Press

Mitchell, W.J. (1994) When is seeing believing? *Scientific American*, Feb 1994.

Rivest, R.L. (1992) RFC 1321: MD5 message digest algorithm

Schneier, B. (1994) Applied Cryptography. *John Wiley and Sons.*

7 BIOGRAPHY

Dietmar Storck has earned a Dipl.-Inform. from the Technical University Darmstadt in 1993. Since then he is employed at the Fraunhofer Institute of Computer Graphics. He is a member of the department „Security Technology for Graphic- and Communication-Systems". His research activities are dealing with encryption of multi-media data and special methods for the security of image data. Further interests includeaccess control and billing of multimedia services via the internet.

32
How to Protect Multimedia Applications

Eckhard Koch
Fraunhofer Institute for Computer Graphics
Wilhelminenstr.7, 64283 Darmstadt, Germany,
phone: +49-6151-155147; fax: +49-6151-155444
e-mail: ekoch@igd.fhg.de

Abstract
New kinds of application and the convolution of different media types require a specific approach to fulfill the arising security demands. In principle, the security of a multimedia system can be supplied on the level of the network, the application and the data. This paper concentrates on the security of the application and data level, decribes various requirements and outlines possible solutions to protect multimedia applications.

Keywords
multimedia applications and data, security platform, application programming interfaces, use control, copyright protection

1 INTRODUCTION

The most significant advancements in cryptography were made about 20 years ago by developing the public key cryptography. On the one hand, the key management, especially the distribution of secret keys, was simplified, and, one the other hand, it was made possible to set digital signatures necessary for electronic business transactions. The main application areas were confined to performing authentication as well as to ensuring confidence, integrity, and non-repudiation via simple point-to-point communication.

In the field of multimedia communication and information services the development has been enforced during the last years. New kinds of applications and services have been developed like groupwork applications (CSCW), broadcasting services (Pay-TV, video on demand, interactive TV, hypermedia information services (WWW)). On the other side, the information and communication technology of today - in contrast to the former, mainly text-oriented computer communication - features a combined use and a merging of all sorts of media types. When developing these new multimedia applications and constructing the system architectures the solution of basic technical problems was in the foreground and thus the integration of security functionality was mostly neglected.

Now these systems and applications are leaving the research community and enter mainstream commercial applications, the lack of appropriate security techniques is becoming more and more an obstacle to the real-world applicability of multimedia systems and applications.

This paper is mainly describing the requirements which come forth when security solutions are added to the existing applications and the possible solutions. Furthermore, the specific problems and solutions of the protection of multimedia data are presented.

2 REQUIREMENTS

In principle, security of a system can be supplied on the level of
- the network and the respective protocols
- the application and
- the data

These levels, however, do not exclude but complement each other. Several security provisions may be done on the network and respective protocol level. Some security requirements, however, cannot be met on this level, like the deliberate setting of a digital signature, the authentication of the actual user, the specific cryptographic treatment of images, video etc. or the protection of the intellectual property of multimedia data (copyright protection).

For these application-, data-, and user-specific requirements expedient and adequate solutions can only be found on the application and data levels.

2.1 Application Level

In order to avoid different security solutions for every specific application and application protocol and to remain flexible towards the use of new applications and services the use of a generic security platform seems reasonable for the integration of security on the application level. This security platform should meet the following requirements (Gehrke and Koch, 1995):
- independence from application,
- independence from security technology,
- independence from communication,
- compatibilty and interoperability,
- media-specific operations.

The last issue offers the possibity to differentiate between media types and to apply specific operations to protect multimedia data.

2.2 Data Level

The main objective in introducing this level of protection is to have some means of control and protection of multimedia data after secure transmission and delivery by any kind of application. This is mainly caused by the needs of any information and content provider who hesitates to offer valuable data without any control on the use and any technical means to protect their intellectual property.

From the technical point of view two requirements can be identified. Firstly, after the secure delivery of digital information to an authorized user it would be neccessary to control the operations the user is allowed to perform. They can cover a wide range of allowed actions, e.g. the right to view, display, manipulate, or copy the multimedia data. Secondly, the multimedia data have to be marked or labeled in some way that allows to identify the ownership, the rights etc. to the data after distribution. This information (mark) has to be embedded into the multimedia data, in a way that the mark is perceptually invisible, unremovable, unalterable and furthermore survives processing which does not seriously reduce the quality of the multimedia data.

3 SECURING MULTIMEDIA APPLICATIONS

To fulfill the various requirements outlined above a security plattform has to offer an application-independent Application Programming Interface (API). This could be the Generic Security Service API (GSS-API, RFC 1508,1509) or a similar one. Using such kind of API allows to plug the security platform in different kinds of application. Furthermore, due to the nature of this API the security of the system is independent of the transport layer and specific communication protocols, because the communication is handled by the application itself.

The requirement to distinguish between different kind of multimedia data can be fulfilled by introducing the concept of filters. The filter has to be integrated into the application and passes the different media to the security platform in order to process them seperately.

To be independent of the security technology means to be able to use any security algorithm and method, hard-, soft- or firmware. This can be reached by the use of a technology-independent API which has to be implemented between the application-independent API and the crypto modules.

One of the critical issues - not only in the field of security - is the interoperability between different security platforms, methods and concepts. Interoperability can be obtained to a certain degree by implementing various quasi-standard security modules and data formats below the technology-independent API. Surely, it has to be indicated by each communication partner which technology he supports. But describing this with a formal security policy allows to negotiate a common configuration of the security platform on sender and receiver side.

The intense demand in such flexible security architecture has led to a first implementation of a platform for secure multimedia applications (Krannig, 1995). This security platform is based on an object-oriented approach which facilitates the compliance with the requirements.

4 PROTECTING MULTIMEDIA DATA

Control of the usage of multimedia data on receiver side is difficult to introduce and to perform from the legal, social and technical point of view. A specific rendering device (filter) has to be installed on user side

- to control which rights (e.g. print, display, copy) to which piece of information the user has and
- to account, monitor and trace the usage

On service provider side the information has to be prepared for controlled distribution, which means the multimedia data have to be marked, encrypted, signed and encapsulated. On the user side only the specific device can handle and use the prepared and protected information. This device can be implemented in software and/or hardware and can be integrated as plug-in module in the user's application (e.g. a WWW browser).

Once in possession (e.g. by means of authorized copying) of the information there is no further protection possible, except for marking or labeling the multimedia data. A digital (water-)mark can discourage illicit copying and dissemination of (proprietary) information by making misuse traceable and by providing evidence of misbehavior. By embedding information about the ownership, rights, etc. a proof of ownership is possible, by embedding information about the authorized recipient the distribution path can be tracked.

Meanwhile various watermarking schemes are under development (Koch and Zhao, 1995, Macq and Quisquater, 1995). A key issue is to mark the data in a secure and robust way.

5 CONCLUSION

Today security is not a stand-alone solution. Security always has to be linked to the corresponding applications and the data which have to be protected. Therefore security on application and data level will play an increasing role in securing already existing and future multimedia applications.

6 REFERENCES

Gehrke, M. and Koch, E. (1995) A security platform for future telecommunication applications and services. In: *Proc. of the 6th Joint European Networking Conference* (JENC6), Tel Aviv.

Koch, E. and Zhao J. (1995) Towards robust and hidden image copyright labeling. In: *Proc. of IEEE Workshop on Nonlinear Signal and Image Processing*, Neos Marmaras.

Krannig, A. (1996) PLASMA - Platform for secure multimedia applications. In *Communications and Multimedia Security*, IFIP Joint Working Conference TC-6 and TC-11, Essen, to be published.

Macq, B. and Quisquater, J.J. (1995) Cryptology for digital TV broadcasting. In: Proc. of the IEEE, vol. 83, no. 6.

RFC 1508, 1509 (1993), Generic Security Service API.

7 BIOGRAPHY

Eckhard Koch studied physics and received his doctor degree in 1993. After this he joined the Fraunhofer Institute for Computer Graphics as head of the department 'Security Technology for Graphics and Communication Systems'. He is partner in several European projects working on access control for broadcasting services and copyright protection. Besides this, he leads several industrial and national projects in his department. These projects deal with the integration of security services in information and communication systems, the development of cryptographic methods for multimedia data. Special attention is given to data compression, face and voice recognition, copyright protection and digital watermarking.

PART FOUR

Secure Mobile Applications

33
Phone card application and authentication in wireless communications

C. H. Lee, M. S. Hwang and W. P. Yang

Department of Computer and Information Science
National Chiao Tung University, Hsinchu, Taiwan 30050, R.O.C.
e-mail: wpyang@cis.nctu.edu.tw
Fax: 886-35-721490   Tel: 886-35-712121 ext. 56617

Abstract

The Subscriber Identity Module (SIM) concept is widely used in many mobile communication systems. Some limitations and considerations of using SIM significantly reduce the availability and accessibility for the public in the wireless communications. The idea of using phone card for the public telephones is introduced for the mobile communications in this study. This new application is proposed to improve, in the meantime, to contain the merits of SIM. Major technical issues for using phone card includes the location tracking of the user, authentication of the phone card, and the billing scheme. In this study, three authentication approaches are also presented.

Keywords

Authentication; Phone Card.

1   INTRODUCTION

The Subscriber Identity Module (SIM) concept has been widely used in modern mobile communication systems. The SIM plays a major role in the security for both the subscribers and the networks (Mazziotto, 1992). The SIM has several major advantages in mobile communications (Mouly and Pautet, 1992). First, the SIM serves as a key to the communication equipment which charges to the account of SIM for its use. Once the SIM is removed, an equipment is no longer associated with the subscriber and the billing protection can be achieved. Second, the SIM is removable from one equipment to another and facilitates more flexible applications with the same billing account. Third, the SIM-roaming concept may provide the services at a much larger scale between systems based on different radio techniques.

Nevertheless, the requirements of registration for the SIM and the ownership of a specified physical object (the subscriber's card) significantly reduce the availability and accessibility of SIM for the public in the wireless communications. The use of SIM is constrained by having a definite subscribing process and its economic concerns for short-term and more flexible uses of it. Those who use the wireless communications occasionally, wish to pay flexible and limited expenses at certain time and certain places, and hope to run less risk by losing the specific SIM card, will demand more flexibility for the application of mobile phones. As compared to a telephone system, the SIM card is more like a private phone and is not readily accessible to the public on the street. More versatile applications are deemed necessary and the idea of using phone card for the public telephones is thus introduced for the mobile communications in this study.

2 PHONE CARD APPLICATION

The phone card, a new application of mobile phone, is proposed to improve and, in the meantime, to contain the merits of SIM. The phone card is a value-added alternative of SIM for the users. The scenario is that a user buys a phone card from a telecommunication company just as he owns a temporary SIM card and he can use any mobile phone at any place to make outgoing calls and to receive incoming calls with this card. The architecture of wireless mobile communication with phone card is shown in Figure 1.

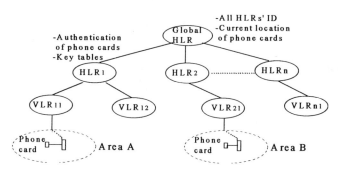

Figure 1   Architecture of wireless communication with phone card.

As illustrated in the Figure 1, the Global Home Location Register (Global HLR) holds all HLR's IDs and current locations of phone cards. The Global HLR plays an important role in location tracking when the user is roaming to different areas, i.e. the card-holder may roam from Area A to Area B. The HLR1 issues a phone card and stores n keys in it. It keeps the key tables for its phone cards and has the responsibility to authenticate the phone cards. The key format of the phone card is drawn as

| HLR.id | Keys ..... | NB (or TU) |

In this format, HLR.id is the global unique identifier who issues the cards. There are n keys generated by HLR and are stored in the card. NB stands for the Number of Bills, and TU is the Time Unit for a call. Both NB and TU can be used as the billing unit, where NB is used to charge by counts and TU is used to charge by time.

Major technical issues for using phone card includes the location tracking of the user, authentication of the phone card, and the billing scheme for Visited Location Register (VLR) (Wilkes, 1995). In the following, the descriptions of these issues are presented.

3 LOCATION TRACKING OPERATION

Location tracking is required when the network attempts to deliver a call to the card holder with a specific mobile phone. Figure 2 denotes the process of location tracking of the phone card. If a card-holder wants to receive an incoming call in a visited area, he may insert the card in any mobile phone and key-ins his Personal ID (PID) to register in this area. The visited HLR_2 will pass his PID to the Global HLR to record the location of this specific user. Whenever and wherever an incoming call is for this PID, Global HLR first locates the user for the call, then the network can set up the trunk for the communication.

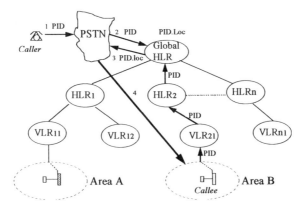

Figure 2   Process of location tracking of phone card.

4 AUTHENTICATION OF PHONE CARD

Authentication of the mobile user is a critical issue of correct billing for wireless communications (ETSI, 1993; Aziz and Diffie, 1994; Beheim, 1994; Beller, et al., 1993; Carlsen, 1994; Hagen, 1992; Lee et al., 1996; Molva et al., 1994). The process of authentication of card can protect both the card holders and the HLRs. The use of phone card in the holder's hometown is much less a problem than it is used out of town. If the holder uses the card in a visited area, visited HLR has to verify the card before a call can be delivered.

Three authentication methods for the phone card are proposed in this article. The three approaches are the Key Table approach, the Signature approach and the Polynomial approach.

4.1 Key table approach

The first authentication approach is Key Table approach, as shown in Figure 3. HLR generates n keys, stores them in the card, and maintains a table in itself. When card-holder wants to make a call at any place, i.e., his home domain or a visited domain, he inserts the card to a mobile phone and key-ins his PID, and then the information of authentication, i.e., (HLR.id, key$_i$, NB), can be routed to his home HLR. HLR checks if the keyin matches to one of the keys in the table.

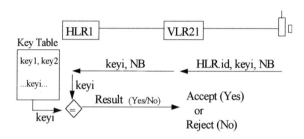

Figure 3 Key Table approach for authentication of phone card.

4.2 Signature approach

The second method is the Signature approach. Though the Key Table approach is simple, there is a problem for the plain key table in HLR to be exposed to and unlawfully used by unauthorized users. To avoid this concern, HLR will maintain a set of keywords instead of the plain keys in the table and stores the signatures of those keywords in the phone card. When a call is to be made, card sends the signature, i.e., Si, to HLR. HLR checks if there is any one of signatures of the keywords matches the Si . Figure 4 shows the Signature approach.

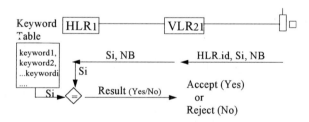

Figure 4 Signature approach for authentication of phone card.

4.3 Polynomial approach

Both the Key Table and Signature approaches require storage space for the table. The third approach is proposed to use a polynomial instead of storing the tables in HLR. First, HLR selects a polynomial,

$$h(x) = (x \oplus k_1) + x\, k_2\, Mod\, P = y, \tag{1}$$

and its own secret keys, k_1 and k_2. HLR calculates n pairs of (x_i, y_i) from the polynomial and then stores those keys in the phone card. When the authentication takes place, HLR checks if the phone card holds the authorized key pair (x_i, y_i). This approach is illustrated in Figure 5.

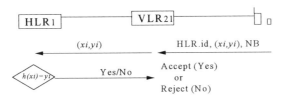

Figure 5 Polynomial approach for the authentication of phone card.

Comparisons are made for the three authentication methods. The Key Table Approach is simple but has to store the tables in HLR. Another drawback is that the card is much easier to be forged if the tables are obtained by unlawful users. The Signature Approach improves the first method by using signature to avoid the exposure of plain keys. However, it does not get rid of the need for the storage space. The Polynomial approach carries the advantages over storing key tables and leaking of sensitive information. One disadvantage in this approach is the problem of replaying. That is, how to avoid the key in phone card from using more than once? Our solution is that a bit map is set up in HLR to record the status of usage for those keys. Whenever a key is used for authentication, HLR sets the bit to 1 in the map. The HLR checks the correctness of the key and then checks if the key is used for more than once. The space for the bit map is considerably small when compared to the space for the Keyword Table in the Signature approach.

5 BILLING SCHEME

The third issue in the phone card application is the billing scheme. Billing scheme is designed to prevent the VLR of other domains from improper service charges. Two common occasions are often seen. One is that a VLR shares the key, which is transmitted from the card, with other VLRs. Another one is that VLR may overcharge for a given call. To avoid the keys being shared by several VLRs, we can check that none of the keys can be used more than once, i.e., to check the replaying. To prevent the overcharging problem, the authentication process must be done again when a TU is expired.

6 CONCLUSION

Solutions for the location tracking, authentication, and the billing scheme make it more feasible for the application of phone card in mobile communications. The application of phone card is valuable for more diversified and extended uses of mobile phone with the merits but reduces the constraints of SIM card. Using a phone card is convenient for a person without carrying his mobile phone and the specific SIM card. Even if the phone card is lost, the financial risk would be reduced more to limited extents than that of losing a SIMs. The phone card application is expected to provide more channels for the public access and enhances the technological achievements of mobile telecommunications.

7 REFERENCES

European Telecommunications Standards Institute (1993) GSM 03.20: Security related network functions.

Aziz, A. and Diffie, W. (1994) Privacy and authentication for wireless local area network. IEEE Personal Communication, 1(1), 25-31.

Beheim, J. (1994) Security first in Europe's mobile communication. Telecom Report International, 17(1), 31-4.

Beller, M. J., Chang L. F. and Yacobi, Y. (1993) Privacy and authentication on a portable communications system. IEEE Journal on Selected Areas in Communication, 11(6), 821-9.

Carlsen, U. (1994) Optimal privacy and authentication on a portable communications system ACM Operation System Review, 16-23.

Hagen, R. (1992) Security requirements and their realization in mobile networks. XIV International Switching Symposium '92, 1, 127-131.

Lee, C. H., Hwang, M. S. and Yang, W. P. (1996) Authentication of mobile users in GSM system," Submitted for publication.

Mazziotto, G. (1992) The Subscriber Identity Module for the European digital cellular system GSM and other mobile communication systems. XIV International Switching Symposium, Yokohama, Japan.

Molva, R., Samfat D. and Tsudile, G. (1994) Authentication of mobile users. IEEE Network, March/April, 26-34.

Mouly M. and Pautet, M. B. (1992) *The GSM System for Mobile Communications.* ISBN: 2-9507190-0-7,.

Wilkes, J. E. (1995) Privacy and authentication needs of PCS. IEEE Personal Communication, 11-5.

8 BIOGRAPHY

Chrissy Chii-Hwa Lee received the BS from National Taiwan University, Taiwan, in 1976 and Master of Computer Science from Texas A&M University, USA, in 1982. She is currently a Ph.D. candidate of Computer and Information Science in National Chiao Tung University,

Taiwan. She joined to work on the projects of C3I System in Chung Shan Institute of Science and Technology (CSIST) under Department of Defense, Republic of China, since 1985. She was also the leader of Management Information Systems of CSIST since 1988. Her research interests include database security, mobile computing, mobile communications and database systems.

Min-Shiang Hwang received the BS in EE from National Taipei Institute of Technology, Taiwan, in 1980; the MS in EE from National Tsing Hua University, Taiwan, in 1988; and the Ph.D. in Computer and Information Science from National Chiao Tung University, Taiwan, in 1995. He was the leader of the Computer Center at Telecommunication Laboratories (TL), Ministry of Transportation and Communications. He was also a project leader for research in computer security at TL since 1990. He is a member of IEEE and ACM. His research interests include cryptography, data security, database systems, and network management.

Wei-Pang Yang received the BS in mathematics from National Taiwan Normal University in 1974, and the MS and Ph.D. from the National Chiao Tung University (NCTU) in 1979 and 1984, respectively, both in computer engineering. He has been on the faculty of the Department of Computer Science and Information Engineering at NCTU, Taiwan, since 1979. In 1988, he joined the Department of Computer and Information Science at NCTU. His research interests include database theory, database security, object-oriented database, image database, and Chinese database retrieval systems. Professor Yang is a senior member of IEEE and a member of ACM.

34

Real-time mobile EFTPOS: challenges and implications of a world first application

S.R. Elliot
School of Information Systems
The University of NSW, Sydney 2052, Australia
Telephone   [+612] 385 4736,   Fax [+612] 661 4062
Email   s.elliot@unsw.edu.au

Abstract

This paper describes the world's first implementation of real-time mobile electronic funds transfer at point of sale (EFTPOS). Keys to the success of this application were: business enthusiasm creating demand-pull; a mobile data packet switching service providing technology-push; an industry and community ready for change; a determined systems integrator; and a small independent taxi co-operative with a record of innovation which was seeking to address its business problems. These problems included cash flow difficulties for owners and zero population growth in its core geographical market. The major benefit sought by the taxi company was an expansion in its non-cash market to include EFTPOS card holders - an expected $1 million in new business in the first full financial year. The implications of this application are likely to be of major significance for other industries and in other countries.

Keywords

Strategic IS, telecommunications technologies, technological innovation, distributed systems, banking, transportation

1   INTRODUCTION

The world's first implementation of real-time mobile electronic funds transfer at point of sale (EFTPOS) occured in Sydney, Australia late in 1994. A small taxi co-operative, with less than 5% of a market dominated by a major competitor and facing zero population growth in its core geographical market, sought to expand its market by direct appeal to customers in the community wanting to use non-cash means of payment. The taxi co-operative provided the initial application, however, it was only one of seven major players necessary to implement mobile EFTPOS. This paper describes the motivations and challenges faced by the different firms; considers how the system operates and the components and protocols necessary to support its operation; looks at the outcomes; and reviews the likely implications of this application for other industries and for other countries.

2 THE INDUSTRY

Each year in Australia an estimated 400 million taxi passenger journeys are taken generating a total revenue of more than $2 billion. The industry provides regular employment for 50 000 drivers and support staff and is made up of some 10 000 small businesses which are mostly organised into co-operatives. The purpose of the co-operatives is to share costs and provide services, particularly the establishment and operation of a radio call service. Apart from a major role in the local public transport industry, taxis play a important part of the tourism industry. Australia has more than 15 000 taxis with some 4800 located in Sydney. (source: Cabcharge).

Payment of taxi fares is primarily cash-based, although non-cash payment means are increasing in importance. Owned by major taxi groups, Cabcharge Australia P/L was formed in 1976 to provide account services to customers. Today, some 85% of the nation's taxis utilise Cabcharge. The Cabcharge system also accepts third party credit cards (including American Express and Diners Club) which do not require authorisation of individual transactions. Cabcharge supports its operations through a 10% service fee on each transaction. In the last financial year, Cabcharge's turnover exceeded $275 million. The taxi industry has for some years, however, recognised the need for an updated non-cash payment system. Many large customers became dissatisfied with paying Cabcharge the 10% surcharge since it could amount to hundreds of thousands of dollars for each customer each year. Drivers and owners were unhappy since they had to wait 5-6 weeks for payment of account fares. The Cabcharge dockets also did not reflect the growing community demand for EFTPOS as a means of payment.

Australia has one of the world's highest levels of EFTPOS use. Since its introduction in 1985, some 12.7 million EFTPOS cards are now being used by the population of 18 million. EFTPOS is used in Australia differently from other countries. Internationally, EFTPOS is predominantly a means of processing credit transactions whereas debit transactions (ie access to a customer's own funds) have proven to be more popular in Australia. The taxi industry was, therefore, in a position where community demand for new payment mechanisms had not been reflected by the industry's offerings. Taxis were not catering for a potential market of customers wanting to pay their fare by EFTPOS, using a debit card. This situation appeared, to several firms, to be an unrealised business opportunity.

3 MOTIVATIONS AND CHALLENGES

This application is characterised by the large number of major players involved: the taxi co-operative (Manly Warringah Cabs - MWC); the telecommunications service provider (Telstra); the manufacturers (Ingenico and Motorola); the EFTPOS switch (MasterCard); the bank (Commonwealth Bank of Australia - CBA); and the systems integrator, Mobile EFT Technologies - MET. The number of major players required was related more to the complexity of the application than the degree of complexity in the technology.

Not surprisingly, business motivations varied. The motivation for the two person systems integration firm, MET, was a commitment to the vision of mobile EFTPOS and a determination to make it work. The acquiring bank, CBA, sought to promote card use and a new business area. MasterCard sought to increase the acceptance of its debit cards. Telstra, facing a recently de-regulated telecommunications marketplace, looked to establish itself in a new market area, and the manufacturers, Motorola and Ingenico, sought future hardware sales.

The initial impetus for EFTPOS by MWC, the taxi co-operative, was to address driver dissatisfaction and to improve owner cash flow. This position quickly moved to an appreciation of business opportunities to attract customers in the existing market and to acquire a new market segment. Existing corporate customers were attracted by the potential to reduce the 10% levy charged by Cabcharge. The new market segment was non-cash, non-corporate customers. Recognising the irresistible movement towards use of credit and debit cards by the community, EFTPOS in taxis could attract more casual customers . Of secondary importance was the tourist market which should grow in the leadup to Sydney's Olympics Games in 2000. Electronic payments with international cards were considered to be increasingly significant in this future.

The challenges, both technical and managerial, were not insignificant although perhaps the biggest challenge was to get all seven major players working together to a common schedule. Telstra's technical challenge was to provide wireline performance within a wireless link. Motorola and Ingenico worked jointly to develop reliable EFTPOS application software on an integrated mobile terminal device experiencing variable power supply. MET specified the terminal device software and hardware, tested and implemented the systems, trained users and generally drove implementation at an operational level. MWC, as a co-operative, needed to convince its taxi owners to spend $2500 for each taxi system - a not insignificant barrier to entry for the co-operative's 225 taxis. CBA established a pathway for mobile EFTPOS transactions into the local and international banking system, certified the wireless terminal device and developed settlement systems. MasterCard's major challenges were in project management of the parties.

4 THE OPERATION

First implemented in December 1994, the operation of mobile EFTPOS in taxis appears identical to EFTPOS use in stationary outlets. The operator (ie the taxi driver) places a debit or credit card in the EFTPOS handset card reader and the customer enters a Personal Identification Number and the amount. Instead of transmitting the transaction by land-line, the mobile EFTPOS transaction request is transmitted via radio and public data network to MasterCard's Australian Processing Centre (APC). Subsequently, the request is treated as any other EFTPOS transaction to be switched to the customer's bank and account, as shown in Figure 1. Advice of approval or denial is returned by the bank to the APC switch and thence back to the taxi. Average message response times are 3-4 seconds for local banks and 6-10 seconds for banks located overseas.

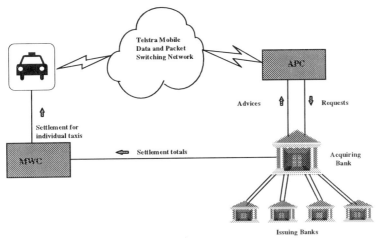

Figure 1 Processing of Mobile EFTPOS transactions.

Settlement with taxi owners is different from other EFTPOS arrangements, although this is not due to the mobility of the application. The taxi company, MWC, is a co-operative of taxi owners rather than a single merchant. The customer's bank (ie the card issuing bank) advises the acquiring bank of transactions to affect settlement with the taxi owners. The acquirer (here the CBA) provides details of transactions to the taxi company. The taxi company separates and amalgamates transactions to individual taxi owners and returns a file of direct credit transactions. Payment is made into taxi owner's accounts within three days of the original transaction. Settlement for a day's EFTPOS transactions is made by the Commonwealth Bank to MWC overnight. This settlement is the source of funding for MWC's direct payment to owners / drivers. In return for this rapid payment system the taxi owners / driver receive only 98% of the fare.

5 PRIMARY COMPONENTS

The wireless terminal device comprises a secure EFTPOS handset and radio modem. The EFTPOS handset selected is an Ingenico PX318 pinpad similar to those used by major retail chains. Use of an established EFTPOS handset assisted in implementation and contributed to customer acceptance. Magnetic cards are read in a dip style rather than by a swipe reader as the dip style allows for the future implementation of contact Smart card technology. The PX318 has a memory capacity of 256KB, sufficient to enable concurrent processing of credit, debit and Smart card applications. The PX pinpad is a secure processor in which all programs, data and internal data transfers are encrypted based on a 64 bit Data Encryption Standard (DES) key. A small impact printer produces a duplicate receipt.

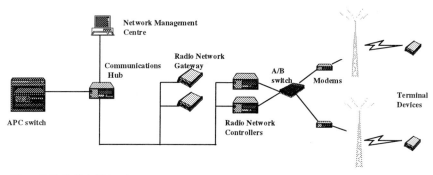

Figure 2 Radio Data Network.

The radio modem is Motorola's RM405i transmitting at 800 MHz with a raw data rate of 19.2 kbs. Located in a small junction box which also houses a power supply adaption circuit connected to the car electronics, the radio modem is connected to the handset and a roof mounted aerial. The power source is the vehicle's 12 volt battery. Substantial software development was necessary for both handset and modem to support this EFTPOS application.

The radio data network infrastructure, Telstra's MobileData service, is based on Motorola's DataTAC 5000 system. As can be seen in Figure 2, the major components of the infrastructure are: radio towers (forming a cellular data network); a radio network gateway (converting protocols, authorising all service access on the network and registering each wireless terminal device); radio network controllers (routing messages between the gateway and wireless terminal devices over particular radio towers); a communications hub (providing protocol conversion between the DataTAC system and the host); and a network management centre (supporting system management functions). Communications between the radio network and MasterCard's APC are via a public data packet switched service. Capacity of the Telstra MobileData service is rated at 100,000 packets per hour.

The component responsible for the biggest operational problem has been the vehicle's power source. Taxis are prone to battery failure since the high daily mileage means that batteries boil most days and last for only a few months. When the car battery is flat the call service radio and EFTPOS are both inoperable, although this is not the worst problem. A failing battery is unable to produce a reliable current. In these circumstances the terminal device may inexplicably cease to function. The EFTPOS handset and radio modem have required modification to operate reliably at a range of currents due to varying battery states. The handsets and radio modems have proven to be very robust despite the additional vibration involved in being mobile. Printers are the biggest day to day headache. Experience has shown that cheap paper does not produce an acceptable copy. Telecommunications black spots due to Sydney's high rise buildings and hilly terrain have also posed a problem, which has been addressed by Telstra installing additional radio towers to improve radio area coverage.

Alternative systems, including analogue radio and analogue and digital telephony were considered but found wanting. The first thought was to utilise the existing radio network for the taxi call service which is based on analogue radio technology. This operates at a slow 12 MHz; has little or no security; and proved to be unreliable for a data transfer service due to radio blackspots and to the volume of radio traffic in peak times. Several companies looked at ways to support mobile EFTPOS services based on use of the mobile GSM digital telephone adapted for EFTPOS transactions. Each on-line transaction would require a call to the authorising body (ie the card issuer) and another call to approve the transaction. Two calls became an expensive component in a small value transaction and the time to establish two calls was excessive. Off-line transactions (ie accumulated in the taxi during the day and transferred by fixed telephone line to the bank each night) were considered but could not adequately support debit payments.

6 PROTOCOLS

At the time of implementation, the only radio frequency transmission protocols available were proprietary standards from major equipment suppliers. Telstra avoided potential incompatability problems by selecting the latest protocol set from a major supplier and building their service capability based exclusively on Motorola's DataTAC 5000 system. As can be seen in Figure 2, communications from the host to the Motorola DataTAC system are via a public data network X.25 protocol, which maps to the ISO Reference Model's physical, data link and network layers. Motorola's Radio Network Gateway 5000 protocol maps to transport, session, presentation and application layers. Motorola's Standard Context Routing protocol supports communication from the host to a number of terminal devices using a single logical X.25 connection. Communications between the various DataTAC internal system components rely on Ethernet, RS-232, HDLC and TCP/IP. Motorola's Native Control Language protocol maps to physical, data link and presentation layers for both the EFTPOS handset and the radio packet modem. Wireless transmissions are supported by Motorola's RD-LAP open protocol, which enables multiple concurrent use of the medium with collision detection, contention and retries.

EFTPOS transactions comply with Australian Standard 2805 which is consistent with the requirements of major local and international debit and credit card owning organisations. All EFTPOS transactions, mobile and stationary, require encryption to the international 64 bit Data Encryption Standard.

7 OUTCOMES

Twelve months after implementation, MWC's anticipated benefits of $1 million in new business in the first full financial year appear conservative. Account work has increased more than 50% with EFTPOS transactions being responsible for the majority of the increase. MWC's owners and drivers have responded favourably to the increased business and rapid settlement of non-cash business. Mobile EFTPOS has proven to be rewarding for MasterCard and financial institutions and to be of no greater risk than for non-mobile EFTPOS transactions. Hardware sales have increased rapidly for both

Ingenico and Motorola and Telstra is seen to provide the nation's most utilised mobile data service.

Nine months after implementation, however, Cabcharge responded with the launch of pilot implementations of rival EFTPOS and stored value card products. The stored value card has been targetted at a broader market being designed for use in trains and buses as well as cabs. This smartcard has not been widely released. Its distribution has been complicated by the international piloting of other, incompatible, smartcard products in Australia by MasterCard and Visa. Cabcharge's EFTPOS system currently processes only credit cards and retains the 10% service fee. There is certain to be contention in the taxi industry while these non-cash payment issues are being resolved.

8 IMPLICATIONS

Many applications of real-time mobile computing have been in specialised and support functions, such as co-ordination of equipment servicing or price checking by sales staff, rather than in mainstream business. This application of mobile computing addresses a core business activity in an international industry. It demonstrates integration of stationary and mobile processors in a time critical mainstream business application and performs reliably and consistently in this role. Initial implications for senior management in other industries are that mobile computing must be considered for core business activities. First and other early adopters of mobile EFTPOS are achieving competitive benefits which will be regarded with interest not only by the taxi industries world-wide but by any local or international industry seeking a transportable, relocatable and flexible non-cash payment system. The existing system could be implemented with little modification in other mobile industries such as couriers and truckers.

Now that the nexus between EFTPOS and a fixed location has been broken new applications abound. The system could be developed further to produce an attache case-sized device complete with 12 volt battery to provide transportable EFTPOS. As much small business is conducted away from fixed premises (eg market stall holders, home gardeners, lawn mowing companies, electricians, plumbers etc) a transportable EFTPOS system could prove a desirable payment system.

The application of most interest, however, is internationally - particularly in Asia. Many Asian countries experience poor communications by land line and unreliable power sources and so cannot adequately support the electronic payment systems necessary for their international tourism industries. The flexibility of mobile EFTPOS is such that a radio data network could be implemented at low cost and a reliable EFTPOS / ATM system made operational within a matter of weeks. A pilot implementation of such a system has already been undertaken.

Further implications of this case are for IS researchers, initially in the development of theory in first adoption of IT. The market leader (totally dominant in the provision of non-cash payment systems to the industry) was aware of the gap between the payment systems provided by the industry and changing community demand. This firm had researched the problem over a period of several years, selected a stored value card solution and had piloted a product. Nonetheless, their planning procedures appear to have taken too long; their stored value card solution did not comply with the standard adopted by the international leaders in the stored value card industry (Eurocard, MasterCard and Visa); and the company appeared to have been caught unaware by the implementation of mobile EFTPOS. Their response, after nine months, consisted of a rival mobile EFTPOS system which retained the 10% surcharge. There appears to be uncertainty at present as to widespread release of the stored value product.

Finally, the case provides significant insights into the mixed motivations present in consortia seeking co-operative advantage and to the empirical realities of strategic IS planning for inter-organisational systems. The innovative forces present in the different firms demonstrate the requirements for integration of innovations research and intentions-based models of IS adoption and diffusion. Neither of these theoretical areas, in isolation, provides a complete explanation of the circumstances resulting in this world's first application of IT.

The contribution of the many who willingly assisted in this project is gratefully acknowledged.

9 REFERENCES

Australian Payment Systems Council, (1995) Annual Report 1994/95. Reserve Bank of Australia, Sydney.

Cabcharge Australia, (1995) Corporate Report.

Kermode R.L., (1995) "Electronic payment system in taxis". *Taxi*. NSW Taxi Council July-August. Vol. 40, Nr. 6, pp 17-21.

Manly Warringah Cabs Co-operative Society Limited, (1994) Chairman's Report, Papers of the Annual General Meeting.

Manly Warringah Cabs Co-operative Society Limited, (1995) Chairman's Report, Papers of the Annual General Meeting.

Motorola, (1994) DataTAC 5000 System - Host Application Programmer's Guide. Motorola Canada. Richmond BC Canada.

Motorola, (1994) DataTAC Open Protocol Specification - Native Control Language Release 1.2. Motorola Canada. Richmond BC Canada.

Motorola, (1995) Personal Messenger 100D: Two-Way Wireless Modem Card - Application Developer's Guide. Motorola Inc. Bothell WA. USA.

NSW Department of Transport, (1995) Annual Report 1994/1995. Sydney.

Yin R.K., (1994) Case Study Research: Design and Methods, 2nd Ed. Sage Publications Inc. Thousand Oaks Calif.

and various technical, publicity and strategic planning details from Cabcharge, Ingenico, Manly Warringah Cabs, MasterCard, MET, Motorola and Telstra,

10 BIOGRAPHY

Since commencing in the computer industry in 1972 Steve Elliot has worked in Australia, Europe and Asia in business, government, education and with the United Nations. He is currently Director of the Information Technology Research Centre at the University of New South Wales in Sydney. Steve has degrees in Economics and Information Systems from University of Sydney and University of Technology, Sydney and a PhD in strategic IS planning from University of Warwick, UK. His major research interest lies in the strategic use and management of IS, particularly telecommunications-based IS.

INDEX OF CONTRIBUTORS

Akaiwa, Y. 111
Alanko, T. 151
Assis Silva, F.M. 181

Belz, C. 98
Bergold, M. 98
Bönigk, J. 167
Brown, M. 69
Buot, T. 119

Canchi, R. 111
Carlier, D. 78
Chan, B.Y.L. 31
Choi, Y. 132
Cordonnier, V. 239

Dassanayake, P. 141
Diehl, N. 9
Dong, C. 267

Eckardt, T. 50
Elliot, S.R. 330

Ha, E. 132
Häckelmann, H. 98
Harjono, H. 219
Hwang, M.S. 323

Kim, C. 132
Kirste, T. 41, 86
Koch, E. 267, 317
Kojo, M. 151
Kottmann, D. 23
Kowalski, B. 245
Krause, S. 181
Kümmel, S. 159

Laamanen, H. 151
Le, M.T. 205
Lee, C.H. 323
Leong, H.V. 31
Leske, J. 11
Loyolla, W. 181

Magedanz, T. 50, 181
Marco, C. 301
Mendes, M. 181
Morillo, P. 301
Mu, Y. 258

Nurkic, I. 277

Perkins, C. 219
Popescu-Zeletin, R. 50

Raatikainen, K. 151
Rabaey, J. 205

Sasaki, H. 3
Schill, A. 159
Schirmer, J. 86
Schumann, K. 159
Si, A. 31
Smith, T. 285
Storck, D. 309
Strack, R. 98, 129

Tienari, M. 151
Trane, P. 78

Ulbricht, C. 50

Varadharajan, V. 258
von Lukas, U. 167

Yang, W.P. 323

Zhao, J. 267
Zheng, Y. 249
Ziegert, T. 159
Zonoozi, M. 141

KEYWORD INDEX

A3   285
A5   285
A8   285
Access control   245
Active mail   86
Adaptation   159
Advanced mobile applications   23
Agent
  roles   181
  skills   181
Agents   181
ALOHA   119
Animation   98
Anonymity   258
Application
  partitioning   41
  programming interfaces   317
  -support systems   23
Asymmetric   277
Authentication   245, 249, 323
  protocols   258

Banking   330
Buffer overhead   132

Cell residence time   141
Color PDP   3
Communications   277
Company card   245
Computational mail   86
Copyright protection   317
CORBA   167
Cryptography   245, 249, 277, 301
Customer control   50

Data
  broadcasting   31
  caching   31
  distribution   159
  models   41
  reduction   167

DCS1800   98
Detail-on-demand   167
Device technology   3
DHCP   219
Digital signature   245
Direct Sequence-Code Division Multiple
    Access   205
Directory service   219
Disconnected operation   159
Discrete logarithm   301
Distributed systems   41, 167, 330
Dynamic configuration   219

Electronic commerce   267
Electronic
  display   3
  markets   181
  payment   267
Encoding   98
Encryption   245, 285

Fault tolerance   78
Flat panel display   3
Flexible information systems   181
Format conversion   50
Frame models   41
Fraud   285
Frequency-domain   309

Global system for mobiles   285
GPRS   11
GSM   11, 98, 277, 285
GSM data service   151

Handoff   132
Handover   141
HSCSD   11
Hybrid approach   258

IN   50
Integrity to digital images   309

Keyword index

Intelligent agents 151
Internet 267
IRMA 111
Itinerant agents 86

Key 277
  distribution 249

LCD 3
Light weight 219

Media Conversion 50
MIME 86
Mobile 277
  agents 86
  clients 151
  communication 167
  computing 23, 41, 78, 159, 167, 181, 219, 249
  data 11
  database 31
  IP 219
  multimedia 69, 98
  security 258
  system architecture 167
  transport system 159
Mobility 69, 141
Modeling 98
Multicast 132
Multimedia 3
  applications and data 317
  mail 86

Nomadic users 151

Object orientation 167
On-line services 98
Oriented-object technology 78

PCS 50
PCSS 50
Personal communications 151, 181
Personalisation 69
Phone card 323
Picocellular network 132
Polling 119
Power control 205
Public key cryptography 245

Queuing systems 159

Random access 119
RDP 219
Remote programming 86
Replication 23
Reservation 119
Resource discovery protocol 219

Search engine 219
Security 249, 267, 277
  platform 317
Service
  interworking 50
  location 219
  Personalization 50
SGMH 132
SLP 219
Smart card 301, 245
Smartcard 277
Strategic IS 330
Synchronization 86

TDMA 111
Technological innovation 330
Telecommunications technologies 330
TFT-LCD 3
Throughput delay 111
TMN 50
Transient fluid approximation 119
Transportation 330
Trusted Third Parties 245

UPT 50
URL 219
URN 219
Use control 317
User mobility 50

Virtual enterprises 181
VRML 98

Wireless 11
  data 119
  network 205
  services 98
World Wide Web (WWW) 98, 267